Functional Molecular Materials Insights—a Themed Issue in Honour of Professor Manuel Almeida on the Occasion of His 70th Birthday

Functional Molecular Materials Insights—a Themed Issue in Honour of Professor Manuel Almeida on the Occasion of His 70th Birthday

Laura C. J. Pereira
Dulce Belo

Laura C. J. Pereira
Department of Nuclear
Sciences and Engineering
University of Lisbon
Lisbon
Portugal

Dulce Belo
Department of Nuclear
Sciences and Engineering
University of Lisbon
Lisbon
Portugal

Editorial Office
MDPI AG
Grosspeteranlage 5
4052 Basel, Switzerland

This is a reprint of articles from the Special Issue published online in the open access journal *Magnetochemistry* (ISSN 2312-7481) (available at: www.mdpi.com/journal/magnetochemistry/special_issues/GET5H14FD2).

For citation purposes, cite each article independently as indicated on the article page online and using the guide below:

Lastname, A.A.; Lastname, B.B. Article Title. *Journal Name* **Year**, *Volume Number*, Page Range.

ISBN 978-3-7258-1796-2 (Hbk)
ISBN 978-3-7258-1795-5 (PDF)
https://doi.org/10.3390/books978-3-7258-1795-5

Cover image courtesy of Laura C. J. Pereira

© 2024 by the authors. Articles in this book are Open Access and distributed under the Creative Commons Attribution (CC BY) license. The book as a whole is distributed by MDPI under the terms and conditions of the Creative Commons Attribution-NonCommercial-NoDerivs (CC BY-NC-ND) license (https://creativecommons.org/licenses/by-nc-nd/4.0/).

Contents

About the Editors vii

Preface ix

Laura C. J. Pereira and Dulce Belo
Functional Molecular Materials Insights
Reprinted from: *Magnetochemistry* **2023**, *10*, 1, doi:10.3390/magnetochemistry10010001 1

José F. Malta, Marta S. C. Henriques, José A. Paixão and António P. Gonçalves
Synthesis and Structural and Magnetic Properties of Polycrystalline GaMo4Se8
Reprinted from: *Magnetochemistry* **2023**, *9*, 182, doi:10.3390/magnetochemistry9070182 5

Hiroyuki Tajima, Takeshi Oda and Tomofumi Kadoya
Nonthermal Equilibrium Process of Charge Carrier Extraction in Metal/Insulator/Organic Semiconductor/Metal (MIOM) Junction
Reprinted from: *Magnetochemistry* **2023**, *9*, 180, doi:10.3390/magnetochemistry9070180 15

Reizo Kato and Takao Tsumuraya
Dirac Cone Formation in Single-Component Molecular Conductors Based on Metal Dithiolene Complexes
Reprinted from: *Magnetochemistry* **2023**, *9*, 174, doi:10.3390/magnetochemistry9070174 24

Vladislav D. Kochev, Seidali S. Seidov and Pavel D. Grigoriev
On the Size of Superconducting Islands on the Density-Wave Background in Organic Metals
Reprinted from: *Magnetochemistry* **2023**, *9*, 173, doi:10.3390/magnetochemistry9070173 43

Ryuhei Oka, Keishi Ohara, Kensuke Konishi, Ichiro Yamane, Toshihiro Shimada and Toshio Naito
Band Structure Evolution during Reversible Interconversion between Dirac and Standard Fermions in Organic Charge-Transfer Salts
Reprinted from: *Magnetochemistry* **2023**, *9*, 153, doi:10.3390/magnetochemistry9060153 59

Adam Berlie, Ian Terry and Marek Szablewski
Driving a Molecular Spin-Peierls System into a Short Range Ordered State through Chemical Substitution
Reprinted from: *Magnetochemistry* **2023**, *9*, 150, doi:10.3390/magnetochemistry9060150 77

Bin Zhang, Yan Zhang, Zheming Wang, Yang Sun, Tongling Liang and Mei Liu et al.
Two One-Dimensional Copper-Oxalate Frameworks with the Jahn–Teller Effect: $[(CH_3)_3NH]_2[Cu(\mu-C_2O_4)(C_2O_4)] \cdot 2.5H_2O$ (I) and $[(C_2H_5)_3NH]_2[Cu(\mu-C_2O_4)(C_2O_4)] \cdot H_2O$ (II)
Reprinted from: *Magnetochemistry* **2023**, *9*, 120, doi:10.3390/magnetochemistry9050120 84

Aritro Sinha Roy, Boris Dzikovski, Dependu Dolui, Olga Makhlynets, Arnab Dutta and Madhur Srivastava
A Simulation Independent Analysis of Single- and Multi-Component cw ESR Spectra
Reprinted from: *Magnetochemistry* **2023**, *9*, 112, doi:10.3390/magnetochemistry9050112 95

Qingyun Wan, Masanori Wakizaka, Haitao Zhang, Yongbing Shen, Nobuto Funakoshi and Chi-Ming Che et al.
A New Organic Conductor of Tetramethyltetraselenafulvalene (TMTSF) with a Magnetic Dy(III) Complex
Reprinted from: *Magnetochemistry* **2023**, *9*, 77, doi:10.3390/magnetochemistry9030077 111

Brett Laramee, Raju Ghimire, David Graf, Lee Martin, Toby J. Blundell and Charles C. Agosta
Superconductivity and Fermi Surface Studies of β''-(BEDT-TTF)$_2$[(H$_2$O)(NH$_4$)$_2$Cr(C$_2$O$_4$)$_3$]·18-Crown-6
Reprinted from: *Magnetochemistry* 2023, 9, 64, doi:10.3390/magnetochemistry9030064 ...... **122**

Janice L. Musfeldt, Zhenxian Liu, Diego López-Alcalá, Yan Duan, Alejandro Gaita-Ariño and José J. Baldoví et al.
Vibronic Relaxation Pathways in Molecular Spin Qubit Na$_9$[Ho(W$_5$O$_{18}$)$_2$]·H$_2$O under Pressure
Reprinted from: *Magnetochemistry* 2023, 9, 53, doi:10.3390/magnetochemistry9020053 ...... **131**

Danica Krstovska, Eun Sang Choi and Eden Steven
Giant Angular Nernst Effect in the Organic Metal $\alpha-(\text{BEDT}-\text{TTF})_2\text{KHg(SCN)}_4$
Reprinted from: *Magnetochemistry* 2023, 9, 27, doi:10.3390/magnetochemistry9010027 ...... **140**

Sergei I. Pesotskii, Rustem B. Lyubovskii, Gennady V. Shilov, Vladimir N. Zverev, Svetlana A. Torunova and Elena I. Zhilyaeva et al.
Effect of External Pressure on the Metal–Insulator Transition of the Organic Quasi-Two-Dimensional Metal κ-(BEDT-TTF)$_2$Hg(SCN)$_2$Br [†]
Reprinted from: *Magnetochemistry* 2022, 8, 152, doi:10.3390/magnetochemistry8110152 ...... **155**

Jean-Paul Pouget
Spin-Peierls, Spin-Ladder and Kondo Coupling in Weakly Localized Quasi-1D Molecular Systems: An Overview
Reprinted from: *Magnetochemistry* 2023, 9, 57, doi:10.3390/magnetochemistry9020057 ...... **165**

Fabio Manna, Mariangela Oggianu, Narcis Avarvari and Maria Laura Mercuri
Lanthanide-Based Metal–Organic Frameworks with Single-Molecule Magnet Properties [†]
Reprinted from: *Magnetochemistry* 2023, 9, 190, doi:10.3390/magnetochemistry9070190 ...... **184**

About the Editors

Laura C. J. Pereira

Laura C. J. Pereira, senior researcher at the Department of Nuclear Engineering and Sciences of the Instituto Superior Técnico (IST), obtained her Ph.D. in Inorganic Chemistry from the Faculty of Science of the University of Lisbon in 1998 under the supervision of Prof. Manuel Almeida. Since then, she has been developing her research in the field of solid-state chemistry and physics, namely in the structure/magnetic property correlations of new advanced materials, such as multifunctional molecular-based compounds with unconventional magnetic and electrical properties, magnetic-oxide-based nanostructures, and intermetallics with 3d- and f- elements (lanthanides and/or actinides). Together with Prof. Manuel Almeida, she launched the Low Temperature and High Magnetic Fields Laboratory (LTHMFL) of the IST. She is currently leading the research group after the retirement of Prof. Manuel Almeida.

Dulce Belo

Dulce Belo completed her Ph.D. degree in Chemistry from Instituto Superior Técnico (IST), Universidade de Lisboa, in 2002. She is a principal researcher of the Solid State Group at IST and has devoted most of her scientific career to the design, synthesis, and physical characterization of molecular materials with unconventional magnetic and transport properties. Among them, she developed pioneering work in the preparation and study of Single-Component Molecular Metals (SCMMs) and is interested in their processing as active components for electronic devices. In 2020, she started a complementary line of research dedicated to Therapeutic Molecular Materials (T-MoMas). As in SCMM, T-MoMas are Transition Metal Bisdithiolate Complexes, which display promising biological activities as both antitumor and antimicrobial agents. This type of material has been extensively used as a molecular building block for magnetic and conductive molecular materials, but it never been studied before in this context. Since 2013, her responsibilities have also included teaching duties as an invited professor at the Chemistry Department of IST, having been annually distinguished with the pedagogical excellence award of IST.

Preface

This Special Issue of *Magnetochemistry*, "Functional Molecular Materials Insights—a Themed Issue in Honour of Professor Manuel Almeida on the occasion of his 70th Birthday", has published fifteen insightful contributions in which eminent researchers from around the globe in the field of molecular materials science gathered to acknowledge and celebrate the notable scientific contributions of Professor Almeida. This reprint is a themed collection of state-of-the-art publications illustrating recent achievements in the development of molecular materials, especially those with unconventional magnetic and transport properties, as well as their potential applications.

We would like to sincerely thank all the authors who contributed to this Special Issue for their dedicated efforts and the outstanding quality of their submissions. We also thank the anonymous reviewers and the editorial team of *Magnetochemistry*, especially Jamie, whose assistance has played a crucial role in preparing this Special Issue.

The development of functional molecular materials has become one of the main challenges for chemists, physics, and materials researchers. Molecular materials are based on well-designed molecular building blocks, prepared by advanced organic/inorganic synthetic methods, and crystal engineering plays a fundamental role in obtaining the desired (nano)structures. Crystal engineering has enabled the development of materials with tuneable chemical and physical solid-state properties. This field has evolved significantly, with a growing number of molecular materials capable of reproducing the different types of properties commonly found in other materials, as well as demonstrating new ones. For this reason, multifunctional materials are potentially useful in technological applications, namely in electronic devices, sensors, and spintronics.

Laura C. J. Pereira and Dulce Belo
Editors

Editorial

Functional Molecular Materials Insights

Laura C. J. Pereira * and Dulce Belo

Centro de Ciências e Tecnologias Nucleares, Departamento de Engenharia e Tecnologias Nucleares, Instituto Superior Técnico, Universidade de Lisboa, 2695-066 Bobadela, Portugal
* Correspondence: lpereira@ctn.tecnico.ulisboa.pt

In the commemorative Special Issue titled "Insights into Functional Molecular Materials—A Themed Collection Honoring Professor Manuel Almeida on His 70th Birthday", eminent researchers from around the globe in the field of molecular materials science come together to acknowledge and celebrate the notable scientific contributions of Professor Almeida.

The topics covered in the published articles exemplify the interdisciplinary nature of this scientific domain, seamlessly integrating preparative chemistry and condensed matter physics in equal proportions. This unique Special Issue features fifteen insightful contributions, and a brief overview of these contributions is presented below.

The first paper, "Synthesis and Structural and Magnetic Properties of Polycrystalline $GaMo_4Se_8$" by J.F. Malta and A.P. Gonçalves from IST, the University of Lisbon (Portugal), and M.S.C. Henriques and J.A. Paixão from the University of Coimbra (Portugal), reports an innovative two-step synthesis of pure polycrystalline $GaMo_4Se_8$, a lacunar spinel belonging to the GaM_4X_8 family. Phase purity and composition are confirmed through XRD and SEM analyses. Magnetic investigations reveal the presence of cycloidal, skyrmionic, and ferromagnetic phases, contributing valuable insights into the compound's magnetic behavior and potential applications (contribution 1).

The second paper by H. Tajima and T. Kadoya from the Japanese Universities of Hyogo and Konan investigates "Nonthermal Equilibrium (NTE) Process of Charge Carrier Extraction in Metal/Insulator/Organic Semiconductor/Metal (MIOM) Junction" with Schottky-type contacts, contrasting it with standard capacitors. Strategies like ohmic contacts or high-mobility organic semiconductors are suggested to mitigate NTE's negative impact in organic field-effect transistors. The NTE process could find applications in OFETs as memory devices and for finding charge injection barriers (contribution 2).

The contribution by R. Kato from the RIKEN Laboratory at Wako, Saitama, and T. Tsumuraya from Kumamoto University in Japan explores "Dirac Cone Formation in Single-Component Molecular Conductors Based on Metal Dithiolene Complexes", using tight-biding models and first-principles density functional theory (DFT) calculations. The tight-binding model predicts the emergence of Dirac cones in the studied systems, which is associated with a stretcher bond type of molecular arrangement (contribution 3).

The paper "On the Size of Superconducting Islands on the Density-Wave Background in Organic Metals" by V.D. Kochev, S.S. Seidov, and P.D. Grigoriev from the National University of Science and Technology "MISiS" (Russia) and the L.D. Landau Institute for Theoretical Physics (Russia) explores spatial inhomogeneity in high-temperature organic superconductors, specifically the transition from superconductivity to a density wave state in quasi-one-dimensional metals with imperfect nesting. External pressure influences this transition by changing electron dispersion. By estimating the size of superconducting islands during this transition, the authors provide insights into spatial heterogeneity in organic superconductors (contribution 4).

In a paper entitled "Band Structure Evolution during Reversible Interconversion between Dirac and Standard Fermions in Organic Charge-Transfer Salts" by R. Oka, K.

Citation: Pereira, L.C.J.; Belo, D. Functional Molecular Materials Insights. *Magnetochemistry* 2024, 10, 1. https://doi.org/10.3390/magnetochemistry10010001

Received: 12 December 2023
Accepted: 14 December 2023
Published: 20 December 2023

Copyright: © 2023 by the authors. Licensee MDPI, Basel, Switzerland. This article is an open access article distributed under the terms and conditions of the Creative Commons Attribution (CC BY) license (https://creativecommons.org/licenses/by/4.0/).

Ohara, K. Konishi, and T. Naito from Ehime University (Japan) and I. Yamane and T. Shimada from Hokkaido University (Japan), organic charge-transfer salts studied, α-D_2I_3 (D = ET, BETS), exhibit temperature-sensitive charge-transfer interactions, transforming flat-bottomed bands into Dirac cones with decreasing temperature. The observed band reshaping under ambient pressure provides insights into the lifecycle of Dirac fermions (DFs). While shedding light on DF systems, the study emphasizes the need for further research to deepen our understanding of the nature of DFs (contribution 5).

"Driving a Molecular Spin-Peierls System into a Short Range Ordered State through Chemical Substitution" explores the effect of the introduction of bromine atoms on quasi-1D spin-Peierls system potassium TCNQ (TCNQ = 7,7,8,8-tetracyanoqunidimethane). The resulting derivative, potassium $TCNQBr_2$, shows evidence of residual magnetism, suggesting short-range and potentially disordered correlations. While magnetic susceptibility data hint at 1D behavior, muon spin spectroscopy reveals a departure from the expected spin-Peierls system behavior, indicating a dominance of short-range magnetic correlations. This study by A. Berlie from the Rutherford Appleton Laboratory, at Harwell Campus in Oxfordshire (UK) and I. Terry and M. Szablewski from Durham University (UK) suggests that the bulky bromine atoms prevent the structural changes required for the system to undergo a spin-Peierls transition (contribution 6).

In paper number seven entitled "Two One-Dimensional Copper-Oxalate Frameworks with the Jahn–Teller Effect: $[(CH_3)_3NH]_2[Cu(\mu-C_2O_4)(C_2O_4)]\cdot 2.5H_2O$ (I) and $[(C_2H_5)_3NH]_2[Cu(\mu-C_2O_4)(C_2O_4)]\cdot H_2O$ (II)", B. Zhang, Y. Sun, T. Liang, M. Liu, and D. Zhu from the Chinese Academy of Sciences and Y. Zhang and Z. Wang from Peking University, China, reported that these salts feature Jahn–Teller-distorted copper ions in an octahedral coordination and that the crystal structure is characterized by hydrogen bonds among ammonium, water, and the copper–oxalate framework, creating a 3D network. Both salts are insulators and exhibit ferromagnetic and weak-ferromagnetic behaviors with no long-range ordering observed above 2 K (contribution 7).

The contribution "A Simulation Independent Analysis of Single- and Multi-Component cw ESR Spectra" by A. S. Roy, B. Dzikovski, and M. Srivastava from Cornell University (USA), D. Dolui and A. Dutta from the Indian Institute of Technology Bombay (India), and O. Makhlynets from Syracuse University (USA) introduces a novel simulation-independent approach for the precise analysis of continuous-wave electron spin resonance (cw ESR) spectra, crucial for understanding free radicals and paramagnetic metal complexes. The method, based on wavelet packet transform, accurately extracts spectral information, overcoming challenges posed by poorly resolved spectra. Applied to various systems, this approach demonstrates consistency and accuracy, even in identifying the features of a 5% minor component in a two-component system. The method's efficacy was validated with well-studied systems (contribution 8).

A paper by Q. Wan, M. Wakizaka, H. Zhang, N. Funakoshi, S. Takaishi, and M. Yamashita from Tohoku University (Japan), Y. Shen from Xi'an Jiaotong University (China), and C.-M. Che from The University of Hong Kong (China), entitled *A New Organic Conductor of Tetramethyltetraselenafulvalene (TMTSF) with a Magnetic Dy(III) Complex*", reports the synthesis of $(TMTSF)_5[Dy(NCS)_4(NO_3)_2]CHCl_3$ using the electrochemical oxidation method. This salt shows a semiconducting behavior with a conductivity of 0.2 S·cm^{-1} at room temperature and an activation energy of 34 meV at ambient pressure. This preliminary study provides information for designing new hybrid materials based on molecular conductors and polyvalent magnetic 4f metal complexes (contribution 9).

The study "Superconductivity and Fermi Surface Studies of β″-(BEDT-TTF)$_2$[(H$_2$O)(NH$_4$)$_2$Cr(C$_2$O$_4$)$_3$]·18-Crown-6" presents radiofrequency penetration depth measurements on a 2D organic superconductor with the largest layer separation between consecutive conduction layers, using a contactless tunnel diode oscillator measurement technique. Measurements reveal its behavior under different orientations to the crystal conduction planes. When parallel to the layers, $Hc2$ is 7.6 T with no signs of inhomogeneous superconductivity. Perpendicular orientation shows Shubnikov–de Haas oscillations, indicating high anisotropy in $Hc2$, particularly a low $Hc2\perp$ of 0.4 T, possibly due to a lower effective mass. The work was developed by B. Laramee, R. Ghimire, and C.C. Agosta from Clark University (Worcester, MA, USA), D. Graf from the National High Magnetic Field Laboratory (Florida State University, USA), and L. Martin and T.J. Blundell from Nottingham Trent University (UK) (contribution 10).

The contribution "Vibronic Relaxation Pathways in Molecular Spin Qubit $Na_9[Ho(W_5O_{18})_2]\cdot 35H_2O$ under Pressure", authored by J.L. Musfeldt from the University of Tennessee (USA), Z. Liu from the University of Illinois Chicago (USA), and D. López-Alcalá, Y. Duan, A. Gaita-Ariño, J.J. Baldoví, and E. Coronado from the University of Valencia (Spain), investigates controlling spectral sparsity and decoherence in a qubit system using diamond anvil cell techniques, infrared spectroscopy, and first-principles calculations. The results suggest that applying negative pressure through chemical means or strain could improve transparency in the spin qubit system, offering potential for better managing decoherence in quantum devices (contribution 11).

The paper "Giant Angular Nernst Effect in the Organic Metal α-$(BEDT-TTF)_2KHg(SCN)_4$" by D. Krstovska from Ss. Cyril and Methodius University (North Macedonia), E.S. Choi from Florida State University (USA), and E. Steven from Jakarta Utara DKI (Indonesia) reports a substantial Nernst effect in the charge density wave state of this organic metal. Momentum relaxation dynamics in the low-field CDW state indicate significant carrier mobility, contributing to the large Nernst signal. However, this effect diminishes in the high-field CDW state, where only phonon drags and electron–phonon interactions contribute to the thermoelectric signal. These findings challenge previous understandings of the complex properties of this organic metal (contribution 12).

The metal–insulator transition in κ-$(BEDT-TTF)_2Hg(SCN)_2Br$ at $T_{MI} \approx 90$ K involves a crystal shift from monoclinic to triclinic. The triclinic phase tends towards a Mott insulating state, causing increased resistance below T_{MI}, which is suppressed by external pressure. This remarkable "Effect of External Pressure on the Metal–Insulator Transition of the Organic Quasi-Two-Dimensional Metal κ-$(BEDT-TTF)_2Hg(SCN)_2Br$" is reported in this paper by S.I. Pesotskii, R.B. Lyubovskii, G. V. Shilov S.A. Torunova, V.N. Zverev, and E.I. Zhilyaeva from the Russian Academy of Sciences (Russia) and E. Canadell from ICMAB-CSIC (Spain). Quantum oscillations align with the calculated Fermi surface for the triclinic phase, explaining the material's behavior around 100 K. Notably, differences are observed in the behaviors of the isostructural salts κ-$(BEDT-TTF)_2Hg(SCN)_2Br$ and κ-$(BEDT-TTF)_2Hg(SCN)_2Cl$ (contribution 13).

"Spin-Peierls, Spin-Ladder and Kondo Coupling in Weakly Localized Quasi-1D Molecular Systems: An Overview", by J.-P. Pouget from Université Paris-Saclay (France), explores magneto-structural properties in quasi-one-dimensional molecular organics with electron–electron correlations, emphasizing spin–charge decoupling and singlet dimer formation. Examples like $(TMTTF)_2X$ Fabre salts and Per_2-$M(mnt)_2$ systems illustrate the spin-Peierls instabilities and the 3D-SP ground states. This study also delves into the unique features of correlated 1D systems, including coexisting orders and soliton nucleation in perturbed spin-Peierls systems (contribution 14).

"Lanthanide-Based Metal–Organic Frameworks with Single-Molecule Magnet Properties" allow for tunable magnetic behaviors through factors like solvent, temperature, and organic linkers. This overview covers synthetic methods and strategies, including redox activity and chirality, for controlling SMM behavior in Ln-MOFs. The discussion also touches on intriguing phenomena like the CISS effect and CPL. The paper was authored by F. Manna, M. Oggianu, and M.L. Mercuri from the University of Cagliari (Italy) and N. Avarvari from the University of Angers (France) (contribution 15).

This issue will provide valuable insights into the rapidly evolving landscape of current research in molecular materials and related studies. We would like to sincerely thank all the authors who contributed to this Special Issue for their dedicated efforts and the outstanding quality of their submissions. Finally, we would like to express our gratitude for the unwavering support and commitment of the editorial team at *Magnetochemistry*, whose assistance has played a crucial role in preparing this Special Issue.

Author Contributions: L.C.J.P. and D.B. contributed to the writing of this editorial. All authors have read and agreed to the published version of the manuscript.

Funding: This research received no external funding.

Conflicts of Interest: The authors declare no conflict of interest.

List of Contributions:

1. Malta, J.F.; Henriques, M.S.C.; Paixão, J.A.; Gonçalves, A.P. Synthesis and Structural and Magnetic Properties of Polycrystalline $GaMo_4Se_8$. *Magnetochemistry* **2023**, *9*, 182.
2. Tajima, H.; Oda, T.; Kadoya, T. Nonthermal Equilibrium Process of Charge Carrier Extraction in Metal/Insulator/Organic Semiconductor/Metal (MIOM) Junction. *Magnetochemistry* **2023**, *9*, 180.
3. Kato, R.; Tsumuraya, T. Dirac Cone Formation in Single-Component Molecular Conductors Based on Metal Dithiolene Complexes. *Magnetochemistry* **2023**, *9*, 174.
4. Kochev, V.D.; Seidov, S.S.; Grigoriev, P.D. On the Size of Superconducting Islands on the Density-Wave Background in Organic Metals. *Magnetochemistry* **2023**, *9*, 173.
5. Oka, R.; Ohara, K.; Konishi, K.; Yamane, I.; Shimada, T.; Naito, T. Band Structure Evolution during Reversible Interconversion between Dirac and Standard Fermions in Organic Charge-Transfer Salts. *Magnetochemistry* **2023**, *9*, 153.
6. Berlie, A.; Terry, I.; Szablewski, M. Driving a Molecular Spin-Peierls System into a Short Range Ordered State through Chemical Substitution. *Magnetochemistry* **2023**, *9*, 150.
7. Zhang, B.; Zhang, Y.; Wang, Z.; Sun, Y.; Liang, T.; Liu, M.; Zhu, D. Two One-Dimensional Copper-Oxalate Frameworks with the Jahn–Teller Effect: $[(CH_3)_3NH]_2[Cu(\mu-C_2O_4)(C_2O_4)]\cdot 2.5H_2O$ (I) and $[(C_2H_5)_3NH]_2[Cu(\mu-C_2O_4)(C_2O_4)]\cdot H_2O$ (II). *Magnetochemistry* **2023**, *9*, 120.
8. Roy, A.S.; Dzikovski, B.; Dolui, D.; Makhlynets, O.; Dutta, A.; Srivastava, M. A Simulation Independent Analysis of Single- and Multi-Component cw ESR Spectra. *Magnetochemistry* **2023**, *9*, 112.
9. Wan, Q.; Wakizaka, M.; Zhang, H.; Shen, Y.; Funakoshi, N.; Che, C.-M.; Takaishi, S.; Yamashita, M. A New Organic Conductor of Tetramethyltetraselenafulvalene (TMTSF) with a Magnetic Dy(III) Complex. *Magnetochemistry* **2023**, *9*, 77.
10. Laramee, B.; Ghimire, R.; Graf, D.; Martin, L.; Blundell, T.J.; Agosta, C.C. Superconductivity and Fermi Surface Studies of β''-$(BEDT-TTF)_2[(H_2O)(NH_4)_2Cr(C_2O_4)_3]\cdot 18$-Crown-6. *Magnetochemistry* **2023**, *9*, 64.
11. Musfeldt, J.L.; López-Alcalá, Z.L.D.; Duan, Y.; Gaita-Ariño, A.; Baldoví, J.J.; Coronado, E. Vibronic Relaxation Pathways in Molecular Spin Qubit $Na_9[Ho(W_5O_{18})_2]\cdot 35H_2O$ under Pressure. *Magnetochemistry* **2023**, *9*, 53.
12. Krstovska, D.; Choi, E.S.; Steven, E. Giant Angular Nernst Effect in the Organic Metal α-$(BEDT-TTF)_2KHg(SCN)_4$. *Magnetochemistry* **2023**, *9*, 27.
13. Pesotskii, S.I.; Lyubovskii, R.B.; Shilov, G.V.; Zverev, V.N.; Torunova, S.A.; Zhilyaeva, E.I.; Canadell, E. Effect of External Pressure on the Metal–Insulator Transition of the Organic Quasi-Two-Dimensional Metal κ-$(BEDT-TTF)_2Hg(SCN)_2Br$. *Magnetochemistry* **2023**, *9*, 152.
14. Pouget, J.-P. Spin-Peierls Spin-Ladder and Kondo Coupling in Weakly Localized Quasi-1D Molecular Systems: An Overview. *Magnetochemistry* **2023**, *9*, 57.
15. Manna, F.; Oggianu, M.; Avarvari, N.; Mercuri, M.L. Lanthanide-Based Metal–Organic Frameworks with Single-Molecule Magnet Properties. *Magnetochemistry* **2023**, *9*, 190.

Disclaimer/Publisher's Note: The statements, opinions and data contained in all publications are solely those of the individual author(s) and contributor(s) and not of MDPI and/or the editor(s). MDPI and/or the editor(s) disclaim responsibility for any injury to people or property resulting from any ideas, methods, instructions or products referred to in the content.

Article

Synthesis and Structural and Magnetic Properties of Polycrystalline GaMo₄Se₈

José F. Malta [1,2,*,†], Marta S. C. Henriques [1], José A. Paixão [1] and António P. Gonçalves [2]

[1] CFisUC—Centre for Physics of the University of Coimbra, Department of Physics, University of Coimbra, 3004-516 Coimbra, Portugal
[2] C2TN—Center for Nuclear Sciences and Technologies, Department of Nuclear Sciences and Engineering, Instituto Superior Técnico, University of Lisbon, 2695-066 Bobadela, Portugal
* Correspondence: jfrmalta@gmail.com
† Current address: IT—Instituto de Telecomunicações, Instituto Superior Técnico, Torre Norte, Piso 10, Av. Rovisco Pais, 1, 1049-001 Lisbon, Portugal.

Abstract: GaMo₄Se₈, is a lacunar spinel where skyrmions have been recently reported. This compound belongs to the GaM_4X_8 family, where M is a transition metal (V or Mo) and X is a chalcogenide (S or Se). In this work, we have obtained pure GaMo₄Se₈ in polycrystalline form through an innovative two-step synthetic route. Phase purity and chemical composition were confirmed through the Rietveld refinement of the powder XRD pattern, the sample characterisation having been complemented with SEM analysis. The magnetic phase diagram was investigated using DC (VSM) and AC magnetometry, which disclosed the presence of cycloidal, skyrmionic and ferromagnetic phases in polycrystalline GaMo₄Se₈.

Keywords: skyrmions; Jahn–Teller distortion; lacunar spinel; GaMo₄Se₈

Citation: Malta, J.F.; Henriques, M.S.C.; Paixão, J.A.; Gonçalves, A.P. Synthesis and Structural and Magnetic Properties of Polycrystalline GaMo₄Se₈. *Magnetochemistry* 2022, 9, 182. https://doi.org/10.3390/magnetochemistry9070182

Academic Editor: Joan-Josep Suñol

Received: 21 May 2023
Revised: 4 July 2023
Accepted: 10 July 2023
Published: 12 July 2023

Copyright: © 2022 by the authors. Licensee MDPI, Basel, Switzerland. This article is an open access article distributed under the terms and conditions of the Creative Commons Attribution (CC BY) license (https://creativecommons.org/licenses/by/4.0/).

1. Introduction

The lacunar spinels belonging to the GaM_4X_8 family (M = V, Mo; X = S, Se), crystallise, at room temperature, in the cubic $F\bar{4}3m$ (216) space group. At low temperatures, a polar Jahn–Teller (JT) distortion occurs, which consists of a geometrical distortion of atomic positions departing from cubic into rhombohedral symmetry [1]. This distortion, seeking a lower energy state, takes place via vibronic coupling between the ground and excited states, if the coupling is sufficiently strong [2]. In this case, by the action of the crystal field, it is favourable to split the triply degenerate t_2 orbitals into two sets of orbitals, a_1 and e [3–5]. Such splitting results from the distortion of the structure along the cubic $\langle 111 \rangle$ axis, transforming the structure from $F\bar{4}3m$ to rhombohedral $R3m$. According to the idealised ionic formula, only seven electrons are available for metal–metal bonding in V lacunar spinels, but eleven are available in Mo lacunar spinels. Therefore, the three-fold degenerated highest orbital contains one electron in the V lacunar spinels and five in the Mo lacunar spinels [6].

Figure 1 shows the valence molecular orbital diagrams of both V₄ and Mo₄ metal clusters, as well as the structure distortion from the cubic $F\bar{4}3m$ to the rhombohedral $R3m$ space group. In the crystal structure, the V/Mo and S/Se atoms form a heterocubane arrangement (in red in Figure 1), while Ga-S/Se bonds group these atoms in a tetrahedral arrangement (in green in Figure 1) [6–8]. The rhombohedral angle α in the distorted phase is smaller than 60° in the V lacunar spinels and larger than 60° in the Mo lacunar spinels, which inverts the order of the a_1 and e orbitals resulting from the orbital splitting by the crystal field [6]. This means that when the a_1 orbital energy decreases by -2ε, the two e orbitals increase energy by $+\varepsilon$, and vice-versa. By occupying the molecular orbitals with seven electrons in V lacunar spinels, a stabilisation of -2ε occurs for $\alpha_{rh} < 60°$, and $-\varepsilon$

for $\alpha_{rh} > 60°$. The reverse effect is observed for Mo lacunar spinels, with 11 electrons, for which the stabilisation energy is $-\varepsilon$ for $\alpha_{rh} < 60°$ and -2ε for $\alpha_{rh} > 60°$ [6,9,10].

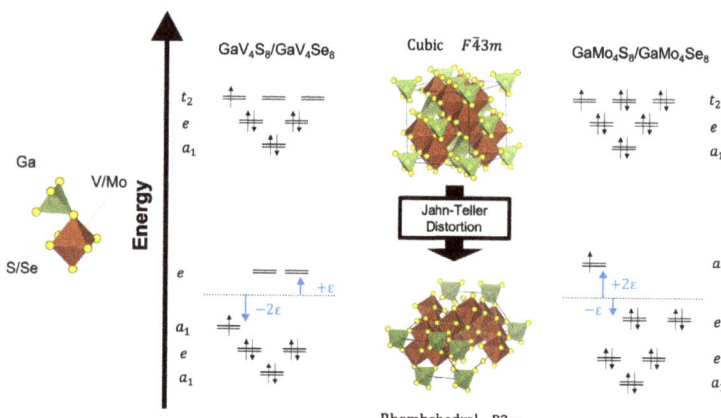

Figure 1. Effect of the Jahn–Teller transition on the molecular orbital diagrams of the V_4 and Mo_4 clusters of atoms.

In the high-temperature phase, the space group of these lacunar spinels, $F\bar{4}3m$, is non-centrosymmetric and non-polar. The V_4 and Mo_4 clusters have a T_d ($\bar{4}3m$) point symmetry that yields a zero DM interaction in the lowest order [11–13]. The JT distortion lowers the T_d ($\bar{4}3m$) point symmetry to that of the polar C_{3v} ($3m$) point group, allowing the emergence of skyrmions, as in the distorted, polar phase the Dzyaloshinskii–Moriya (DM) interaction is non-vanishing at the lowest order of the moments, i.e., the bilinear, two-spin coupling. The low-temperature polar structure can host both ferroelectricity and chiral magnetic structures, including skyrmionic magnetic phases [4]. Skyrmions are mesoscopic swirling spin textures, mostly found in chiral magnets. They are usually stabilised by competition between isotropic Heisenberg ferromagnetic exchange and DM interactions in non-centrosymmetric structures [14–16], but other mechanisms have been disclosed that may also stabilise these spin arrangements [17]. Skyrmionic compounds are attracting much interest as they may find application in high-density data storage systems. In fact, skyrmions can be manipulated via small magnetic fields and even voltages and currents and so may be used to encode binary digits that can both be read and written [18–20]. Thus, provided that they can be stabilised at room temperature, skyrmions may provide efficient digital magnetic recording, due to their small size, high stability and also ease of manipulation, as well as resistive-based memories [21]. Recently, there has been growing interest in 2D topological superconductivity in skyrmionic systems, which may also find a way into interesting applications [22].

The distorted, $R3m$ low-temperature structure of these spinels can host similar spin configurations to those of chiral magnets [23], and indeed it allows the stabilisation of Neél-type skyrmions below the magnetic ordering temperature $T_C < T_{JT}$ in these compounds. As result of the symmetry lowering from the parent cubic structure, the low-temperature rhombohedral phase is microtwinned, with four distinct $\langle 111 \rangle$ domains. By applying a magnetic field, these domains become non-equivalent if the magnetic anisotropy axes produce different angles to the applied magnetic field. As a result, a complex magnetic behaviour arises and more than one magnetic phase may be stabilised depending on the direction of the applied magnetic field [24].

GaV_4S_8 is the most well-known lacunar spinel for which the presence of Neél-type magnetic skyrmions has been confirmed by a variety of experimental techniques [25,26]. GaV_4S_8 is a Mott insulator [27], depicting semiconductor behaviour due to the strong electron–electron interaction. The Jahn–Teller transition occurs at $T_{JT} = 44$ K, and presents a complex magnetic phase diagram below $T_C = 13$ K, where cycloidal and ferromagnetic spin-ordered phases are found, and also a small region where skyrmions can be observed [28]. A similar skyrmionic phase was found in GaV_4Se_8, with an extended region of stability compared to GaV_4S_8 [24,29].

In Mo lacunar spinels, the existence of skyrmions was first suggested to occur in $GaMo_4S_8$ through theoretical studies, namely Monte-Carlo simulations [5]. The experimental evidence of skyrmions was only revealed very recently in such compounds [5,30–32]. $GaMo_4S_8$ is also a Mott insulator, with ferromagnetic and cycloidal spin ordering below $T_C = 19$ K [4,32,33]. Our previous work proved the existence of a cluster spin-glass phase close to T_C in pollycristalline $GaMo_4S_8$ [34]. In the case of $GaMo_4Se_8$, it orders at $T_C = 28$ K and the Jahn–Teller distortion occurs at T_{JT} at 51 K. Recent synchrotron powder diffraction studies on polycrystalline $GaMo_4Se_8$ disclosed that below T_{JT} the rhombohedral $R3m$ phase coexists with a metastable orthorhombic $Imm2$ phase [30,31]. Such metastable orthorhombic $Imm2$ phases can only be detected through Neutron Scattering Techniques at low temperatures, below T_{JT}, and all the related studies are very recent.

In this work, we conducted a thorough study of the $GaMo_4Se_8$ compound, covering a novel synthetic route and structural and magnetic characterisation, to investigate its magnetic properties that provide evidence for the existence of a skyrmionic magnetic phase.

2. Materials and Methods

$GaMo_4Se_8$ was synthesised in polycrystalline form by a simple two-step method, similar to that used in our previous work for $GaMo_4S_8$ [34]. In the first step, the precursor $MoSe_2$ was synthesised by reacting stoichiometric amounts of molybdenum (foil, thickness 0.1 mm, 99.9%, Sigma Aldrich, St. Louis, MO, USA) and selenium (pellets, <5 mm, 99.99%, Sigma Aldrich) with a molar proportion of 1:2 in evacuated sealed quartz tubes. The tube was placed into a vertical furnace and the temperature was slowly increased from room temperature up to 1000 °C at 36 °C/h and kept at that value for 3 days. After this period, the temperature was slowly decreased at the same rate down to room temperature. The second step consisted of reacting the obtained precursor with pure amounts of Ga (99.999%, Sigma Aldrich). Then, the obtained $MoSe_2$ and pure amounts of Ga were both sealed in evacuated quartz tubes under 0.2 atm of argon in a molar proportion of 4:1. These were placed in the furnace for a heating cycle with temperature increasing and decreasing rates identical to those of the first step, but the temperature was held at 1000 °C for one week.

3. Results and Discussion

3.1. Characterisation

To check the purity of the obtained sample, powder X-ray diffraction data were measured on a Bruker AXS D8 Advance diffractometer equipped with a Cu tube (Kα = 1.5418 Å) in Bragg–Brentano geometry. The data were obtained from finely ground powder deposited on a low-background, monocrystalline and off-cut Si sample holder. The presence of pure $GaMo_4Se_8$ without extraneous phases was confirmed by a search/match procedure against the PDF4+ ICDD database, further corroborated by a Rietveld refinement using the Profex software [35]. During refinement, only cell parameters and those related to the assumed Lorentzian size distribution of grain size were allowed to vary, atomic positions and thermal parameters were fixed at the published values obtained from single-crystal XRD, with the stoichiometry kept at nominal values for $GaMo_4Se_8$. Figure 2 shows the measured and calculated X-ray diffractograms. No residual unindexed peaks remained in the difference pattern, and the refinement converged to the final quality factors $R_{wp} = 7.98\%$, $R_{exp} = 5.67\%$, $\chi^2 = 1.96$ and GOF = 1.40. The mean crystallite size refined to the large value of 303(4) nm, close to the upper resolution limit of the diffractometer, showing that the sample was well crystallised. Table 1 shows the crystallographic data from the Rietveld refinement.

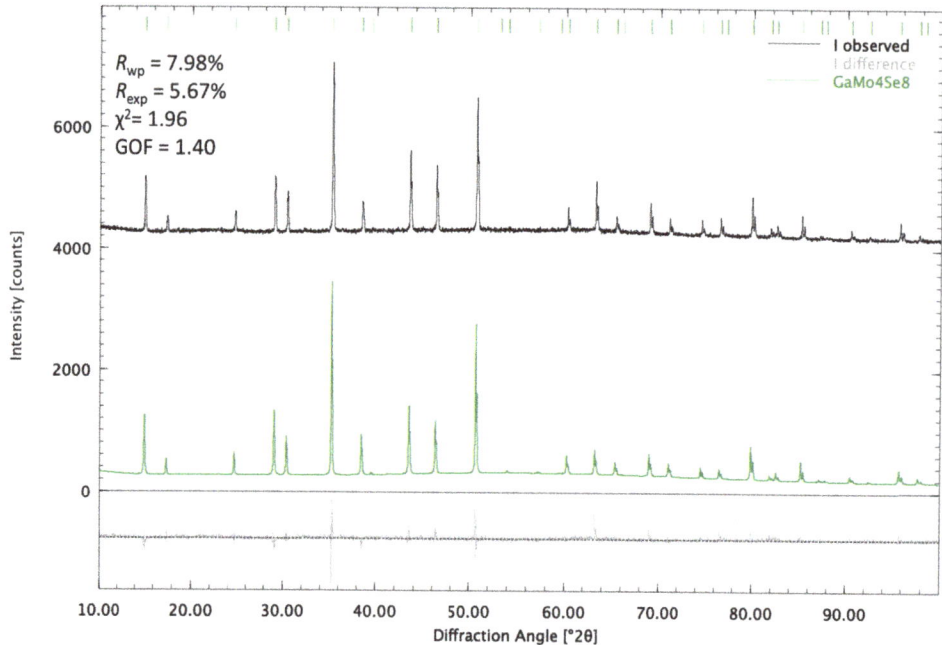

Figure 2. XRD pattern for polycrystalline GaMo$_4$Se$_8$.

Table 1. Crystallographic data (from the Rietveld refinement of powder X-ray data) for GaMo$_4$Se$_8$. Space group $F\bar{4}3m$, $a = 10.1770(1)$ Å. The $4a$ site occupancy for Ga was fixed to the nominal value during refinement.

Atomic Positions	x	y	z	Occupation
Ga (4a)	0	0	0	1
Mo (16e)	0.39925	0.39925	0.39925	0.928(8)
Se1 (16e)	0.63624	0.63624	0.63624	1.00(2)
Se2 (16e)	0.13655	0.13655	0.13655	0.95(2)

The presented data prove that GaMo$_4$Se$_8$ was obtained with no extraneous phases using the present two-step synthetic method, contrasting with other synthetic routes described in the literature where small amounts of the impurity Mo$_3$Ga were always detected [30,31]. A satisfactory Rietveld refinement was obtained (R_{wp} < 10%), and the residual differences between the observed and calculated intensities occurring in the strongest peaks can be ascribed to texture effects and the large particle size in the obtained powder. The Mo$_3$Ga impurity tends to be formed when we react directly stoichiometric amounts of Ga, Mo and Se by the conventional flux method [30,31,36]. Thus, our synthetic route consisting of reacting Ga with the intermediate MoSe$_2$ by avoiding the formation of such impurities affords higher-purity samples of molybdenum lacunar spinels. To investigate in detail the grain size distribution, the morphology of the obtained GaMo$_4$Se$_8$ sample was examined by Scanning Electron Microscopy (SEM) and the composition by energy-dispersive X-ray spectroscopy (EDS). The images were obtained with a 5 keV electron beam using a detector of secondary electrons at a working distance of 8 mm (Figure 3a). The obtained EDS spectra and the average composition are depicted in Figure 3b and Table 2, respectively.

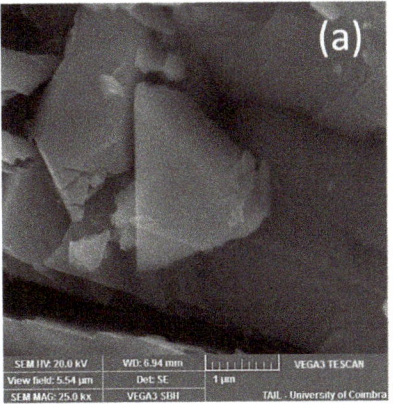

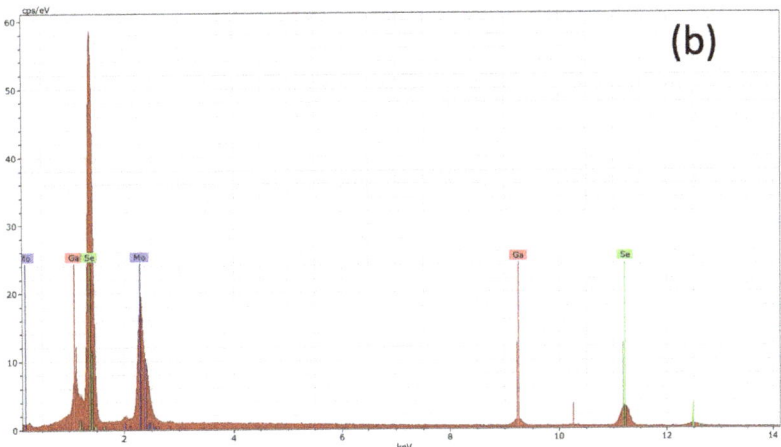

Figure 3. (a) SEM image and (b) EDS spectra for polycristalline GaMo$_4$Se$_8$.

Table 2. Average composition of GaMo$_4$Se$_8$ determined by EDS.

Compound	Ga	Mo	Se
GaMo$_4$Se$_8$	0.974 ± 0.005	4.2 ± 0.2	7.7 ± 0.2

The examined polycrystalline samples had a grain size in the range 300 nm–1 μm, confirming the large value already found in the Rietveld refinement of the powder XRD data. Only Ga, Mo and Se elements can be detected in the EDS spectra. The average composition, obtained from the EDS analysis, is close to that expected for GaMo$_4$Se$_8$. No further purification or heat treatment were applied to this sample, which was used as-synthesized for the subsequent magnetisation studies.

3.2. Magnetisation Studies

The thermomagnetic $M(T)$ measurements were performed on a Quantum Design Dynacool PPMS system equipped with a VSM option. The sample, weighing 3.8 mg, was packed in a clean teflon sample holder. The ZFC (zero-field-cooled) and FC (field-cooled) thermomagnetic curves, measured under an applied magnetic field of 500 Oe, are depicted in Figure 4a, where the anomaly related with T_C can be observed at 27.5 K. A plot of the inverse magnetic susceptibility of the ZFC curve, $\chi^{-1} = H/M$, as a function of temperature is seen in Figure 4b, showing a clear signature of the Jahn–Teller transition, T_{JT}, occuring at 51 K.

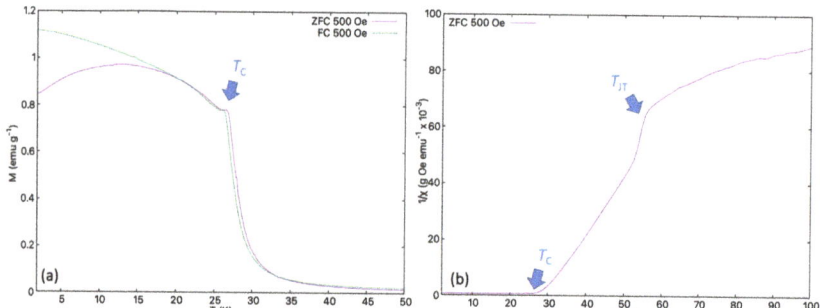

Figure 4. (a) Magnetisation $M(T)$ curves in ZFC (zero-field cooling) and FC (field cooling) for polycrystalline GaMo$_4$Se$_8$ measured with $H = 500$ Oe and (b) inverse magnetic susceptibility $\chi^{-1} = H/M$ curves calculated from the ZFC measurement for polycrystalline GaMo$_4$Se$_8$, measured with an applied magnetic field of 500 Oe. The arrows point to the magnetic ordering temperature, $T_C = 27.5$ K and the Jahn–Teller transition, $T_{JT} = 51$ K.

The skyrmionic magnetic phase found in these type of compounds occurs close to T_C, stabilised under the action of an applied magnetic field. It is known that a signature of the skyrmion phase in GaMo$_4$Se$_8$ can be found in $M(H)$ measurements, better depicted as a change in the derivative of the $M(H)$ curve when the phase boundary of the region where skyrmions exist is crossed. Thus, and in order to unveil the presence of the skyrmionic phase, accurate $M(H)$ measurements with a small H step of 5 Oe and long VSM integration times were performed, covering the temperature between 2 K and 28 K, where skyrmions are expected to exist in GaMo$_4$Se$_8$.

The obtained $M(H)$ curves are presented in detail in Figure 5a. The derivative of the data, $dM(H)/dH$, was calculated by a standard centred three-point numerical method followed by a smoothing filter of the type `acsplines` to reduce numerical noise [37] and it is presented in Figure 5b.

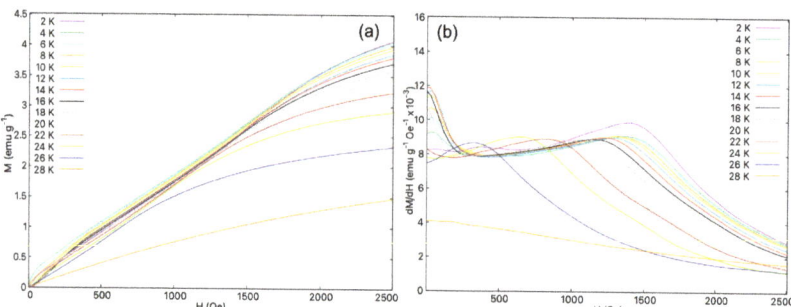

Figure 5. (a) $M(H)$ and (b) dM/dH curves obtained for GaMo$_4$Se$_8$.

From the extensive set of $M(H)$ curves covering the temperature range between 2 and 28 K, it is possible to plot the magnetic phase diagram based on the $dM(H)/dH$ curves. The anomalies representing the transition between cycloidal, skyrmionic and ferromagnetic phases can be detected in such curves. For a better representation of the magnetic phase diagram of GaMo$_4$Se$_8$, the dM/dH data are represented in a coloured pseudo 3-D plot, shown in Figure 6. This plot was produced using the `pm3d` interpolation algorithm implemented in `gnuplot` [37].

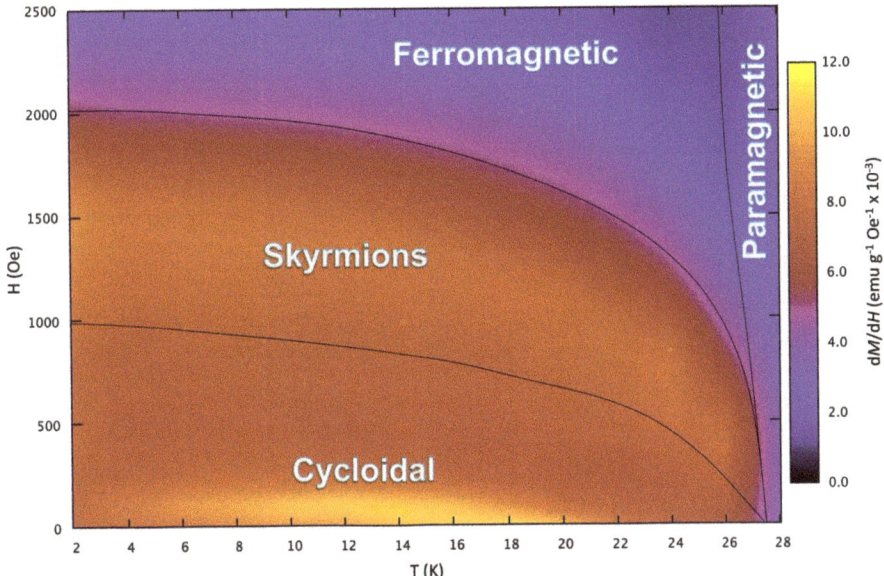

Figure 6. Magnetic phase diagram obtained from the numerative dM/dH curves for $GaMo_4Se_8$.

On the magnetic phase diagram plot in Figure 6, the phase boundaries determined from the more pronounced dM/dH anomalies of the curves in Figure 5 were drawn as guides to the eye, showing in the range of temperatures from 2 K up to T_C the presence of cycloidal, skyrmionic and ferromagnetic phases.

3.3. AC Susceptibility Measurements

To complement the data obtained from the $dM(H)/dH$ curves, AC susceptibility measurements were performed using the ACMS II option of the PPMS system. The data were collected on a small pellet of $GaMo_4Se_8$ (obtained by pressing 10 mg of fine powder) that was glued with a thin layer of GE cryogenic varnish onto a low-background quartz sample holder. The measurements were performed with an AC magnetic field of 5 Oe amplitude in the temperature range 2–28 K, as a function of the applied DC magnetic field up to 2500 Oe. The real (χ') and imaginary (χ'') AC susceptibility components as a function of magnetic field are shown in Figure 7a,b, respectively.

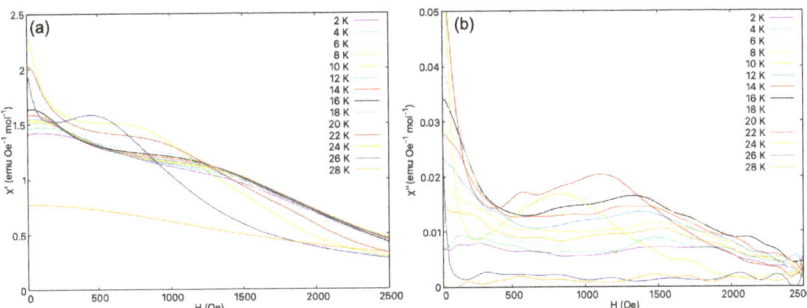

Figure 7. (a) χ' and (b) χ'' components of the AC magnetic susceptibility as function of the DC magnetic field, up to 2500 Oe, for temperatures between 2 K and 28 K. H_{AC} = 5 Oe, f = 1000 Hz.

Both the χ' and χ'' components of the AC susceptibility curves are similar to the obtained dM/dH curves presented in Figure 5. The anomalies related to the magnetic phase transitions can be observed in such curves. The results of the measured curves are depicted in graphical form in Figure 8, obtained from the χ' component of AC susceptibility, with the different magnetic phases represented, using the pm3d interpolation algorithm implemented in gnuplot [37]. The boundaries were determined from the χ'' component, superimposed on the colour map representing the values of $\chi'(H, T)$.

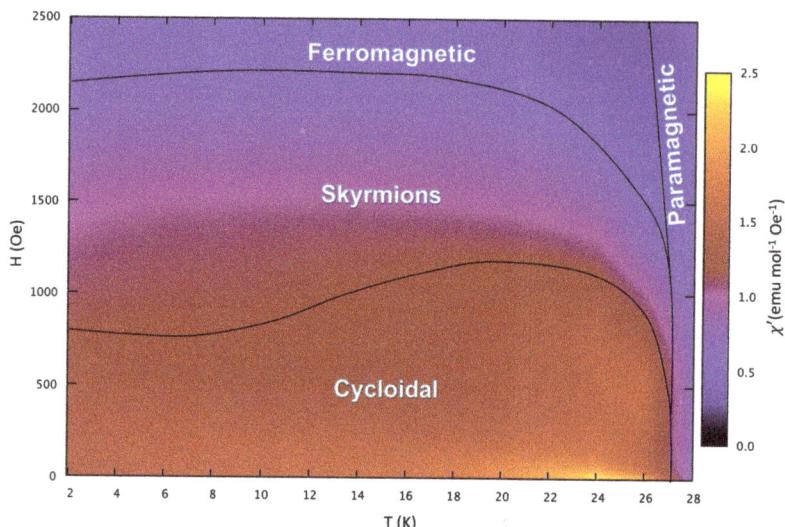

Figure 8. Magnetic phase diagram obtained from the χ' component of AC susceptibility measurements for GaMo$_4$Se$_8$.

Similar to the dM/dH map, presented in Figure 6, the magnetic phases can be depicted in the (H, T) map of the AC susceptibility χ' component and the phase boundaries determined from the χ'' component. From both the VSM and AC susceptibility measurements, the skyrmion phase was found to be present in a large temperature range, from to 2 K up to 27.5 K, where the ferromagnetic/paramagnetic transition occurs. From these data, we can conclude that in GaMo$_4$Se$_8$ the cycloidal and skyrmion regions span more extended temperature regions compared to those observed in other lacunar spinels.

4. Conclusions

In this work, we have accomplished the synthesis of high-purity GaMo$_4$Se$_8$ though a two-step chemical route similar to that used for the synthesis of GaMo$_4$S$_8$. From both VSM $M(H)$ and AC susceptibility measurements, we have shown that the skyrmionic magnetic phase is stabilised under an applied magnetic field in a large temperature range, from 2 K up to T_C, in a polycrystalline sample. These measurements do not show any anomaly that could be associated with the onset of a metastable orthorhombic phase, which has been suggested by synchrotron radiation studies to co-exist with the main, low-temperature, rhombohedral phase. This point, and the details of the magnetic structures and their evolution with the applied magnetic field, deserve to be further investigated by microscopic techniques such as neutron scattering.

Author Contributions: J.F.M.: synthesis, characterisation, physical property measurements, analysis of the results, investigation and writing. M.S.C.H.: physical property measurements, investigation and writing. J.A.P.: physical property measurements, analysis of the results, investigation, writing and funding. A.P.G.: investigation and writing. All authors have read and agreed to the published version of the manuscript.

Funding: José F. Malta's PhD grant was supported by FCT-Fundação para a Ciência e a Tecnologia through the ChemMat PhD programme. Access to the TAIL-UC facility supported by the QREN-Mais Centro programme ICT_2009_02_012_1890 is gratefully acknowledged. This work was partially supported by funds from FEDER (Programa Operacional Factores de Competitividade COMPETE) and from FCT under the projects UIDB/FIS/04564/2020, UIDP/FIS/ 04564/2020 and PTDC/FIS-MAC/32229/2017.

Institutional Review Board Statement: Not applicable.

Informed Consent Statement: Not applicable.

Data Availability Statement: Not applicable.

Conflicts of Interest: The authors declare no conflict of interest.

References

1. Reinen, D. The Jahn-Teller Effect in Solid State Chemistry of Transition Metal Compounds. *J. Solid State Chem.* **1979**, *27*, 71–85. [CrossRef]
2. Barone, P.; Yamauchi, K.; Picozzi, S. Jahn-Teller distortions as a novel source of multiferroicity. *Phys. Rev. B* **2015**, *92*, 014116. [CrossRef]
3. Nikolaev, S.A.; Solovyev, I.V. Skyrmionic order and magnetically induced polarization change in lacunar spinel compounds GaV$_4$S$_8$ and GaMo$_4$S$_8$: A comparative theoretical study. *Phys. Rev. B* **2020**, *102*, 014414. [CrossRef]
4. Wang, Y.; Puggioni, D.; Rondinelli, J.M. Assessing exchange-correlation functional performance in the chalcogenide lacunar spinels GaM$_4$Q$_8$ (M = Mo, V, Nb, Q = S, Se). *Phys. Rev. B* **2019**, *100*, 115149. [CrossRef]
5. Zhang, H.M.; Chen, J.; Barone, P.; Yamauchi, K.; Dong, S.; Picozzi, S. Possible emergence of a skyrmion phase in ferroelectric GaMo$_4$S$_8$. *Phys. Rev. B* **2019**, *99*, 214427. [CrossRef]
6. Pocha, R.; Johrendt, D.; Pottgen, R. Electronic and Structural Instabilities in GaV$_4$S$_8$ and GaMo$_4$S$_8$. *Chem. Mater.* **2000**, *12*, 2882–2887. [CrossRef]
7. Powell, A.V.; McDowall, A.; Szkoda, I.; S. Knight, K.; Kennedy, B.J.; Vogt, T. Cation Substitution in Defect Thiospinels: Structural and Magnetic Properties of GaV$_4$xMoxS$_8$. *Chem. Mater.* **2007**, *19*, 5035–5044. [CrossRef]
8. Bichler, D.; Slavik, H.; Johrendt, D. Low Temperature Crystal Structures and Magnetic Properties of the V4 Cluster Compounds. *Z. Naturforsch.* **2009**, *64*, 915–921. [CrossRef]
9. Geirhos, K.; Krohns, S.; Nakamura, H.; Waki, T.; Tabata, Y.; Kézsmárki, I.; Lunkenheimer, P. Orbital-order driven ferroelectricity and dipolar relaxation dynamics in multiferroic GaMo$_4$S$_8$. *Phys. Rev. B* **2018**, *98*, 224306. [CrossRef]
10. Kim, H.S.; Im, J.; Han, M.J.; Jin, H. Spin-orbital entangled molecular j_{eff} states in lacunar spinel compounds. *Nat. Commun.* **2014**, *5*, 3988. [CrossRef]
11. Ado, I.A.; Qaiumzadeh, A.; Brataas, A.; Titov, M. Chiral ferromagnetism beyond Lifshitz invariants. *Phys. Rev. B* **2020**, *101*, 161403. [CrossRef]
12. Ado, I.A.; Tchernyshyov, O.; Titov, M. Noncollinear Ground State from a Four-Spin Chiral Exchange in a Tetrahedral Magnet. *Phys. Rev. Lett.* **2021**, *127*, 127204. [CrossRef]
13. Petersen, T.; Prodan, D.; Geirhos, K.; Nakamura, H.; Kézsmárki, I.; Hozoi, L. Dressed j_{eff}-1/2 objects in mixed-valence lacunar spinel molybdates. *Sci. Rep.* **2023**, *13*, 2411. [CrossRef]
14. Cheong, S.W.; Mostovoy, M. Multiferroics: A magnetic twist for ferroelectricity. *Nat. Mater.* **2007**, *6*, 13–20. [CrossRef] [PubMed]
15. Versteeg, R.B.; Vergara, I.; Schäfer, S.D.; Bischoff, D.; Aqeel, A.; Palstra, T.T.M. Grüninger, M.; Van Loosdrecht, P.H. Optically probed symmetry breaking in the chiral magnet Cu$_2$OSeO$_3$. *Phys. Rev. B* **2016**, *94*, 094409. [CrossRef]
16. Freimuth, F.; Bamler, R.; Mokrousov, Y.; Rosch, A. Phase-space Berry phases in chiral magnets: Dzyaloshinskii-Moriya interaction and the charge of skyrmions. *Phys. Rev. B* **2013**, *88*, 214409. [CrossRef]
17. Khanh, N.D.; Nakajima, T.; Yu, X.; Gao, S.; Shibata, K.; Hirschberger, M.; Yamasaki, Y.; Sagayama, H.; Nakao, H.; Peng, L.; et al. Nanometric square skyrmion lattice in a centrosymmetric tetragonal magnet. *Nat. Nanotechnol.* **2020**, *15*, 444–449. [CrossRef]
18. Romming, N.; Hanneken, C.; Menzel, M.; Bickel, J.E.; Wolter, B.; von Bergmann, K.; Kubetzka, A.; Wiesendanger, R. Writing and Deleting Single Magnetic Skyrmions. *Science* **2013**, *341*, 636–639. [CrossRef] [PubMed]
19. Hagemeister, J.; Romming, N.; von Bergmann, K.; Vedmedenko, E.Y.; Wiesendanger, R. Stability of single skyrmionic bits. *Nat. Commun.* **2015**, *6*, 8455. [CrossRef]
20. Finocchio, G.; Büttner, F.; Tomasello, R.; Carpentieri, M.; Kläui, M. Magnetic skyrmions: From fundamental to applications. *J. Phys. D Appl. Phys.* **2016**, *49*, 423001. [CrossRef]

21. Büttner, F.; Lemesh, I.; Beach, G.S.D. Theory of isolated magnetic skyrmions: From fundamentals to room temperature applications. *Sci. Rep.* **2018**, *8*, 4464. [CrossRef]
22. Mascot, E.; Bedow, J.; Graham, M.; Rachel, S.; Morr, D.K. Topological superconductivity in skyrmion lattices. *npj Quantum Mater.* **2021**, *6*, 6. [CrossRef]
23. Tokura, Y.; Kanazawa, N. Magnetic Skyrmion Materials. *Chem. Rev.* **2020**, *121*, 2857–2897. [CrossRef]
24. Bordács, S.; Butykai, A.; Szigeti, B.G.; White, J.S.; Cubitt, R.; Leonov, A.O.; Widmann, S.; Ehlers, D.; von Nidda, H.A.K.; Tsurkan, V.; et al. Equilibrium Skyrmion Lattice Ground State in a Polar Easy-plane Magnet. *Sci. Rep.* **2017**, *7*, 7584. [CrossRef]
25. Ruff, E.; Widmann, S.; Lunkenheimer, P.; Tsurkan, V.; Bordács, S.; Kézsmárki, I.; Loidl, A. Multiferroicity and skyrmions carrying electric polarization in GaV_4S_8. *Sci. Adv.* **2015**, *1*, e150091. [CrossRef] [PubMed]
26. Butykai, Á.; Bordács, S.; Kézsmárki, I.; Tsurkan, V.; Loidl, A.; Döring, J.; Neuber, E.; Milde, P.; Kehr, S.C.; Eng, L.M. Characteristics of ferroelectric-ferroelastic domains in Néel-type skyrmion host GaV_4S_8. *Sci. Rep.* **2017**, *7*, 44663. [CrossRef]
27. Yadava, C.; Nigamb, A.K.; Rastogia, A.K. Thermodynamic properties of ferromagnetic Mott-insulator GaV_4S_8. *Phys. B Condens. Matter* **2008**, *403*, 1474–1475. [CrossRef]
28. Kézsmárki, I.; Bordács, S.; Milde, P.; Neuber, E.; Eng, L.M.; White, J.S.; Rønnow, H.M.; Dewhurst, C.D.; Mochizuki, M.; Yanai, K.; et al. Néel-type skyrmion lattice with confined orientation in the polar magnetic semiconductor GaV_4S_8. *Nat. Mater.* **2015**, *14*, 1116–1122. [CrossRef] [PubMed]
29. Gross, B.; Philipp, S.; Geirhos, K.; Mehlin, A.; Bordács, S.; Tsurkan, V.; Leonov, A.; Kézsmárki, I.; Poggio, M. Stability of Néel-type skyrmion lattice against oblique magnetic fields in GaV_4S_8 and GaV_4Se_8. *Phys. Rev. B* **2020**, *102*, 104407. [CrossRef]
30. Schueller, E.C.; Kitchaev, D.A.; Zuo, J.L.; Bocarsly, J.D.; Cooley, J.A.; der Ven, A.V.; Wilson, S.D.; Seshadri, R. Structural evolution and skyrmionic phase diagram of the lacunar spinel $GaMo_4Se_8$. *Phys. Rev. Mater.* **2020**, *4*, 064402. [CrossRef]
31. Routledge, K.; Vir, P.; Cook, N.; Murgatroyd, P.A.E.; Ahmed, S.J.; Savvin, S.N.; Claridge, J.B.; Alaria, J. Mode Crystallography Analysis through the Structural Phase Transition and Magnetic Critical Behavior of the Lacunar Spinel $GaMo_4Se_8$. *Chem. Mater.* **2021**, *33*, 5718–5729. [CrossRef] [PubMed]
32. Butykai, Á.; Geirhos, K.; Szaller, D.; Kiss, L.; Balogh, L.; Azhar, M.; Garst, M.; DeBeer-Schmitt, L.; Waki, T.; Tabata, Y.; et al. Squeezing the periodicity of Néel-type magnetic modulations by enhanced Dzyaloshinskii-Moriya interaction of 4d electrons. *npj Quantum Mater.* **2022**, *7*, 26. [CrossRef]
33. Rastogi, A.K.; Berton, A.; Chaussy, J.; Tournier, R. Itinerant Electron Magnetism in the Mo_4 Tetrahedral Cluster Compounds $GaMo_4S_8$, $GaMo_4Se_8$, and $GaMo_4Se_4Te_4$. *J. Low Temp. Phys.* **1983**, *52*, 539–557. [CrossRef]
34. Malta, J.F.; C. Henriques, M.S.; Paixão, J.A.; P. Gonçalves, A. Evidence of a cluster spin-glass phase in the skyrmion-hosting $GaMo_4S_8$ compound. *J. Mater. Chem. C* **2022**, *10*, 12043–12053. [CrossRef]
35. Doebelin, N.; Kleeberg, R. Profex: A graphical user interface for the Rietveld refinement program BGMN. *J. Appl. Crystallogr.* **2015**, *48*, 1573–1580. [CrossRef] [PubMed]
36. Querré, M.; Corraze, B.; Janod, E.; Besland, M.P.; Tranchant, J.; Potel, M.; Cordier, S.; Bouquet, V.; Guilloux-Viry, M.; Cario, L. Electric pulse induced resistive switching in the narrow gap Mott insulator $GaMo_4S_8$. *Key Eng. Mater.* **2014**, *617*, 135–140. [CrossRef]
37. Gnuplot 5.2: An Interactive Plotting Program. 2019. Available online: http://gnuplot.sourceforge.net/ (accessed on 14 March 2023).

Disclaimer/Publisher's Note: The statements, opinions and data contained in all publications are solely those of the individual author(s) and contributor(s) and not of MDPI and/or the editor(s). MDPI and/or the editor(s) disclaim responsibility for any injury to people or property resulting from any ideas, methods, instructions or products referred to in the content.

Communication

Nonthermal Equilibrium Process of Charge Carrier Extraction in Metal/Insulator/Organic Semiconductor/Metal (MIOM) Junction

Hiroyuki Tajima [1,*], Takeshi Oda [1] and Tomofumi Kadoya [2]

1. Graduate School of Science, University of Hyogo, 3-2-1 Kohto, Kamigori-cho 678-1297, Japan
2. Department of Chemistry, Konan University, 8-9-1 Okamoto, Higashinada, Kobe 658-8501, Japan; tkadoya@konan-u.ac.jp
* Correspondence: tajima@sci.u-hyogo.ac.jp

Abstract: This paper presents the concept and experimental evidence for the nonthermal equilibrium (NTE) process of charge carrier extraction in metal/insulator/organic semiconductor/metal (MIOM) capacitors. These capacitors are structurally similar to metal/insulator/semiconductor/(metal) (MIS) capacitors found in standard semiconductor textbooks. The difference between the two capacitors is that the (organic) semiconductor/metal contacts in the MIOM capacitors are of the Schottky type, whereas the contacts in the MIS capacitors are of the ohmic type. Moreover, the mobilities of most organic semiconductors are significantly lower than those of inorganic semiconductors. As the MIOM structure is identical to the electrode portion of an organic field-effect transistor (OFET) with top-contact and bottom-gate electrodes, the hysteretic behavior of the OFET transfer characteristics can be deduced from the NTE phenomenon observed in MIOM capacitors.

Keywords: hysteresis; bias stress effect; organic semiconductors; organic field-effect transistors; metal/insulator/organic semiconductor/metal; non-thermal equilibrium process

Citation: Tajima, H.; Oda, T.; Kadoya, T. Nonthermal Equilibrium Process of Charge Carrier Extraction in Metal/Insulator/Organic Semiconductor/Metal (MIOM) Junction. *Magnetochemistry* 2023, 9, 180. https://doi.org/10.3390/magnetochemistry9070180

Academic Editors: Laura C. J. Pereira and Dulce Belo

Received: 28 May 2023
Revised: 28 June 2023
Accepted: 7 July 2023
Published: 11 July 2023

Copyright: © 2023 by the authors. Licensee MDPI, Basel, Switzerland. This article is an open access article distributed under the terms and conditions of the Creative Commons Attribution (CC BY) license (https://creativecommons.org/licenses/by/4.0/).

1. Introduction

Organic semiconductors (OSs) exhibit very low conductivities without chemical doping; however, with the injection of dopants they exhibit high conductivities. This characteristic has been employed in the fabrication of organic field-effect transistors (OFETs), which has been extensively studied in recent years. These devices are used not only for practical applications but also for the determination of mobility in OSs [1–3]. Figure 1a shows the typical structure of a top-contact bottom-gate-type OFET in which the source (S)/drain (D) electrodes are formed on an OS film fabricated on a film of insulator (INS) and gate (G) electrodes. Figure 1b shows a capacitor composed of metal 1 (M1)/INS/OS/metal 2 (M2). The metal/insulator/organic semiconductor/metal (MIOM) capacitor is identical to the electrode portion of the OFET.

MIOM capacitors are equivalent to metal/INS/semiconductor/(metal) (MIS) capacitors found in standard textbooks [4] when OS/M2 contacts form an ohmic junction. However, when the OS/M2 contact forms a Schottky junction, charge carrier extraction based on nonthermal equilibrium (NTE) processes frequently occurs. This behavior is significantly different from that of the MIS capacitors. This NTE charge extraction process can be one of the origins of the hysteresis behavior in the transfer characteristics of OFETs [5–11]. However, detailed experiments on MIOM capacitors have not been conducted yet. In this study, we theoretically and experimentally describe a model of the NTE process of charge extraction [12,13]. This model was derived from several studies based on accumulated charge measurements (ACM) [12–19] to determine the electron and hole injection barriers in OS films. The ACM allowed us to estimate the potential distribution in the OS based on experimental data. The obtained values of the injection barriers were approximately consistent with those obtained using ultra violet photoelectron spectroscopy [20,21], photoemission yield spectroscopy [22,23], inverse photoelectron spectroscopy [24], and low-energy inverse photoelectron spectroscopy [25]. Therefore, we believe that the validity of the model has been confirmed in previous studies on ACM.

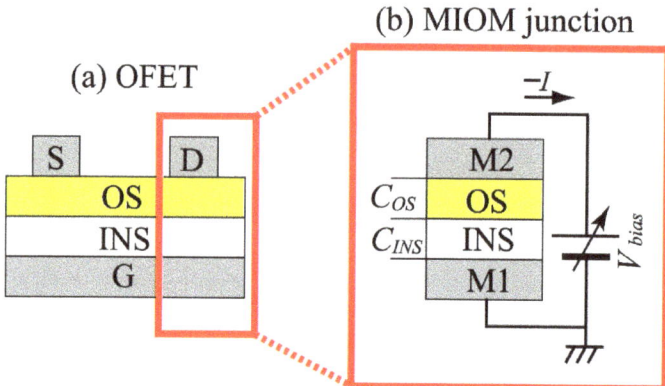

Figure 1. (**a**) Structure of an OFET. Source (S), drain (D), and gate (G) electrodes. Layers of an OS and INS. (**b**) Structure of an MIOM capacitor. M1 and M2 are metal electrodes. The components of the two electrodes are not always the same. Note that the MIOM capacitor is identical to the electrode portion of the OFET. The panel (**b**) was reproduced with permission [13]. Copyright © 2021, AIP Publishing.

2. Concept of NTE Process of Charge Extraction

Figure 2 shows the calculations of potential distribution in the OS layer when a bias voltage, V_{bias}, is applied to an MIOM capacitor, with a hole injection barrier of 0.4 eV for the Schottky junction between the OS and M2. These calculations assume that there are no charge-capturing dopants in the OSs. This assumption is considered reasonable for OSs in which doping is difficult. Figure 2a–d show the potential diagrams of the thermal equilibrium (TE) state obtained by solving the Poisson–Boltzmann equation [13]. As there is a Schottky barrier at the OS/M2 interface, no charge is injected when V_{bias} is low (Figure 2b). The device operates as a capacitor with a series capacitance (C_0) of the OS layer (C_{OS}) and INS layer (C_{INS}), that is, $C_0 = C_{INS}C_{OS}/(C_{INS} + C_{OS})$. At this stage, the total accumulated charge Q_{total} is the same as the interface charge Q_0 between the OS and M2, that is, $Q_{total} = Q_0 = C_0 V_{bias}$. With an increase in V_{bias}, the gap between the Fermi level of the M2 electrode and the highest occupied molecular orbital (HOMO) near the INS decreased and holes were injected into the OS layer (Figure 2c,d). Consequently, the OS layer near the INS layer exhibited band bending owing to the accumulated holes. The degree of band bending increased as the amount of hole injection increased. However, this band bending is limited to the vicinity of the INS/OS interface, and the approximate shape of the potential curve in the OS layer is dominated by Q_0 at the OS/M2 interface. The HOMO level of the OS at the OS/M2 interface was pinned by the Fermi level of M2 and did not exceed this level. Thus, once the holes were injected, most of the voltage applied to the capacitor decreased within the insulator layer.

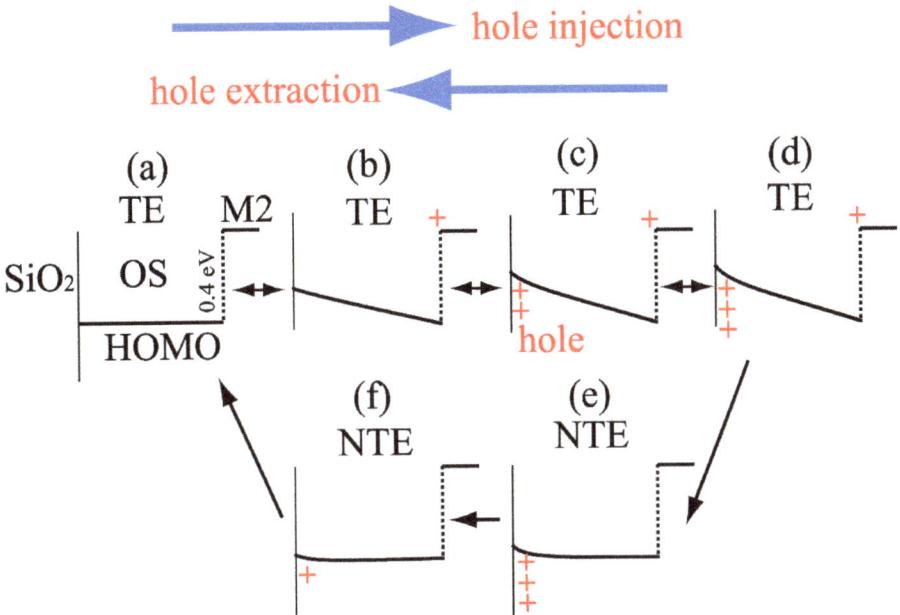

Figure 2. Schematic potential curve to explain the hole injection and extraction processes in the MIOM junction, where a Schottky contact of 0.4 eV is formed between the HOMO of the OS and M2 electrode. The layers of M1 and INS (SiO$_2$) and the potential curve of LUMO are omitted for simplicity in the figures. A flat band state is shown in (**a**). A positive bias voltage is applied to the M2 electrode. The potential curves (**a**–**d**) are obtained for the TE state. The curves (**e**,**f**) are realized in the NTE state when the charges at the OS/M2 interface are extracted before the holes near the INS/OS boundary are extracted.

Furthermore, we considered the charge extraction from this state by decreasing V_{bias}. If charge extraction occurs in the TE, it is the reverse of the charge injection case, with charge extraction at the INS/OS interface and then at the OS/M2 interface. However, as the electric field applied to the sample is still directed from the M2 electrode to the INS side, whether this TE reverse process occurs depends on carrier diffusion. If carrier diffusion is unlikely, the electric field in the OS, which is proportional to Q_0, decreases rapidly before the charge is extracted from the INS/OS interface. This is the case for the NTE model [13]. When Q_0 is slightly reversed, the charge at the INS/OS interface immediately begins to move because of the influence of the electric field. Therefore, the holes near the INS/OS interface are continuously extracted, whereas Q_0 remains near zero, as shown in Figure 2e,f. Consequently, the initial state is achieved. In the NTE model, the potential in the NTE state is semiempirically reproduced based on the potential calculation in the TE state, considering the series of processes described above. States (e) and (f) are not in the TE but are kept metastable by the electric field. If an ohmic junction is formed at the OS/M2 interface, no such metastable state occurs, because the electric field near the OS/M2 interface is negligible. In other words, the formation of such a metastable state is unique to MIOM capacitors with Schottky junctions at the OS/M2 interface.

In the following section, experimental evidence supporting this model is presented.

3. Experimental Evidences of NTE Charge Extraction

3.1. Displacement Current Measurement

The simplest experimental technique for detecting the NTE process during charge extraction is the displacement current measurement. Figure 3a shows the experimental setup. In this experiment, holes (or electrons) were injected through the Schottky barrier at the OS/M2 interface by applying a positive (or negative) V_{bias} at V_{off} for a long period. The injected holes (or electrons) were extracted using voltage sweeps.

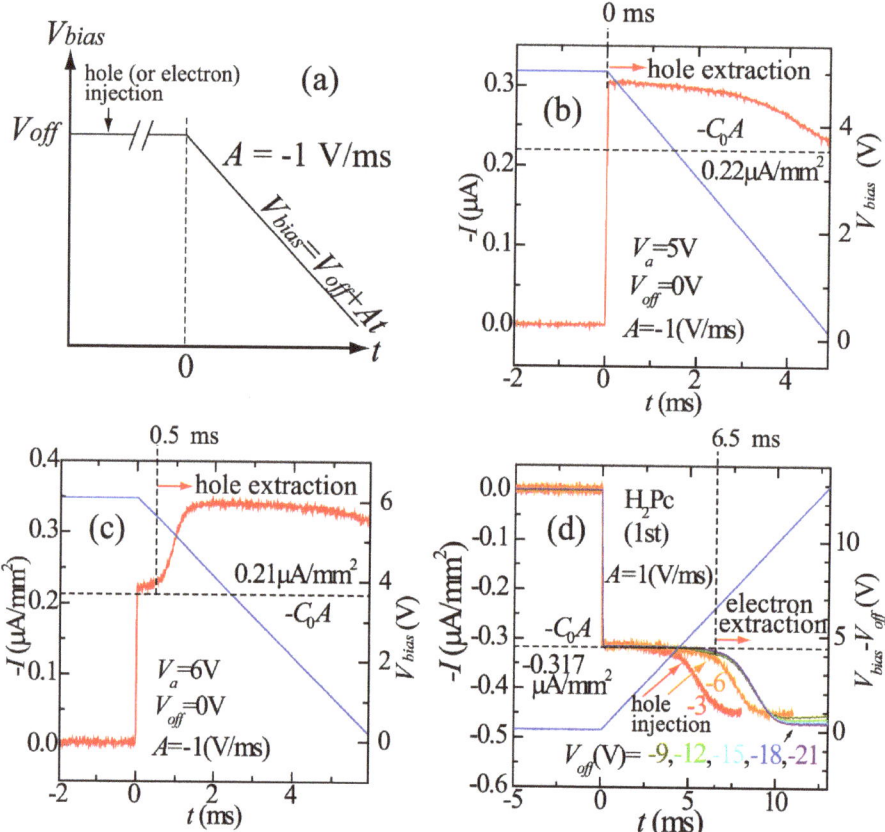

Figure 3. Displacement current (I) measurements as a function of time (t) to investigate the NTE process in an MIOM capacitor. (**a**) Schematic of the applied bias voltage (V_{bias}) pattern. V_{bias} is maintained at V_{off} for a long time to inject holes (or electrons) into the OS and then decreased at a rate of A above $t = 0$ ms. (**b**) Measurement for n-Si/SiO$_2$ (100 nm)/pentacene (50 nm)/Au. The value of I at $t = 0$ ms exceeds $-C_0A$, suggesting that TE-type hole extraction occurs in this sample. (**c**) Measurement for n-Si/SiO$_2$ (100 nm)/H$_2$Pc (52 nm)/Au. The value of I below $t = 0.5$ ms approximately coincides with $-C_0A$, suggesting that NTE-type hole extraction occurs in this sample. (**d**) Measurement for p-Si/SiO$_2$(70 nm)/H$_2$Pc (44 nm)/Au. For $V_{off} = -3$ and -6 V, electron injection does not occur and the changes in I around $t = 3.8$ ms ($V_{off} = -3$ V) and 5.7 ms ($V_{off} = -6$ V) are due to the injection of holes. For $V_{off} = -9, -12, -15, -18$, and -21 V, electron injection occurs. The changes in I around $t = 6.5$ ms are due to the extraction of electrons. Hole injection subsequently occurs after the extraction. (See the text and Figure 4 for details.). Panels (**b**,**c**) were reproduced and modified with permissions [13]. Copyright © 2021, AIP Publishing. Panel (**d**) was reproduced with permission [12]. Copyright © 2023, Elsevier B. V.

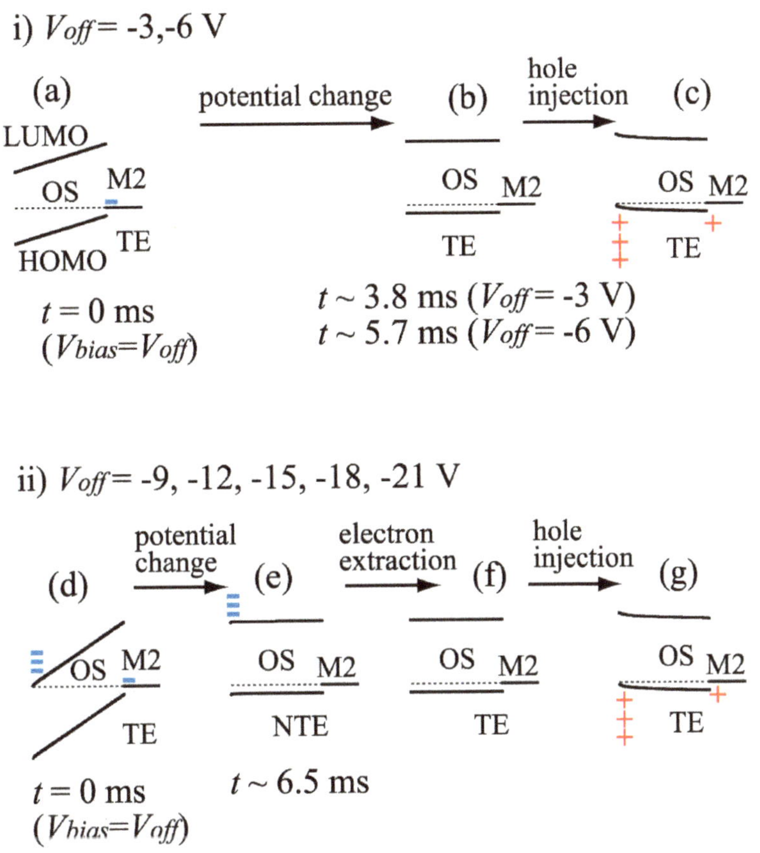

Figure 4. Potential diagram to explain the displacement current of p-Si/SiO$_2$/H$_2$Pc/Au shown in Figure 3d. The layers of M1 and INS (SiO$_2$) are omitted for simplicity in the figure. At V_{off} = −3 and −6 V, V_{bias} = V_{off} is insufficient to inject electrons (**a**). The threshold of the decrease in displacement current is due to the hole injection from M2 to the OS (**b**,**c**). At V_{off} = −9, −12, −15, −18, and −21 V, the time dependence of the displacement current is almost the same, and the displacement current at t < 6.5 ms is approximately $-C_0 A$. This is the direct evidence indicating electron injection by V_{off} (**d**) and the NTE process of electron extraction (**d**)→(**e**). As the voltage drops in the OS layer at V_{bias} = V_{off} are almost the same owing to the electron accumulation near the INS/OS interface, the thresholds of the current decrease are also almost the same. After the electrons are extracted (**e**)→(**f**), holes are subsequently injected (**f**)→(**g**).

Figure 3b shows a typical example of the displacement current used for hole extraction during the TE [13]. The sample used was n-Si/SiO$_2$ (100 nm)/pentacene (50 nm)/Au. A sample behaved as a series capacitor if no charge was injected into (or extracted from) the OS layer. In that case, the displacement current is given by $-I = -C_0 dV_{bias}/dt$, which is indicated by a dotted line in the figure. Here, C_0 is the series capacitance of the INS and OS layers, as previously mentioned. If a current was observed above the dotted line, the accumulated charge near the INS/OS interface was extracted. In the pentacene sample (Figure 3b), the charge accumulated near the INS/OS interface was immediately extracted when the applied positive bias was reduced. This is direct evidence of the TE-type charge extraction in this sample.

Figure 3c shows a typical example of the displacement current used for hole extraction during the NTE. The sample was n-Si/SiO$_2$ (100 nm)/metal-free phthalocyanine (H$_2$Pc,

52 nm)/Au. The displacement current was close to $C_0 dV_{bias}/dt$ up to approximately $t = 0.5$ ms, and then increased sharply. This confirms that NTE charge extraction occurred, indicating that only the electric field in the OS layer changed in the first stage and that the holes injected at $V_{bias} = V_{off}$ were extracted after the electric field inside the sample almost disappeared.

Currently, it is unclear whether charge carriers were extracted via the TE or NTE processes. However, considering the model shown in Figure 2, we believe that TE charge extraction was enhanced by diffusion processes and tended to occur when the carrier density or mobility was high. For the pentacene/Au sample, TE charge extraction was observed in many of the tested samples [13,15]; however, NTE charge extraction was also observed in some samples [18].

Figure 3d shows the electron extraction experiment for p-Si/SiO$_2$ (70 nm)/H$_2$Pc (44 nm)/Au [12]. Care must be taken when analyzing the data because the hole injection barrier was considerably lower than the electron injection barrier in H$_2$Pc/Au. Figure 4 shows a potential diagram derived from the TE and NTE models to explain Figure 3d. In the case of $V_{off} = -3$ and -6 V, an electric field was generated in the OS layer, but no electrons were injected into the OS. A negative interfacial charge Q_0 was generated at the OS/M2 interface because of the electric field (Figure 4a). When V_{bias} increased in the positive direction, Q_0 decreased and eventually reversed, followed by hole injection. That is, a displacement current reflecting the change in Q_0 was observed initially. Subsequently, a displacement current owing to hole injection was observed (Figure 4b,c).

In the case of $V_{off} \leq -9$ V, electron injection at $V_{bias} = V_{off}$ occurred (Figure 4d). In this V_{off} region, the time at which the displacement current began to decrease was almost constant at 6.5 ms (Figure 4e). This is because the voltage drop across the OS became constant when the electrons were injected. Therefore, it was possible to determine whether the electrons were injected by V_{off} based on the obtained experimental data. Although electrons were injected by V_{off}, the displacement current at the beginning of the voltage sweep was by $C_0 dV_{bias}/dt$. This indicated that the injected electrons were not extracted at the beginning of the voltage sweep, suggesting that NTE electron extraction occurred in this sample.

3.2. Accumulated Charge Measurement

Another method to investigate NTE charge extraction is ACM [12,13]. Changes in the accumulated charge are observed in this experiment. As the theoretical calculation of the accumulated charge is much easier than that of the displacement current, the observed data are analyzed more easily in this experiment than in the displacement current measurement.

Figure 5a shows the experimental scheme for the ACM. In this experiment, the states $V_{bias} = V_{off}$ and $V_{bias} = V_a + V_{off}$ were alternately created to determine the amount of accumulated change $Q_{acc}(V_a) = Q_{total}(V_a + V_{off}) - Q_{total}(V_{off})$. Here, $Q_{total}(V_{bias})$ is the amount of charge accumulated at the junction when the applied voltage is maintained at V_{bias} for a long period. In the scheme shown in Figure 5a, long-term displacement current integration is necessary to determine $Q_{acc}(V_a)$. However, as it is affected by the current amplifier offsets and other factors, $Q_{acc}(V_a)$ is determined with high accuracy using a method called the voltage oscillation technique [14]. From the $Q_{acc}(V_a)$ obtained thus, the parameters ΔQ and V_{OS} were obtained using the following equations:

$$\Delta Q = Q_{acc} - C_0 V_a \quad (1)$$

$$V_{OS} = V_a - Q_{acc}/C_{INS} \quad (2)$$

These two parameters are defined as follows. The first is ΔQ, where $C_0 V_a$ is the accumulated charge at the OS/M2 interface; therefore, this parameter is zero when no charge is injected (or extracted) by V_a into the OS and shifts from zero when charge is injected (or extracted). In addition, V_{OS} indicates the voltage drop inside the OS caused by V_a according to Gauss's law. A feature of this experiment is that it is not dynamic, but static, and the experimentally obtained Q_{acc} and V_a as well as ΔQ and V_{OS} are easy to compare with theoretical calculations that assume a quasistatic state.

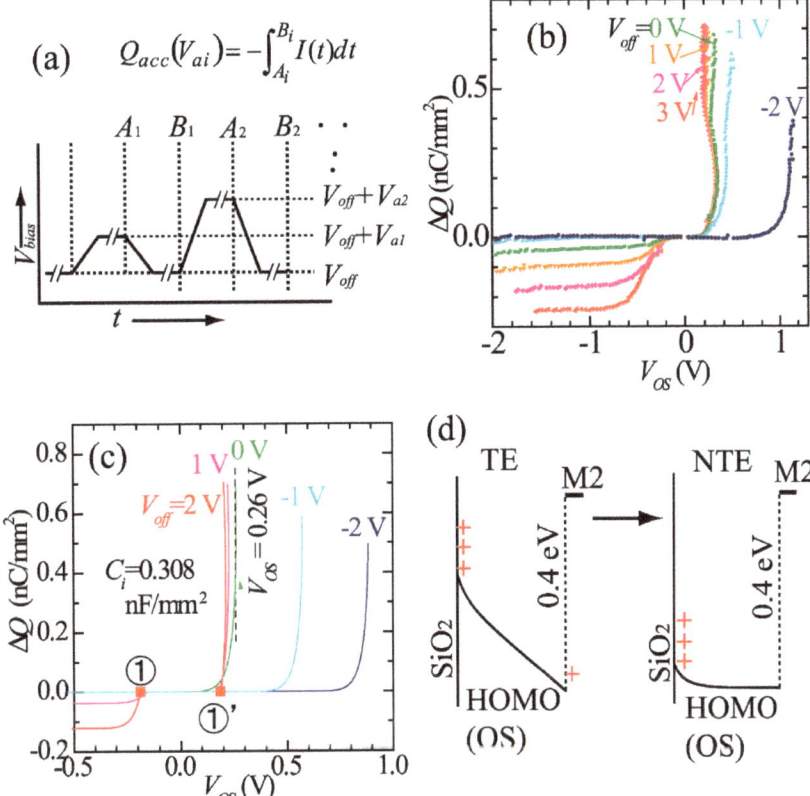

Figure 5. Experimental evidence of the NTE model for charge extraction based on ACM. (**a**) Schematic of applied bias voltage (V_{bias}) as a function of time (t). V_{bias} is a combination of the constant voltage V_{off} and the alternating voltage V_a. Q_{acc} as a function of V_a is obtained by integrating the displacement current from A_i to B_i. From the obtained dataset of Q_{acc} and V_a, ΔQ and V_{OS} are obtained using Equations (1) and (2) in the text. ΔQ indicates whether charges are injected into the OS layer, and V_{OS} is the potential change within the OS layer caused by V_a. (**b**) Experimentally obtained ΔQ–V_{OS} plot for hole injection in the MIOM capacitor of n-Si/SiO$_2$ (100 nm)/H$_2$Pc (52 nm)/Au. (**c**) Calculation of ΔQ–V_{OS} plot based on the NTE model. (**d**) Change in the potential curves associated with the point ① in (**c**). The layers of M1, INS (SiO$_2$) as well as the potential curve of LUMO are omitted in the figures for simplicity. Panel (**a**) was reproduced with permission [12]. Copyright © 2023, Elsevier B. V. Panels (**b**,**c**) were reproduced with permissions [13]. Copyright © 2021, AIP Publishing.

Figure 5b shows the experimental data for ΔQ vs. V_{OS} for n-Si/SiO$_2$/H$_2$Pc/Au during hole injection. The calculations of the ΔQ–V_{OS} plots, assuming the NTE model, are shown in Figure 5c. The variation in the potential curve corresponding to point ① is shown in Figure 5d. This is the point where ΔQ shifts from zero, and according to calculations, a similar potential curve change occurs at point ①'. For OSs that follow the NTE model, the ΔQ–V_{OS} plot allows us to determine the charge injection barrier at the OS/M2 interface experimentally. In the case of H$_2$Pc/Au, the hole injection barrier with approximately 0.2 eV was obtained from the plot [13].

4. Conclusions

The NTE charge extraction described in this study is a phenomenon specific to an MIOM capacitor, where a Schottky barrier exists between the OS/M2 contacts and where the mobility of the OS is not very high. Several MIOM capacitors satisfy the above two conditions; consequently, the NTE process of charge extraction appears to be a common phenomenon in both MIOM capacitors and OFETs. In the case of MIS capacitors and inorganic field-effect transistors, where ohmic contacts are formed between the semiconductor and the metal electrode, this NTE process does not occur. This process in OFETs is detrimental to transistor operation. Considering the mechanism of the process, the fabrication of ohmic contacts between the OS/M2 interface or the application of an OS with high mobility are the most effective ways to avoid this process. However, the NTE process may be useful in the application of OFETs, as memory devices. In addition, the NTE phenomenon can be used to determine the charge injection (extraction) barrier by means of ACM [12,13].

Author Contributions: This study was conceptualized by H.T. The formal analysis and methodology were provided by H.T. and T.O. Data curation was conducted by T.K. All authors have read and agreed to the published version of the manuscript.

Funding: This study was supported by a Grant-in-Aid for Scientific Research (Grant no. 21K05009).

Data Availability Statement: Data will be made available upon request.

Conflicts of Interest: The authors declare no conflict of interest.

References

1. Horowits, G. Organic Field-effect transistors. *Adv. Mater.* **1999**, *10*, 365–377. [CrossRef]
2. Sirringhaus, H. Organic Field-effect transistors. The pass beyond amorphous silicon. *Adv. Mater.* **2014**, *26*, 1319–1335. [CrossRef] [PubMed]
3. Lamport, Z.A.; Haneef, H.F.; Anand, S.; Waldrip, M.; Jurchescu, O.D. Tutorial: Organic field-effect transistors: Material, structure and operation. *J. Appl. Phys.* **2018**, *8*, 071101. [CrossRef]
4. Sze, S.M.; Ng, K.K. *Physics of Semiconductor Devices*, 3rd ed.; John Wiley & Sons, Inc.: Hoboken, NJ, USA, 2007.
5. Kim, S.; Yoo, H.; Choi, J. Effects of charge traps on hysteresis in organic field-effect transistors and their charge trap cause analysis through causal inference techniques. *Sensor* **2023**, *23*, 2265. [CrossRef] [PubMed]
6. Egginger, M.; Bauer, S.; Schwdiauer, R.; Neugebauer, H.; Sariciftci, N.S. Current versus gate voltage hysteresis in organic field effect transistors. *Mon. Chem.* **2009**, *140*, 735–750. [CrossRef]
7. Park, Y.; Baeg, K.; Kim, C. Solution-processed nonvolatile organic transistor memory based on semiconductor blend. *ACS Appl. Mater. Interfaces* **2019**, *11*, 8327–8336. [CrossRef] [PubMed]
8. Wang, W.V.; Zhang, Y.; Li, X.; Chen, Z.; Wu, Z.; Zhang, L.; Lin, Z.; Zhang, H. High performance nonvolatile organic field-effect transistor memory devices based on pyrene diimide derivative. *InfoMat* **2021**, *3*, 814–822. [CrossRef]
9. Yu, T.; Liu, Z.; Wang, Y.; Zhang, L.; Hou, S.; Wan, Z.; Yin, J.; Gao, X.; Wu, L.; Xia, Y.; et al. Deep-trap dominated degradation of the endurance characteristics in OFET memory with polymer charge-trapping layer. *Sci. Rep.* **2023**, *13*, 5865. [CrossRef]
10. Park, S.; Kim, S.H.; Choi, H.H.; Kang, B.; Cho, K. Recent advances in the bias stress stability of organic transistors. *Adv. Funct. Mater.* **2020**, *30*, 1904590. [CrossRef]
11. Hu, Y.; Zheng, L.; Li, J.; Huang, Y.; Wang, Z.; Lu, X.; Yu, L.; Wang, S.; Sun, Y.; Ding, S.; et al. Organic phase-change memory transistor based on an organic semiconductor with reversible molecular conformation transition. *Adv. Sci.* **2023**, *10*, 2205694. [CrossRef]
12. Oda, T.; Yamaguchi, K.; Kadoya, T.; Tajima, H. Measurement of electron injection barriers in OS/Au (OS = phthalocyanine and pentacene) using accumulated charge measurement. *Org. Electron.* **2023**, *120*, 106827. [CrossRef]
13. Tajima, H.; Kadoya, T.; Yamaguchi, K.; Omura, Y.; Oda, T.; Ogino, A. Thermal and non-thermal equilibrium processes of charge extraction in accumulated charge measurement (ACM). *J. Appl. Phys.* **2021**, *130*, 195501. [CrossRef]
14. Tajima, H.; Miyao, F.; Mizukoshi, M.; Sato, S. Determination of charge injection barrier using the displacement current measurement technique. *Org. Electron.* **2016**, *34*, 193–199. [CrossRef]
15. Kadoya, T.; Otsuka, M.; Ogino, A.; Sato, S.; Yokomatsu, T.; Maenaka, K.; Yamada, J.; Tajima, H. Estimation of charge-Injection barriers at the metal/Pentacene interface through accumulated charge measurement. *J. Phys. Chem. C* **2017**, *121*, 2882–2888. [CrossRef]
16. Tajima, H.; Yoshida, K.; Sato, S.; Kadoya, T.; Yamada, J. Estimation of the charge injection barrier at a metal/organic semiconductor interface based on accumulated charge measurement: The effect of offset bias voltages. *J. Phys. Chem. C* **2017**, *121*, 14725–14730. [CrossRef]

17. Tajima, H.; Yasukawa, N.; Nakatani, H.; Sato, S.; Kadoya, T.; Yamada, J. Estimation of hole injection barrier at the poly-3(hexylthiophene)/metal interface using accumulated charge measurement. *Org. Electron.* **2017**, *51*, 162–167. [CrossRef]
18. Tanimura, T.; Tajima, H.; Ogino, A.; Miyamoto, Y.; Kadoya, T.; Komino, T.; Yokomatsu, T.; Maenaka, K.; Ikemoto, Y. Accumulated charge measurement using a substrate with a restricted-bottom-electrode structure. *Org. Electron.* **2019**, *74*, 251–257. [CrossRef]
19. Shimomoto, S.; Kadoya, T.; Tanimura, T.; Maenaka, K.; Yokomatsu, T.; Komino, T.; Tajima, H. Accumulated charge measurement: Control of the interfacial depletion layer by offset voltage and estimation of band gap and electron injection barrier. *J. Phys. Chem. C* **2021**, *125*, 1990–1998. [CrossRef]
20. Ishii, H.; Sugiyama, K.; Ito, E.; Seki, K. Energy Level Alignment and Interfacial Electronic Structures at Organic/Metal and Organic/Organic Interfaces. *Adv. Mater.* **1999**, *11*, 605–625. [CrossRef]
21. Hill, I.G.; Milliron, D.; Schwartz, J.; Kahn, A. Organic semiconductor interfaces: Electronic structure and transport properties. *Appl. Surf. Sci.* **2000**, *166*, 354. [CrossRef]
22. Adachi, C.; Oyamada, T.; Nakajima, Y. *Data Book on Work Function of Organic-Thin Films*, 2nd ed.; CMC International: Tokyo, Japan, 2006.
23. Kirihata, H.; Uda, M. Externally quenched air counter for low-energy electron emission measurements. *Rev. Sci. Instrum.* **1981**, *52*, 68–70. [CrossRef]
24. Zahn, D.R.T.; Gavrila, G.N.; Georgoi, M. The transport gap of organic semiconductors studied using the combination of direct and inverse photoemission. *Chem. Phys.* **2006**, *325*, 99–112. [CrossRef]
25. Yoshida, H. Near-ultraviolet inverse photoemission spectroscopy using ultra-lowenergy electrons. *Chem. Phys. Lett.* **2012**, *180*, 539–540.

Disclaimer/Publisher's Note: The statements, opinions and data contained in all publications are solely those of the individual author(s) and contributor(s) and not of MDPI and/or the editor(s). MDPI and/or the editor(s) disclaim responsibility for any injury to people or property resulting from any ideas, methods, instructions or products referred to in the content.

Article

Dirac Cone Formation in Single-Component Molecular Conductors Based on Metal Dithiolene Complexes

Reizo Kato [1,*] and Takao Tsumuraya [2,3,*]

1. Condensed Molecular Materials Laboratory RIKEN, Saitama 351-0198, Japan
2. Priority Organization for Innovation and Excellence (POIE), Kumamoto University, 2-39-1 Kurokami, Kumamoto 860-8555, Japan
3. Magnesium Research Center, Kumamoto University, 2-39-1 Kurokami, Kumamoto 860-8555, Japan
* Correspondence: reizo@riken.jp (R.K.); tsumu@kumamoto-u.ac.jp (T.T.)

Citation: Kato, R.; Tsumuraya, T. Dirac Cone Formation in Single-Component Molecular Conductors Based on Metal Dithiolene Complexes. *Magnetochemistry* **2023**, *9*, 174. https://doi.org/10.3390/magnetochemistry9070174

Academic Editor: Laura C. J. Pereira

Received: 11 June 2023
Revised: 25 June 2023
Accepted: 26 June 2023
Published: 6 July 2023

Copyright: © 2023 by the authors. Licensee MDPI, Basel, Switzerland. This article is an open access article distributed under the terms and conditions of the Creative Commons Attribution (CC BY) license (https://creativecommons.org/licenses/by/4.0/).

Abstract: Single-component molecular conductors exhibit a strong connection to the Dirac electron system. The formation of Dirac cones in single-component molecular conductors relies on (1) the crossing of HOMO and LUMO bands and (2) the presence of nodes in the HOMO–LUMO couplings. In this study, we investigated the possibility of Dirac cone formation in two single-component molecular conductors derived from nickel complexes with extended tetrathiafulvalenedithiolate ligands, [Ni(tmdt)$_2$] and [Ni(btdt)$_2$], using tight-biding models and first-principles density-functional theory (DFT) calculations. The tight-binding model predicts the emergence of Dirac cones in both systems, which is associated with the stretcher bond type molecular arrangement. The DFT calculations also indicate the formation of Dirac cones in both systems. In the case of [Ni(btdt)$_2$], the DFT calculations, employing a vdW-DF2 functional, reveal the formation of Dirac cones near the Fermi level in the nonmagnetic state after structural optimization. Furthermore, the DFT calculations, by utilizing the range-separated hybrid functional, confirm the antiferromagnetic stability in [Ni(btdt)$_2$], as observed experimentally.

Keywords: Dirac electron systems; single-component molecular conductors; metal dithiolene complexes; tight-binding model; first-principles DFT calculation

1. Introduction

The conventional molecular conductors are categorized as multi-component systems wherein each molecule provides only one type of frontier molecular orbital (HOMO or LUMO) to form the conduction band, where HOMO and LUMO denote highest occupied molecular orbital and lowest unoccupied molecular orbital, respectively. The concept of a single-component molecular metal is based on a multi-orbital system in which more than two molecular orbitals (in this case, HOMO and LUMO) in the same molecule contribute to electronic properties. In a single-component molecular metal, the fully occupied HOMO band and the empty LUMO band overlap, and electron transfer between them induces a partially filled state. This idea was confirmed by the observation of electron and hole pockets in an ambient-pressure single-component molecular (semi)metal [Ni(tmdt)$_2$] (tmdt = trimethylenetetrathiafulvalenedithiolate: Scheme 1) through the detection of quantum oscillations at the begging of this century [1–4]. Since this breakthrough, various single-component molecular conductors have been developed by extending π conjugating systems to reduce the HOMO-LUMO energy gap, or by applying high pressure to increase the band width for each band. In particular, metal dithiolene complexes with a planar central core have played an important role due to their small HOMO-LUMO energy gap [5,6].

On the other hand, the Dirac electron system, typically observed in graphene, exhibits neither partially filled bands, like in metals, nor band gaps, like in insulators. The energy band structure of the Dirac electron system possesses a unique feature known as the Dirac cone, where two conical surfaces with linear dispersion meet at a single point called the Dirac point in momentum space. A system in which the Dirac point forms extended lines or loops in the Brillouin zone is referred to as a nodal line semimetal, and has garnered significant attention due to the possibility of topologically nontrivial states [7,8].

The discovery of the nodal line semimetal state in a single-component molecular conductor [Pd(dddt)$_2$] (dddt = 5,6-dihydro-1,4-dithiin-2,3-dithiolate: Scheme 1) under high pressure has revealed a significant connection between single-component molecular conductors and the Dirac electron system [9]. The Dirac cone formation in [Pd(dddt)$_2$] can be understood using a tight-binding model based on extended Hückel molecular orbital calculations [10]. This mechanism relies on (1) the crossing of the HOMO and LUMO bands and (2) the presence of nodes in the HOMO–LUMO couplings, which favor the emergence of Dirac cones located near the Fermi level. Subsequently, an ambient-pressure nodal line semimetal based on a single-component molecular conductor [Pt(dmdt)$_2$] (dmdt = dimethyltetrathiafulvalenedithiolate: Scheme 1) was reported [11]. Our analysis using tight-binding band calculations indicated that this system exemplifies the mechanism proposed by us in a textbook manner [12]. An important aspect is the molecular arrangement (Figure 1) associated with the symmetry of the frontier molecular orbitals in [Pt(dmdt)$_2$], which satisfies the requirements for the Dirac cone formation. Within the bc plane, [Pt(dmdt)$_2$] molecules exhibit a stretcher bond pattern wherein each molecule overlaps with the molecule above and below it by half (Figure 1a). Transfer integrals labeled p and c are associated with major intermolecular interactions and form a two-dimensional conducting network (we tentatively call this network "layer"). Dirac cones are described on the k_y-k_z plane, and nodal lines extend along the k_x direction. As the HOMO has ungerade (odd) symmetry and the LUMO has gerade (even) symmetry (Figure 1b), the main HOMO–HOMO and LUMO–LUMO couplings (p and c) yield transfer integrals with opposite signs (Table 1), facilitating the crossing of the HOMO and LUMO bands, which is the first requirement for Dirac cone formation. In this work, we focus on two single-component molecular conductors derived from metal complexes with extended tetrathiafulvalenedithiolate ligands, [Ni(tmdt)$_2$] and [Ni(btdt)$_2$] (btdt = benzotetrathiafulvalenedithiolate: Scheme 1) [13], that exhibit molecular arrangements similar to that in [Pt(dmdt)$_2$]. The former is a (semi)metal, and the latter is a narrow-gap semiconductor with high conductivity at room temperature. We explore the possibility of Dirac cone formation in these compounds by means of tight-biding models and first-principles density functional theory (DFT) calculations.

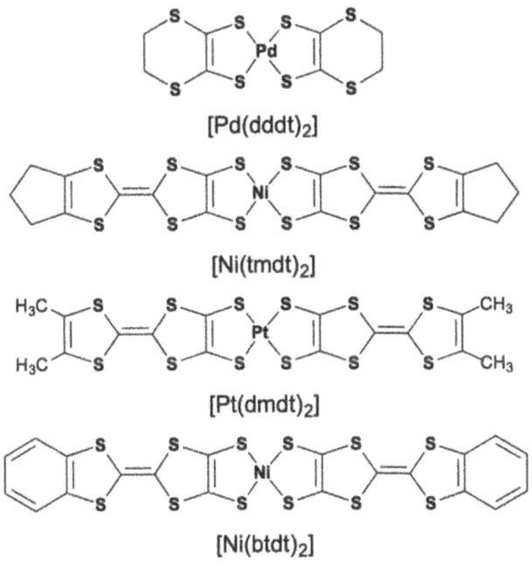

Scheme 1. Component molecules of single-component molecular conductors.

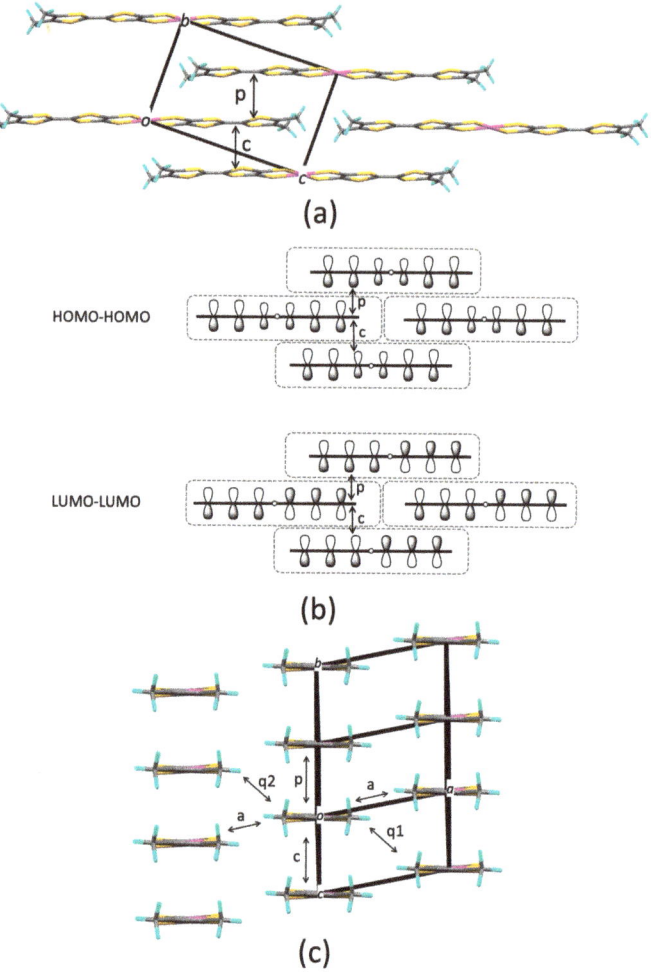

Figure 1. Crystal structure of [Pt(dmdt)$_2$]: (**a**) Molecular arrangement within the *bc* plane, (**b**) HOMO–HOMO and LUMO–LUMO couplings in the stretcher bond arrangement of the [Pt(dmdt)$_2$] molecule, where each frontier molecular orbital is represented using the p orbital of sulfur atom, (**c**) end-on projection (molecular arrangement viewed from the long molecular axis). Each element is identified by color as follows: S, yellow; C, gray; H, blue; Pt, magenta.

Table 1. HOMO–HOMO (H-H), LUMO–LUMO (L-L), and HOMO–LUMO (H-L) intermolecular transfer integrals (*t*) in [Pt(dmdt)$_2$] (meV).[1]

Transfer Integral	H-H	L-L	H-L
t_p	53.4	−49.8	51.7
t_c	67.1	−62.9	64.9
t_a	−6.2	−6.5	0.3
t_{q1}	8.2	−7.4	7.8
t_{q2}	8.2	−7.7	7.9

[1] See Figure 1c. Transfer integrals in this table are used in Equations (2)–(4) with subscripts H-H, L-L, and H-L, which represent HOMO–HOMO, LUMO–LUMO, and HOMO–LUMO couplings, respectively (for example, t_{pH-H} means a transfer integral t_p between HOMO and HOMO).

2. Models and Methods

2.1. Tight-Binding Model

We used published atomic coordinates data [1,13] for our calculations. In order to make a comparison with [Pt(dmdt)$_2$] easier, each unit cell (a_0, b_0, c_0) was transformed to a new cell (a, b, c), as follows:

$a = a_0$, $b = b_0$, $c = a_0 + c_0$ for [Ni(tmdt)$_2$]

$a = a_0$, $b = a_0 + b_0$, $c = c_0$ for [Ni(btdt)$_2$].

In the tight-binding model, we consider one HOMO band and one LUMO band, because the unit cell contains only one molecule. Calculations of molecular orbitals and intermolecular overlap integrals (S) between HOMOs and LUMOs undertaken via the extended Hückel method were carried out using reported sets of semi-empirical parameters for Slater-type atomic orbitals and valence shell ionization potentials for H [14], C [14], Ni [14] and S [15]. Intermolecular transfer integrals, t (eV), were estimated using the equation $t = -10\, S$. The band energies $E(k)$ (k is given by $k = k_x a^* + k_y b^* + k_z c^*$ in terms of the reciprocal lattice vectors a^*, b^*, and c^*) are obtained as eigenvalues of the following 2×2 Hermitian matrix.

$$\mathbf{H}(k) = \begin{pmatrix} h_{\text{H-H}} & h_{\text{H-L}} \\ h_{\text{H-L}} & h_{\text{L-L}} \end{pmatrix} \quad (1)$$

For [Pt(dmdt)$_2$], the matrix elements are given by

$$h_{\text{H-H}} = 2\,[t_{p\text{H-H}} \cos k(b+c) + t_{c\text{H-H}} \cos kc + t_{a\text{H-H}} \cos ka + t_{q1\text{H-H}} \cos k(a+c) + t_{q2\text{H-H}} \cos k(-a+b+c)] \quad (2)$$

$$h_{\text{H-L}} = 2i\,[t_{p\text{H-L}} \sin k(b+c) + t_{c\text{H-L}} \sin kc + t_{a\text{H-L}} \sin ka + t_{q1\text{H-L}} \sin k(a+c) + t_{q2\text{H-L}} \sin k(-a+b+c)] \quad (3)$$

$$h_{\text{L-L}} = \Delta + 2\,[t_{p\text{L-L}} \cos k(b+c) + t_{c\text{L-L}} \cos kc + t_{a\text{L-L}} \cos ka + t_{q1\text{L-L}} \cos k(a+c) + t_{q2\text{L-L}} \cos k(-a+b+c)], \quad (4)$$

where Δ is an energy gap between HOMO and LUMO (0.25 eV for [Pt(dmdt)$_2$]). Matrix elements $h_{\text{H-H}}$, $h_{\text{L-L}}$, and $h_{\text{H-L}}$ are associated with HOMO–HOMO, LUMO–LUMO, and HOMO–LUMO couplings, respectively.

2.2. DFT Calculations

We utilized various methods in our first-principles DFT calculations. To plot the Dirac cones in [Ni(tmdt)$_2$] and [Ni(btdt)$_2$], we performed band structure calculations using an all-electron full-potential linearized plane wave (FLAPW) method implemented in QMD-FLAPW12 [16,17]. The exchange-correlation functional employed was the generalized gradient approximation by Perdew, Burke, and Ernzerhof (GGA-PBE) [18]. We have found that a cut-off energy for plane waves of 20 Ry is sufficient for calculating the band structure. The cut-off energy for electron densities is assumed to be 213 Ry. For [Ni(tmdt)$_2$] and [Ni(btdt)$_2$], we employed uniform k-point meshes of $6 \times 6 \times 4$ and $8 \times 8 \times 4$, respectively. Since the c axis direction is longer than the a and b directions in these crystal structures, the reciprocal lattice vector is shorter, allowing for a reduced number of k-point meshes along the c direction. To calculate three-dimensional band dispersion of the Dirac cone, we used a high-density k-mesh, similar to our previous works [19,20].

For [Ni(btdt)$_2$] at ambient pressure, we performed structural relaxation using the van der Waals density functional (vdW-DF2) [21–23]. Specifically, we chose the vdW-DF2-b86r functional proposed by Hamada [24]. The calculations were carried out using the Quantum Espresso v.6.8 [25]. We employed the ultrasoft pseudopotential method with plane-wave basis sets. Ultrasoft pseudopotentials established by Garrity, Bennett, Rabe, and Vanderbilt (GBRV) were used [26]. During the structural relaxations, a uniform k-point mesh was set as $4 \times 4 \times 2$. When performing structural optimization with stress tensors, it is crucial to use a fixed number of plane waves based on the initial lattice parameters rather than a constant cut-off energy, as the latter can introduce significant errors in the calculated stress tensors [27]. To minimize errors in the calculated stress tensors, we used higher cut-off energies for plane waves, specifically, 60 Ry. The cut-off energy for electron densities was set to 488 Ry.

To investigate the stability of antiferromagnetic (AFM) ordering and the possible realization of the semiconducting phase in [Ni(btdt)$_2$], we employed a hybrid functional approach using the exchange-correlation functional set out by Heyd, Scuseria, and Ernzerhof (HSE06) [28,29]. The HSE06 calculations were performed by use of the Vienna Ab initio Simulation Package (VASP) [30–32], which is also based on the pseudopotential technique employing the projected augmented plane wave (PAW) method [33,34]. In the HSE06 calculations, we first obtained a converged charge density through the self-consistent calculations within GGA. Subsequently, self-consistent hybrid functional calculations were performed using the GGA charge density as the initial state. A k-point mesh of 3 × 4 × 2 was employed for the calculations. The cut-off energy for plane waves was set to 36.75 Ry. The range-separation parameter in the HSE06 calculations was 0.2 Å^{-1}, and 25% of the exact exchange was mixed with the GGA exchange for short-range interactions.

3. Results

3.1. [Ni(tmdt)$_2$]

3.1.1. Tight-Binding Model

Figure 2 illustrates the crystal structure of [Ni(tmdt)$_2$]. The molecular arrangement in the bc plane exhibits the stretcher bond pattern. The end-on projection indicates zigzag face-to-face stacking, and the molecular pair labeled p exhibits the strongest HOMO–HOMO and LUMO–LUMO couplings (Table 2). However, this does not imply a simple "dimerization", because the [Ni(tmdt)$_2$] molecules stack "uniformly" along the $b + c$ direction (corresponding to p) and along the c direction (corresponding to c), as depicted in Figure 2a. Table 2 demonstrates that conducting layers parallel to the bc plane are strongly interconnected through interlayer transfer integrals q1 and a, which imports three-dimensional characteristics of the electronic structure. This differs from the case of [Pt(dmdt)$_2$] that fundamentally possesses a two-dimensional character. For most transfer integrals, including t_p and t_c, the HOMO–HOMO and LUMO–LUMO couplings exhibit opposite signs.

Table 2. HOMO–HOMO (H-H), LUMO–LUMO (L-L), and HOMO–LUMO (H-L) intermolecular transfer integrals (t) in [Ni(tmdt)$_2$] (meV). [1]

Transfer Integral	H-H	L-L	H-L
t_p	72.0	−68.8	70.5
t_c	33.9	−26.7	30.1
t_a	−32.8	−33.1	0.6
t_{q1}	46.9	−46.9	46.9
t_{q2}	3.6	−3.2	3.4
t_r	2.8	−2.6	2.7

[1] See Figure 2b. Transfer integrals in this table are used in Equations (5)–(7) with subscripts H-H, L-L, and H-L, that represent HOMO–HOMO, LUMO–LUMO, and HOMO–LUMO couplings, respectively.

The matrix elements of $\mathbf{H}(k)$ for [Ni(tmdt)$_2$] are given by

$$h_{\text{H-H}} = 2\left[t_{p\text{H-H}}\cos k(b+c) + t_{c\text{H-H}}\cos kc + t_{a\text{H-H}}\cos ka + t_{q1\text{H-H}}\cos k(c-a) + t_{q2\text{H-H}}\cos k(a+b+c) + t_{r\text{H-H}}\cos k(-a+b+c)\right] \quad (5)$$

$$h_{\text{H-L}} = 2i\left[t_{p\text{H-L}}\sin k(b+c) + t_{c\text{H-L}}\sin kc + t_{a\text{H-L}}\sin ka + t_{q1\text{H-L}}\sin k(c-a) + t_{q2\text{H-L}}\sin k(a+b+c) + t_{r\text{H-L}}\sin k(-a+b+c)\right] \quad (6)$$

$$h_{\text{L-L}} = \Delta + 2\left[t_{p\text{L-L}}\cos k(b+c) + t_{c\text{L-L}}\cos kc + t_{a\text{L-L}}\cos ka + t_{q1\text{L-L}}\cos k(c-a) + t_{q2\text{L-L}}\cos k(a+b+c) + t_{r\text{L-L}}\cos k(-a+b+c)\right], \quad (7)$$

where Δ is 0.15 eV. Figures 3a and 4 display the band dispersion and Fermi surface obtained from the tight-binding model. On the k_y-k_z plane ($k_x = 0$), a crossing of the HOMO and LUMO bands occurs at (k_x, k_y, k_z) = (0.0, ±0.578, ±0.210) (marked as N in Figure 3a), resulting in the emergence of symmetrical Dirac cones around the Γ point (0.0, 0.0, 0.0) (Figure 5). The Dirac points reside below the Fermi level, resulting in electron pockets. Due to the strong interlayer interactions among the conducting networks parallel to the bc plane, the Dirac points move within the three-dimensional wave vector space depending on k_x and form a loop. The reciprocal unit cell contains one nodal loop, which is nearly planar. The loop exhibits energy variations, which gives rise to electron pockets and hole pockets, indicating the presence of a nodal line semi-metal (Figure 6). Compared to [Pt(dmdt)$_2$], however, the deviation of the band energy from the Fermi energy is relatively large, resulting in a system with a large Fermi surface (Figure 4).

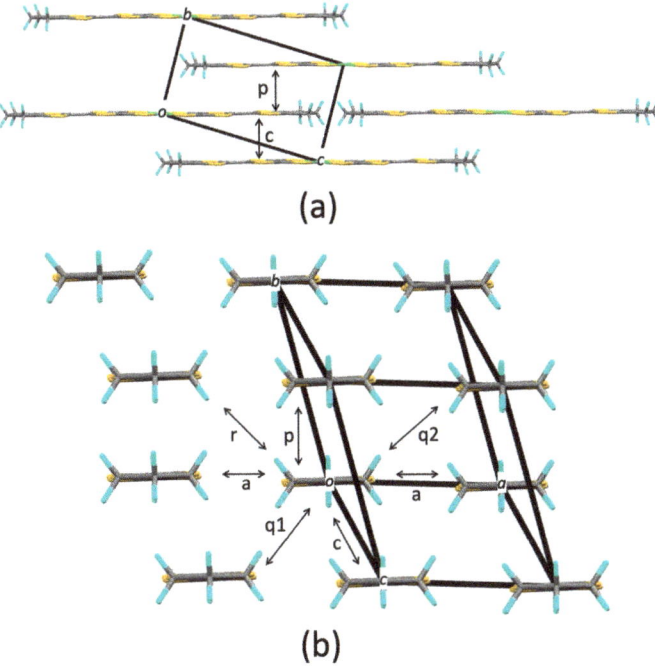

Figure 2. Crystal structure of [Ni(tmdt)$_2$]: (**a**) Molecular arrangement within the *bc* plane, (**b**) end-on projection. Each element is identified by color as follows: S, yellow; C, gray; H, blue; Ni, green.

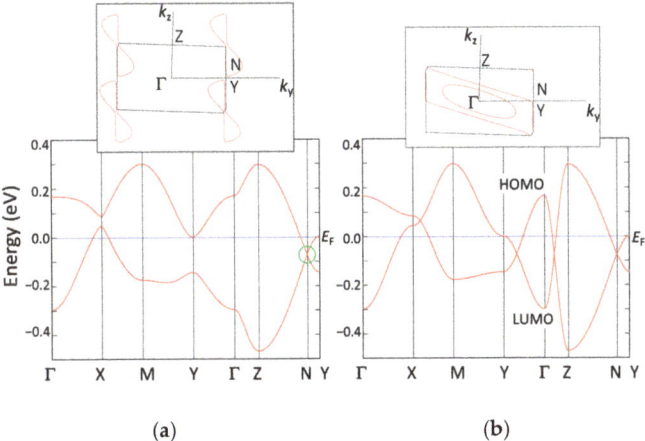

Figure 3. Band dispersion and Fermi surface (on the $k_x = 0$ plane) of [Ni(tmdt)$_2$] (**a**) with HOMO–LUMO couplings (pristine), (**b**) without HOMO–LUMO couplings ($h_{H-L} = 0$). Γ = (0,0,0), X = (1/2,0,0), M = (1/2,1/2,0), Y = (0,1/2,0), Z = (0,0,1/2), and N = (0,0.578,0.210), in units of the reciprocal lattice vectors. N denotes a position of the Dirac point.

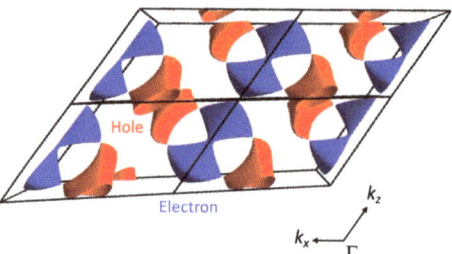

Figure 4. Fermi surface of [Ni(tmdt)$_2$] viewed from the k_y axis. This figure is plotted with FermiSurfer version 2.1 [35].

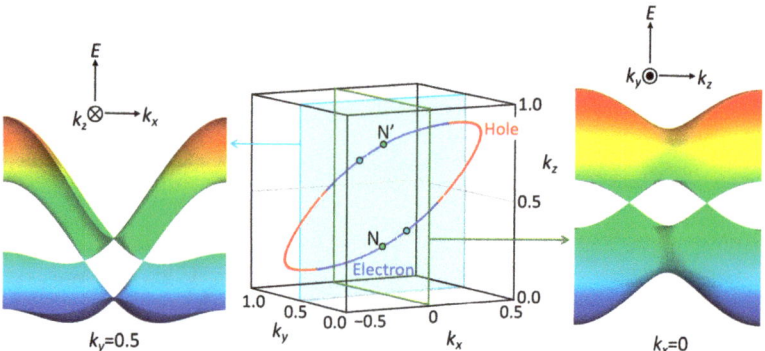

Figure 5. Loop formed by Dirac points (Nodal line) in the three-dimensional wave vector space and a pair of Dirac cones on the k_x = 0 and k_y = 0.5 planes in [Ni(tmdt)$_2$]. The hole-like characteristic is indicated in red and the electron-like characteristic in blue.

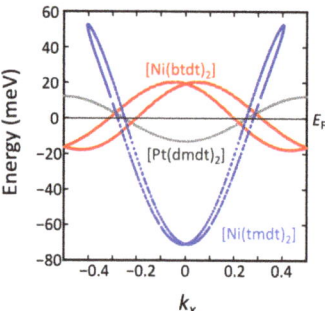

Figure 6. k_x dependence of the band energy at the Dirac point calculated with tight-binding models.

3.1.2. First-Principles DFT Calculations

Figure 7 indicates the DFT band structure of [Ni(tmdt)$_2$] calculated with the GGA-PBE functional. In the k_x = 0 plane, due to the space inversion symmetry, a pair of Dirac points exists at (k_x, k_y, k_z) = (0, 0.685, 0.265) and (0, 0.315, −0.265). The Dirac point is located 65 meV lower than the Fermi level. The tight-binding band structure in Figure 3a is generally in agreement with the DFT band structure, e.g., hall-like characteristics near the X point and electron-like characteristics near the N point.

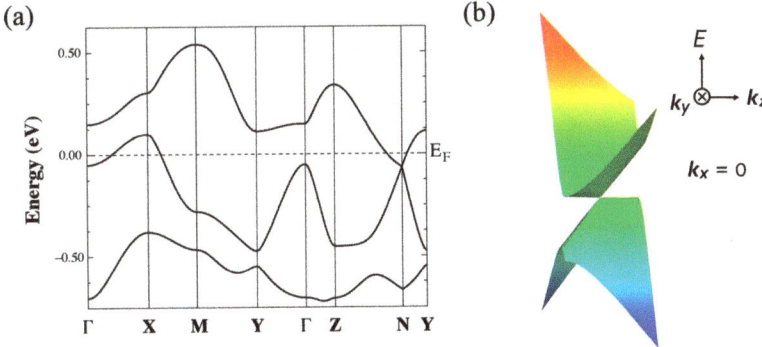

Figure 7. (a) Band structure of [Ni(tmdt)$_2$] calculated with GGA-PBE. Γ = (0,0,0), X = (1/2,0,0), M = (1/2,1/2,0), Y = (0,1/2,0), Z = (0,0,1/2), and N = (0,0.685,0.265). N represents a position of the Dirac point. (b) Dirac cone on the $k_x = 0$ plane.

3.2. [Ni(btdt)$_2$]

3.2.1. Tight-Binding Model

As shown in Figure 8a, the molecular arrangement of [Ni(btdt)$_2$] within the *bc* plane exhibits the stretcher bond pattern. The interplanar distance in the molecular pair labeled c (4.24 Å) is significantly longer than that in the molecular pair p (3.01 Å), leading to a considerable reduction in intermolecular interaction c. In contrast, in [Pd(dmdt)$_2$] and [Ni(tmdt)$_2$], the corresponding interplanar distances are nearly equivalent (3.54 Å and 3.53 Å in [Pd(dmdt)$_2$], 3.46 Å and 3.58 Å in [Ni(tmdt)$_2$]). The main HOMO–HOMO and LUMO–LUMO couplings are observed in the molecular pairs labeled p and q1 (Table 3, Figure 8b). Both transfer integrals t_p and t_{q1} exhibit HOMO–HOMO and LUMO–LUMO couplings with opposite signs. Notably, t_p is opposite in sign to t_p in [Pt(dmdt)$_2$] and [Ni(tmdt)$_2$].

Table 3. HOMO–HOMO (H-H), LUMO–LUMO (L-L), and HOMO–LUMO (H-L) intermolecular transfer integrals (*t*) in [Ni(btdt)$_2$] (meV). [1]

Transfer Integral	H-H	L-L	H-L
t_p	−59.8	51.0	−55.5
t_c	6.6	2.5	1.7
t_a	7.9	9.6	0.2
t_{q1}	44.8	−24.5	34.2
t_{q2}	0.6	−0.6	0.6
t_s	−1.7	1.5	−1.6

[1] See Figure 8b. Transfer integrals in this table are used in Equations (8)–(10) with subscripts H-H, L-L, and H-L, that represent HOMO–HOMO, LUMO–LUMO, and HOMO–LUMO couplings, respectively.

The matrix elements of **H**(**k**) for [Ni(btdt)$_2$] are given by

$$h_{\text{H-H}} = 2\,[t_{\text{pH-H}}\cos k(b+c) + t_{\text{cH-H}}\cos kc + t_{\text{aH-H}}\cos ka + t_{\text{q1H-H}}\cos k(c+a) + t_{\text{q2H-H}}\cos k(-a+b+c) + t_{\text{sH-H}}\cos k(b+2c)] \quad (8)$$

$$h_{\text{H-L}} = 2i\,[t_{\text{pH-L}}\sin k(b+c) + t_{\text{cH-L}}\sin kc + t_{\text{aH-L}}\sin ka + t_{\text{q1H-L}}\sin k(c+a) + t_{\text{q2H-L}}\sin k(-a+b+c) + t_{\text{sH-L}}\sin k(b+2c)] \quad (9)$$

$$h_{\text{L-L}} = \Delta + 2\,[t_{\text{pL-L}}\cos k(b+c) + t_{\text{cL-L}}\cos kc + t_{\text{aL-L}}\cos ka + t_{\text{q1L-L}}\cos k(c+a) + t_{\text{q2L-L}}\cos k(-a+b+c) + t_{\text{sL-L}}\cos k(b+2c)], \quad (10)$$

where Δ is 0.15 eV. The band dispersion and Fermi surface (Figures 9a and 10) obtained from the tight-binding model indicate a semimetallic electronic structure where hole and electron pockets align alternately along the $a^* + b^* - c^*$ direction. The Dirac points emerge at $(k_x, k_y, k_z) = (0.0, \pm 0.118, \pm 0.264)$ (marked as N in Figure 8a) on the k_y-k_z plane ($k_x = 0$) and $(\pm 0.308, \pm 0.404, 0.0)$ on the k_x-k_y plane ($k_z = 0$). In the three-dimensional reciprocal lattice, the Dirac point forms a line along the $a^* + b^* - c^*$ direction (Figure 11). The energy at the Dirac point fluctuates around the Fermi level along the line (Figure 6), resulting in the presence of hole and electron pockets.

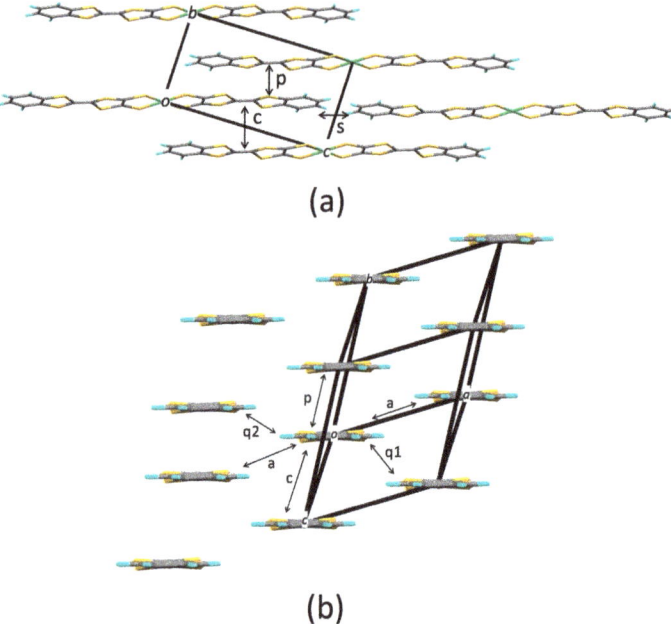

Figure 8. Crystal structure of [Ni(btdt)$_2$]: (**a**) Molecular arrangement within the *bc* plane, (**b**) end-on projection. Each element is identified by color as follows: S, yellow; C, gray; H, blue; Ni, green.

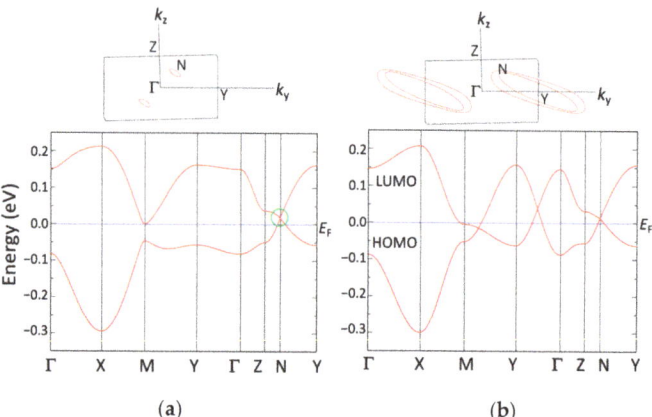

Figure 9. Band dispersion and Fermi surface (on the $k_x = 0$ plane) of [Ni(btdt)$_2$] (**a**) with HOMO–LUMO couplings (pristine), (**b**) without HOMO–LUMO couplings ($h_{H-L} = 0$). Γ = (0,0,0), X = (1/2,0,0), M = (1/2,1/2,0), Y = (0,1/2,0), Z = (0,0,1/2), and N = (0,0.118,0.264), in units of the reciprocal lattice vectors. N denotes a position of the Dirac point.

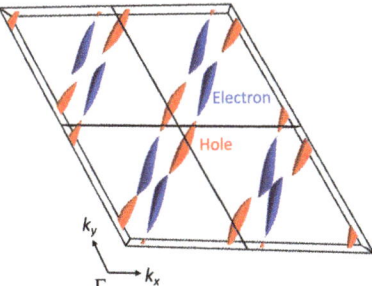

Figure 10. Fermi surface of [Ni(btdt)2] viewed from the k_z axis. This figure is plotted with FermiSurfer version 2.1 [33].

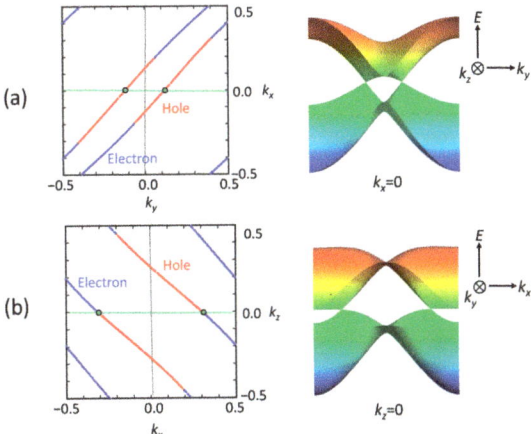

Figure 11. Nodal line projected on the k_x-k_y (a) and k_x-k_z (b) planes and a pair of Dirac cones on the $k_x = 0$ and $k_z = 0$ planes in [Ni(btdt)$_2$]. The hole-like characteristic is indicated in red and the electron-like characteristic in blue.

3.2.2. First-Principles DFT Calculations

Next, we describe the structural and electronic properties of [Ni(btdt)$_2$] as calculated using the first-principles method. Figure 12a shows the band structure for the experimental crystal structure with the original unit cell vectors, indicating a metallic character. Unlike the tight-binding model, the DFT band structure does not exhibit a Dirac cone.

As mentioned in the previous subsection, the experimental crystal structure, determined using powder X-ray diffraction analysis with constrained conditions, has a much shorter interplanar distance (~3.01 Å) compared to other single-component metal dithiolene complexes such as [Ni(tmdt)$_2$] [1] and [Ni(hfdt)$_2$] (hfdt = bis(trifluoromethyl) tetrathiafulvalenedithiolate) (~3.40 Å) [36]. Therefore, we performed structural relaxation for the lattice parameters and internal coordinates using a vdW-DF2 functional (vdw-df2-b86r) [28]. The vdW-DF method, proposed by Dion et al. [21], incorporates a semi-local exchange-correlation functional and a nonlocal correlation functional to account for dispersion interactions. In single-component molecular crystals, dispersion interactions are significant at ambient and low-pressure conditions. We anticipate that structural relaxation with the vdW-DF functional will yield an efficient determination of structural properties.

Figure 12b displays the band structure for the structure with relaxed atomic positions using the vdw-df2-b86r functional (Appendix A). The band gap remains unopened, and a massless Dirac cone emerges at the Fermi level. The band at the Y point has shifted lower, and there is a significant separation between the top and lower bands of the valence band. A new unit cell described in Section 2.1 for the optimized structure has the lattice parameters $a = 6.36$, $b = 6.98$, $c = 14.46$ Å, $\alpha = 107.53$, $\beta = 78.34$, and $\gamma = 79.80°$, while those for the experimental structure are $a = 6.720$, $b = 7.925$, $c = 12.563$ Å, $\alpha = 90.41$, $\beta = 93.62$, and $\gamma = 59.85°$. The DFT-optimized structure, where the interplanar distances are adjusted to ordinary values (~3.47 Å), appears to be a reasonable structure. However, despite using the DFT-optimized structure to simulate the XRD pattern, the resulting simulated pattern differs from the pattern obtained through the powder X-ray diffraction method. Hence, we consider the DFT-optimized structure as a model structure.

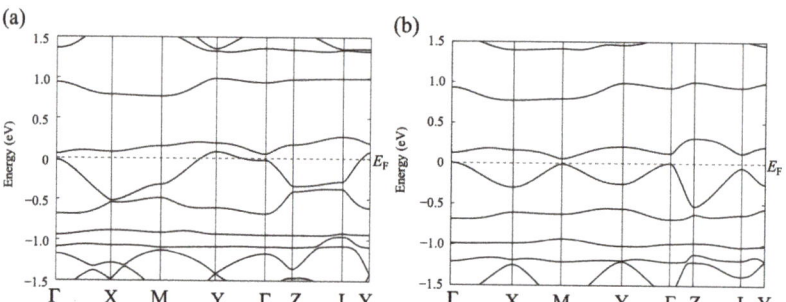

Figure 12. (a) Band structure for the experimental structure of [Ni(btdt)$_2$] calculated with GGA-PBE functional. (b) Band structure for the optimized structure calculated with a vdW-DF2 functional (vdw-df2-b86r). These band structures are calculated based on the original unit cell vectors. $\Gamma = (0,0,0)$, X = (1/2,0,0), M = (1/2,1/2,0), Y = (0,1/2,0), Z = (0,0,1/2), and L = (0,1/2,1/2), in units of the reciprocal lattice vectors.

Figure 13a illustrates the band structure for the new unit cell of the optimized structure. The Dirac points are located at specific points in k-space, represented by the coordinates $(k_x, k_y, k_z) = (\pm 0.3120, \mp 0.014, 0)$. Figure 13b and Figure 13c depict the Dirac cones observed in the k_x-k_y and k_x-k_z planes, respectively. It should be added that the tight-binding band calculation based on the DFT-optimized structure also indicates the presence of Dirac cones at $(k_x, k_y, k_z) = (\pm 0.414, 0, \mp 0.004)$.

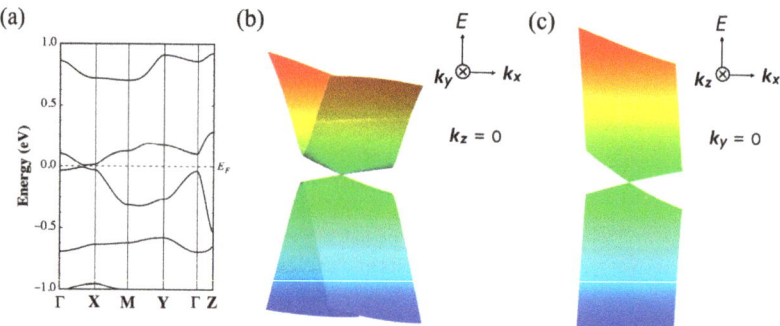

Figure 13. (a) Band dispersion of nonmagnetic calculations in [Ni(btdt)$_2$] for the crystal structure optimized with a vdW-DF2 functional (vdw-df2-b86r). $\Gamma = (0,0,0)$, X = (1/2,0,0), M = (1/2,1/2,0), Y = (0,1/2,0), Z = (0,0,1/2), and L = (0,1/2,1/2), for the new unit cell. Band structure close to the Dirac point on (b) the k_x-k_y and (c) k_x-k_z planes. These electronic structures are calculated with the GGA-PBE functional using the all-electron FLAPW method.

4. Discussion
4.1. Dirac Cone Formation in Single-Component Molecular Conductors

The tight-binding models have revealed that all three single-component molecular conductors with the stretcher bond type molecular arrangement, [Pd(dmdt)$_2$], [Ni(tmdt)$_2$], and [Ni(btdt)$_2$], form Dirac cones. The emergence of the Dirac cone is determined by satisfying the following conditions simultaneously in $\mathbf{H}(\mathbf{k})$ [12]:

$$h_{\text{H-H}} = h_{\text{L-L}} \quad (11)$$

$$h_{\text{H-L}} = 0. \quad (12)$$

Equation (11) represents the HOMO–LUMO band crossing, and Equation (12) indicates a node of the HOMO–LUMO coupling. When the HOMO–LUMO coupling $h_{\text{H-L}}$ is set to zero, the HOMO and LUMO bands intersect, resulting in a large Fermi surface as shown in Figures 3b and 9b. Notably, the energy at the intersection is situated near the Fermi level, because the HOMO is fully occupied and the LUMO is entirely empty in the isolated molecule state. Figure 14 illustrates that the intersection forms a (green) loop on the k_y-k_z plane ($k_x = 0$) for [Ni(tmdt)$_2$] and [Ni(btdt)$_2$]. In real systems, however, the HOMO–LUMO coupling induces band hybridization and maintains a gap between the two bands, known as band repulsion. An exception occurs at the node of the HOMO–LUMO coupling. In Figure 14, the node ($h_{\text{H-L}} = 0$) is represented by the orange line, and the intersection point with the green loop gives the Dirac point where Equations (11) and (12) are satisfied.

This mechanism operates as long as the HOMO and LUMO bands cross each other. The band crossing arises due to the small energy gap between the HOMO and LUMO (Δ), leading to the emergence of Dirac cones within a specific range of the Δ value ($\Delta < 0.62$ eV for [Ni(tmdt)$_2$] and $\Delta < 0.37$ eV for [Ni(btdt)$_2$], respectively). Furthermore, the curvature of the two bands is also a crucial factor for the band crossing. Bands with opposite curvatures tend to intersect each other. The band curvature is associated with the sign of the transfer integrals. The stretcher bond type molecular arrangement facilitates HOMO–HOMO and LUMO–LUMO transfer integrals with opposite signs. In Figure 15, the transfer integrals in the face-to-face stacking arrangement are mapped as a function of the offset from the center of the molecule (Pt or Ni) for the [Pt(dmdt)$_2$] and [Ni(tmdt)$_2$] molecules. The maps for both compounds are very similar to each other. In the region of Figure 15, there are two (positive or negative) peaks around $s = 0$ Å and three (positive or negative) peaks around $s = 1.75$ Å. The transfer integrals p are situated around $s = 0$ Å and the transfer integrals c are located around $s = 0$ Å ([Pt(dmdt)$_2$]) or 1.75 Å ([Ni(tmdt)$_2$]). Importantly, the HOMO–HOMO and LUMO–LUMO transfer integrals exhibit opposite signs almost everywhere.

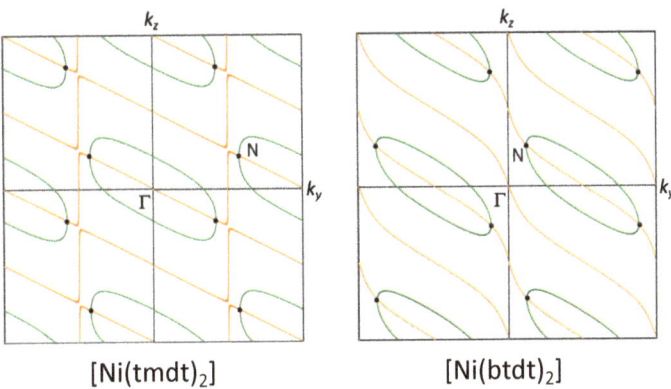

Figure 14. Intersection of the HOMO and LUMO bands ($h_{\text{H-H}} = h_{\text{L-L}}$: green loop), node of the HOMO–LUMO coupling ($h_{\text{H-L}} = 0$: orange line), and the Dirac points (black dots) on the $k_x = 0$ plane (See text).

The [Ni(btdt)$_2$] molecule, with the terminal benzene rings, provides a different pattern in the maps within the $L > 14$ Å region (Figure 16). Nevertheless, even in this case, the HOMO–HOMO and LUMO–LUMO transfer integrals maintain opposite signs.

As observed above, although the face-to-face stacking mode in the stretcher bond type molecular arrangement displays several variations, each mode tends to yield significantly strong HOMO–HOMO and LUMO–LUMO couplings with opposite signs, thereby inducing band crossing. In the early stages of research on single-component molecular conductors, the presence of crossing band structures was considered undesirable for achieving a metallic state due to the band repulsion, and a parallel band structure composed of HOMO and LUMO bands with the same curvature was preferred [5]. This viewpoint, however, is applicable only to one-dimensional systems. Currently, single-component molecular metals with two- or three-dimensional crossing band structures are not uncommon and can be understood within the framework of the conventional tight-binding band picture. The truly unique aspect of single-component molecular conductors with crossing band structure is the emergence of the Dirac electron system, which represents a new direction for the study of functional single-component molecular conductors.

Unfortunately, both [Ni(tmdt)$_2$] and [Ni(btdt)$_2$] do not clearly exhibit the nature of the Dirac electron system. In [Ni(tmdt)$_2$], the (semi)metallic nature associated with the large Fermi surface dominates electronic properties. This is because the energy at the Dirac point is largely shifted away from the Fermi level (Figure 6). On the other hand, for [Ni(btdt)$_2$], we considered the effects of antiferromagnetic (AFM) spin order, as demonstrated in the following subsection.

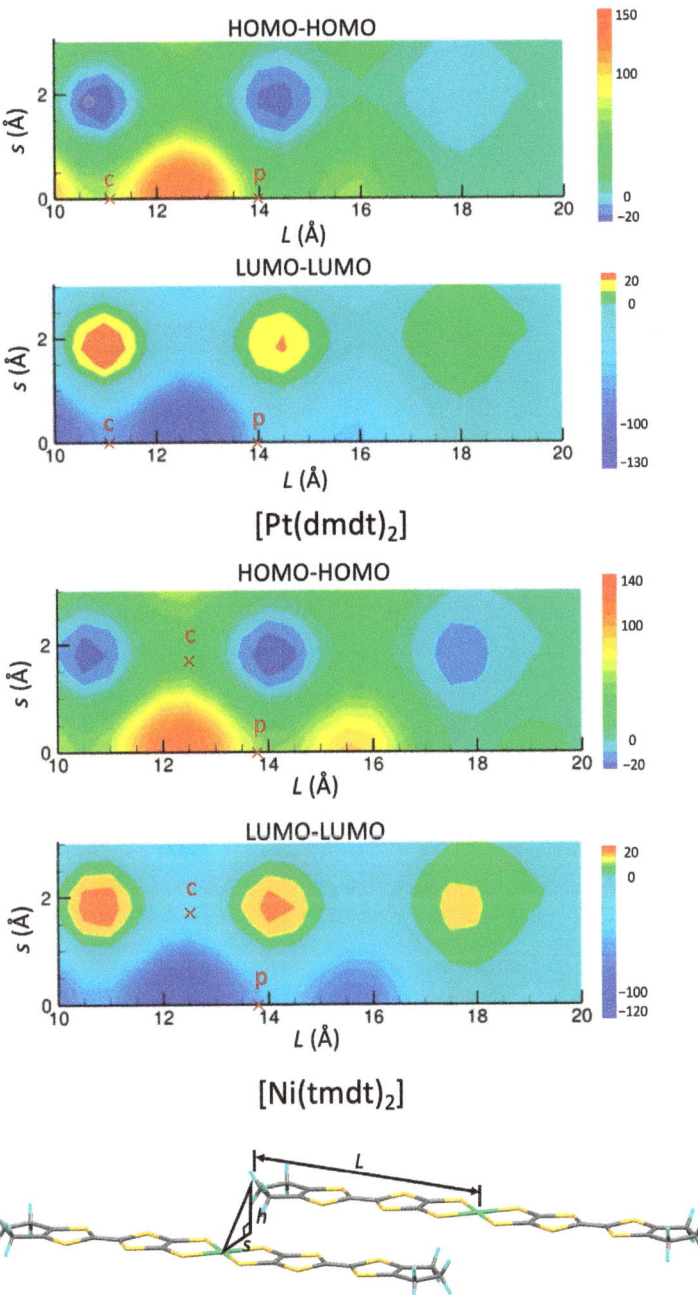

Figure 15. Map of HOMO–HOMO and LUMO–LUMO transfer integrals (meV) between two molecules in the face-to-face stacking arrangement for [Pt(dmdt)$_2$] and [Ni(tmdt)$_2$], as a function of a shift along the long molecular axis (L) and a shift along the short molecular axis (s). Interplanar distance h is fixed to 3.5 Å. The symbols p and c indicate transfer integrals, $t_{\text{pH-H}}$, $t_{\text{pL-L}}$, $t_{\text{cH-H}}$, and $t_{\text{cL-L}}$.

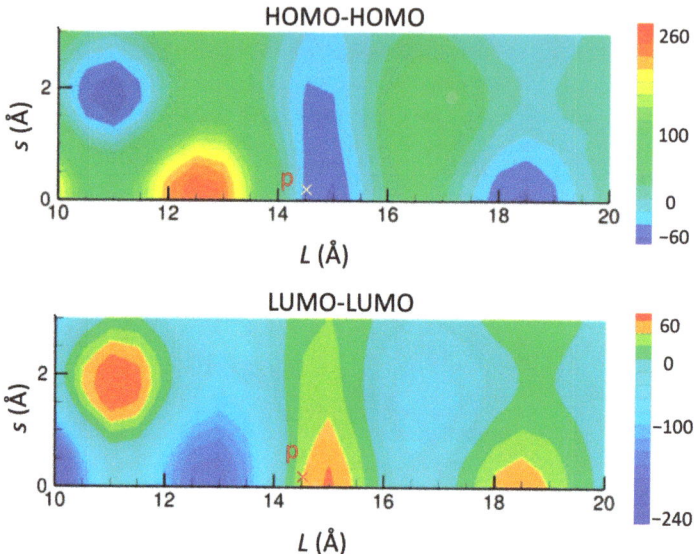

Figure 16. Map of HOMO–HOMO and LUMO–LUMO transfer integrals (meV) between two molecules in the face-to-face stacking arrangement for [Ni(btdt)$_2$], as a function of a shift along the long molecular axis (L) and a shift along the short molecular axis (s). Interplanar distance h is fixed to 3.0 Å. The symbol p indicates transfer integrals $t_{\text{pH-H}}$ and $t_{\text{pL-L}}$ (we do not consider $t_{\text{cH-H}}$, and $t_{\text{cL-L}}$ that are much smaller due to the longer interplanar distance).

4.2. Spin Ordering in [Ni(btdt)$_2$]: Antiferromagnetic HSE06 Calculations

To consider the AFM order, the magnetic unit cell needs to contain a minimum of two [Ni(btdt)$_2$] molecules, unlike the single molecule found in the unit cell of the experimental structure. To construct a magnetic cell, the original lattice vector of the crystal structure optimized using the vdW-DF2 functional is transformed using a rotation matrix consisting of (1, −1, 0), (1, 1, 0), and (0, 0, 1), with an origin shift of (0.5, 0.5, 0). The resulting lattice parameters are a = 6.98, b = 13.38, c = 14.46 Å, α = 110.43, β = 107.53, and γ = 69.32°.

We have found that a range-separated hybrid functional developed by Heyd, Scuseria, and Ernzerhof (HSE) is effective in describing the electronic states of molecular charge order systems [37,38]. The hybrid functional method calculates the exchange-correlation energy by combining the exact Fock exchange with the exchange energy functional from GGA.

As shown in Figure 17a, a finite band gap of 0.034 eV is obtained from spin-polarized HSE06 calculations. In the HSE calculations, the magnetic moments at the Ni site and (btdt)$_2$ ligands are ±0.05 and ±0.18 μ_B, respectively. Thus, the total magnetic moment per [Ni(btdt)$_2$] molecule is 0.23 μ_B. The magnetic moment of the ligand is determined by the summation of the magnetic moments of the C and S atoms belonging to the ligands. In contrast, the GGA calculations yield a magnetic moment of ±0.02 μ_B at the Ni site and ±0.06 μ_B for the ligands. A small band gap of approximately 0.01 eV is opened within the spin-polarized GGA, as plotted in Figure 17b.

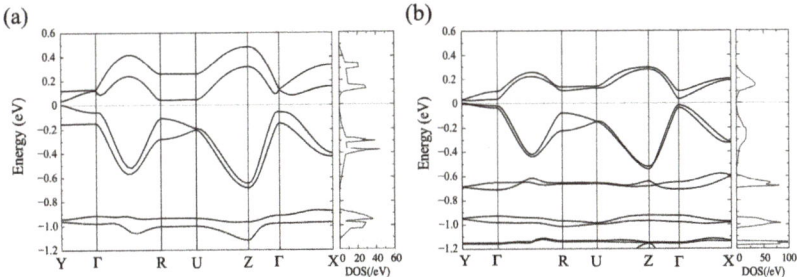

Figure 17. Band dispersion and total DOS for the AFM state in [Ni(btdt)$_2$] calculated with (**a**) HSE06 and (**b**) GGA-PBE functionals. The zero energy at the longitudinal axis is set at the top of the valence bands. Y = (0,1/2,0), Γ = (0,0,0), R = (1/2,1/2,1/2), U = (1/2,0,1/2), Z = (0,0,1/2), and X = (1/2,0,0), in units of the reciprocal lattice vectors.

Next, we discuss the AFM pattern of spin ordering. Figure 18 illustrates the spin densities in the AFM state of [Ni(btdt)$_2$]. As shown in Figure 18a, the spins alternate in a pattern along the b axis of the magnetic unit cell. The spin densities are mainly localized on the four S atoms closest to the Ni atom, as depicted in Figure 18b. The presence of the d_{xz} orbital of the Ni atom indicates that the LUMO of the [Ni(btdt)$_2$] molecule contributes to the spin densities (see supporting materials of Ref. [13]). We note that these spin density and DOS analyses have been widely used in various other studies that describe the electronic structure of materials [39,40].

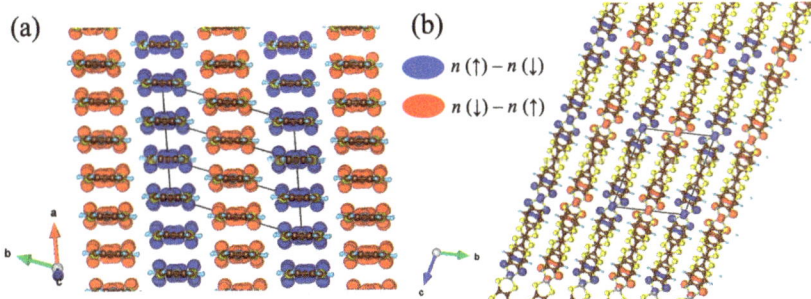

Figure 18. Contour plots of the spin densities of an AFM state in [Ni(btdt)$_2$]. (**a**) ab and (**b**) bc planes. The isosurface level of the spin density is set as 0.00043. The figures are plotted using VESTA software version 3 [41]. n (↑) and n (↓) are the spin-up and spin-down densities calculated with a HSE06 functional.

In the experiments, the magnetic susceptibility above 25 K follows the Curie-Weiss law, with the constants C (0.52 emu K mol^{-1}) and θ (−28.63 K), indicating that intermolecular antiferromagnetic coupling is the dominant factor [13]. We obtain a stable AFM state through the spin-polarized HSE calculations using the structure optimized by the vdW-DF2 functional. On the other hand, using the experimental structure [13], we can stabilize an AFM state with a different ordering pattern from Figure 18, which was found to be metallic. Although the HSE functional includes a portion of the exact exchange, it does not account for the correlation effect. In the future, a semiconducting band structure can be achieved using methods that consider the correlation effect, such as GW approximation and dynamical mean field theory calculations.

5. Conclusions

Both the tight-binding model and the DFT calculation have indicated the possibility of Dirac cone formation in two single-component molecular conductors derived from metal complexes with extended tetrathiafulvalenedithiolate ligands, [Ni(tmdt)$_2$] and [Ni(btdt)$_2$]. Despite various band calculations performed on these systems, the formation of Dirac cones has never been reported before. This is because the close relationship between single-component molecular conductors and the Dirac electron system has not been recognized. Our proposed mechanism suggests that the stretcher bond type molecular arrangements favor the formation of Dirac cones, providing an important clue for the development of new Dirac electron systems. The manifestation of the Dirac electron system depends primarily on the proximity of the Dirac point to the Fermi level. Furthermore, the study of [Ni(btdt)$_2$] has revealed the importance of antiferromagnetic spin order effects. As observed in [Ni(btdt)$_2$], the results obtained from the tight-binding model and DFT calculations do not always agree. Although resolving this discrepancy remains a future challenge, this work has demonstrated that combining the strengths of both methods is a powerful tool for the design of functional molecular materials.

Author Contributions: Conceptualization, R.K. and T.T.; methodology, R.K. and T.T.; formal analysis, R.K. and T.T.; investigation, R.K. and T.T.; writing—original draft preparation, R.K. and T.T.; writing—review and editing, R.K. All authors have read and agreed to the published version of the manuscript.

Funding: This research was funded by a Grant-in-Aid for Scientific Research (JP19K21860) from the Japan Society for the Promotion of Science (JSPS) and JST, CREST Grant Number JPMJCR2094, Japan. This work was performed under the GIMRT Program of the Institute for Materials Research (IMR), Tohoku University (Proposal No. 202012-RDKGE-0034 and 202212-RDKGE-0048).

Institutional Review Board Statement: Not applicable.

Informed Consent Statement: Not applicable.

Data Availability Statement: The data are available from corresponding authors.

Acknowledgments: The study's computations were mainly conducted using the computer facilities of MASAMUNE at IMR, Tohoku University, Japan.

Conflicts of Interest: The authors declare no conflict of interest.

Appendix A. Crystal Data of [Ni(btdt)$_2$] Optimized with the vdw-df2-b86r Functional

Space group: $P\bar{1}$
Lattice constants: $a_o = 6.36$, $b_o = 8.57$, $c_o = 14.46$ Å, $\alpha_o = 113.28$, $\beta_o = 78.34$, $\gamma_o = 126.73°$
Fractional atomic coordinates:

Atom	x	y	z
Ni	0.0000	0.0000	0.0000
S1	0.2206	−0.0388	0.0741
S2	−0.1432	0.1154	0.1400
S3	0.3365	0.0884	0.2961
S4	−0.0134	0.2300	0.3580
S5	0.4602	0.2102	0.5300
S6	0.1176	0.3598	0.5927
C1	0.1804	0.0637	0.1994
C2	0.5499	0.3109	0.7322
C3	0.0192	0.1324	0.2288
C4	0.1948	0.1925	0.3948
C5	0.2495	0.2466	0.4943
C6	0.4262	0.3049	0.6600

Atom	x	y	z
C7	0.2691	0.3799	0.6900
C8	0.2393	0.4651	0.7923
C9	0.3672	0.4743	0.8641
C10	0.5199	0.3961	0.8342
H1	0.1208	0.5274	0.8155
H2	0.3477	0.5446	0.9444
H3	0.6189	0.4020	0.8906
H4	0.6678	0.2480	0.7085

References

1. Tanaka, H.; Okano, Y.; Kobayashi, H.; Suzuki, W.; Kobayashi, A. A Three-Dimensional Synthetic Metallic Crystal Composed of Single-Component Molecules. *Science* **2001**, *291*, 285–287. [CrossRef] [PubMed]
2. Rovira, C.; Novoa, J.J.; Mozos, J.-L.; Ordejon, P.; Canadell, E. First-principles study of the neutral molecular metal Ni(tmdt)$_2$. *Phys. Rev. B* **2002**, *65*, 081104. [CrossRef]
3. Tanaka, H.; Tokumoto, M.; Ishibashi, S.; Graf, D.; Choi, E.S.; Brooks, J.S.; Yasuzuka, S.; Okano, Y.; Kobayashi, H.; Kobayashi, A. Observation of Three-Dimensional Fermi Surfaces in a Single-Component Molecular Metal, [Ni(tmdt)$_2$]. *J. Am. Chem. Soc.* **2004**, *126*, 10518–10519. [CrossRef] [PubMed]
4. Seo, H.; Ishibashi, S.; Okano, Y.; Kobayashi, H.; Kobayashi, A.; Fukuyama, H.; Terakura, K. Single-Component Molecular Metals as Multiband π–d Systems. *J. Phys. Soc. Jpn.* **2008**, *77*, 023714. [CrossRef]
5. Kobayashi, A.; Fujiwara, E.; Kobayashi, H. Single-Component Molecular Metals with Extended-TTF Dithiolene Ligands. *Chem. Rev.* **2004**, *104*, 5243–5264. [CrossRef]
6. Velho, M.F.G.; Silva, R.A.L.; Belo, D. The quest for single component molecular metals within neutral transition metal complexes. *J. Mater. Chem. C* **2021**, *9*, 10591–10609. [CrossRef]
7. Murakami, S. Phase transition between the quantum spin Hall and insulator phases in 3D: Emergence of a topological gapless phase. *New J. Phys.* **2007**, *9*, 356. [CrossRef]
8. Fang, C.; Weng, H.; Dai, X.; Fang, Z. Topological nodal line semimetals. *Chin. Phys. B* **2016**, *25*, 117106. [CrossRef]
9. Kato, R.; Cui, H.; Tsumuraya, T.; Miyazaki, T.; Suzumura, Y. Emergence of the Dirac Electron System in a Single-Component Molecular Conductor under High Pressure. *J. Am. Chem. Soc.* **2017**, *139*, 1770–1773. [CrossRef]
10. Kato, R.; Cui, H.; Minamidate, T.; Yeung, H.H.-M.; Suzumura, Y. Electronic Structure of a Single-Component Molecular Conductor [Pd(dddt)$_2$] (dddt = 5,6-dihydro-1,4-dithiin-2,3-dithiolate) under High Pressure. *J. Phys. Soc. Jpn.* **2020**, *89*, 124706. [CrossRef]
11. Zhou, B.; Ishibashi, S.; Ishii, T.; Sekine, T.; Takehara, R.; Miyagawa, K.; Kanoda, K.; Nishibori, E.; Kobayashi, A. Single-component molecular conductor [Pt(dmdt)$_2$]—A three-dimensional ambient-pressure molecular Dirac electron system. *Chem. Commun.* **2019**, *55*, 3327–3330. [CrossRef]
12. Kato, R.; Suzumura, Y. A Tight-binding Model of an Ambient-pressure Molecular Dirac Electron System with Open Nodal Lines. *J. Phys. Soc. Jpn.* **2020**, *89*, 044713. [CrossRef]
13. Wen, H.-R.; Li, C.-H.; Song, Y.; Zuo, J.-L.; Zhang, B.; You, X.-Z. Synthesis and Magnetic Properties of a Highly Conducting Neutral Nickel Complex with a Highly Conjugated Tetrathiafulvalenedithiolate Ligand. *Inorg. Chem.* **2007**, *46*, 6837–6839. [CrossRef]
14. Summerville, R.H.; Hoffmann, R. Tetrahedral and other M$_2$L$_6$ transition metal dimers. *J. Am. Chem. Soc.* **1976**, *98*, 7240–7254. [CrossRef]
15. Clementi, E.; Roetti, C. Basis Functions and Their Coefficients for Ground and Certain Excited States of Neutral and Ionized Atoms, Z ≤ 54. *At. Data Nucl. Data Tables* **1974**, *14*, 177. [CrossRef]
16. Wimmer, E.; Krakauer, H.; Weinert, M.; Freeman, A.J. Full-potential self-consistent linearized-augmented-plane-wave method for calculating the electronic structure of molecules and surfaces: O$_2$ molecule. *Phys. Rev. B* **1981**, *24*, 864–875. [CrossRef]
17. Weinert, M. Solution of Poisson's equation: Beyond Ewald-type methods. *J. Math. Phys.* **1981**, *22*, 2433–2439. [CrossRef]
18. Perdew, J.P.; Burke, K.; Ernzerhof, M. Generalized Gradient Approximation Made Simple. *Phys. Rev. Lett.* **1996**, *77*, 3865. [CrossRef]
19. Tsumuraya, T.; Kato, R.; Suzumura, Y. Effective Hamiltonian of Topological Nodal Line Semimetal in Single-Component Molecular Conductor [Pd(dddt)$_2$] from First-Principles. *J. Phys. Soc. Jpn.* **2018**, *87*, 113701. [CrossRef]
20. Kitou, S.; Tsumuraya, T.; Sawahata, H.; Ishii, F.; Hiraki, K.; Nakamura, T.; Katayama, N.; Sawa, H. Ambient-pressure Dirac electron system in the quasi-two-dimensional molecular conductor α-(BETS)$_2$I$_3$. *Phys. Rev. B* **2021**, *103*, 035135. [CrossRef]
21. Dion, M.; Rydberg, H.; Schröder, E.; Langreth, D.C.; Lundqvist, B.I. Van der Waals Density Functional for General Geometries. *Phys. Rev. Lett.* **2004**, *92*, 246401. [CrossRef] [PubMed]
22. Thonhauser, T.; Cooper, V.R.; Li, S.; Puzder, A.; Hyldgaard, P.; Langreth, D.C. Van der Waals density functional: Self-consistent potential and the nature of the van der Waals bond. *Phys. Rev. B* **2007**, *76*, 125112. [CrossRef]
23. Sabatini, R.; Küçükbenli, E.; Kolb, B.; Thonhauser, T.; Girnocoli, S. Structural evolution of amino acid crystals under stress from a non-empirical density functional. *J. Phys. Condens. Matter* **2021**, *24*, 424209. [CrossRef] [PubMed]
24. Hamada, I. van der Waals density functional made accurate. *Phys. Rev. B* **2014**, *89*, 121103(R). [CrossRef]

25. Giannozzi1, P.; Andreussi, O.; Brumme, T.; Bunau, O.; Nardelli, M.B.; Calandra, M.; Car, R.; Cavazzoni, C.; Ceresoli, D.; Cococcioni, M.; et al. Advanced capabilities for materials modelling with Quantum ESPRESSO. *J. Phys. Cond. Matter* **2017**, *29*, 465901. [CrossRef]
26. Garrity, K.F.; Bennett, J.W.; Rabe, K.M.; Vanderbilt, D. Pseudopotentials for high-throughput DFT calculations. *Comput. Mater. Sci.* **2014**, *81*, 446–452. [CrossRef]
27. Miyazaki, T.; Ohno, T. First-principles study of pressure effects on the molecular solids $(CH_3)_4X[M(dmit)_2]_2$ (X = N, P and M = Ni, Pd). *Phys. Rev. B* **2003**, *68*, 035116. [CrossRef]
28. Heyd, J.; Scuseria, G.E.; Ernzerhof, M. Hybrid functionals based on a screened Coulomb potential. *J. Chem. Phys.* **2003**, *118*, 8207–8215, Erratum in *J. Chem. Phys.* **2006**, *124*, 219906. [CrossRef]
29. Heyd, J.; Scuseria, G.E. Efficient hybrid density functional calculations in solids: Assessment of the Heyd–Scuseria–Ernzerhof screened Coulomb hybrid functional. *J. Chem. Phys.* **2004**, *121*, 1187–1192. [CrossRef]
30. Kresse, G.; Hafner, J. Ab initio molecular dynamics for liquid metals. *Phys. Rev. B* **1993**, *47*, 558–561. [CrossRef]
31. Kresse, G.; Hafner, J. Ab initio molecular-dynamics simulation of the liquid-metal–amorphous-semiconductor transition in germanium. *Phys. Rev. B* **1994**, *49*, 14251–14269. [CrossRef]
32. Kresse, G.; Furthmüller, J. Efficient iterative schemes for ab initio total-energy calculations using a plane-wave basis set. *Phys. Rev. B* **1996**, *54*, 11169–11186. [CrossRef]
33. Blöchl, P.E. Projector augmented-wave method. *Phys. Rev. B* **1994**, *50*, 17953–17979. [CrossRef]
34. Kresse, G.; Joubert, D. From ultrasoft pseudopotentials to the projector augmented-wave method. *Phys. Rev. B* **1999**, *59*, 1758–1775. [CrossRef]
35. Kawamura, M. FermiSurfer: Fermi-surface viewer providing multiple representation schemes. *Comp. Phys. Commun.* **2019**, *239*, 197. [CrossRef]
36. Cui, H.B.; Kobayashi, H.; Ishibashi, S.; Sasa, M.; Iwase, F.; Kato, R.; Kobayashi, A. A Single-Component Molecular Superconductor. *J. Am. Chem. Soc.* **2014**, *136*, 7619–7622. [CrossRef]
37. Tsumuraya, T.; Seo, H.; Miyazaki, T. First-principles study of the charge ordered phase in κ−D_3(Cat-EDT-TTF/ST)$_2$: Stability of π-electron deuterium coupled ordering in hydrogen-bonded molecular conductors. *Phys. Rev. B* **2020**, *101*, 045114. [CrossRef]
38. Tsumuraya, T.; Seo, H.; Miyazaki, T. First-Principles Study on the Stability and Electronic Structure of the Charge-Ordered Phase in α-(BEDT-TTF)$_2$I$_3$. *Crystals* **2021**, *11*, 1109. [CrossRef]
39. Lv, Z.; Xu, H.; Xu, W.; Peng, B.; Zhao, C.; Xie, M.; Lv, X.; Gao, Y.; Hu, K.; Fang, Y.; et al. Quasi-Topological Intercalation Mechanism of $Bi_{0.67}NbS_2$ Enabling 100 C Fast-Charging for Sodium-Ion Batteries. *Adv. Energy Mater* **2023**, 2300790. [CrossRef]
40. Xiao, W.; Kiran, G.K.; Yoo, K.; Kim, J.-H.; Xu, H. The Dual-Site Adsorption and High Redox Activity Enabled by Hybrid Organic-Inorganic Vanadyl Ethylene Glycolate for High-Rate and Long-Durability Lithium–Sulfur Batteries. *Small* **2023**, *19*, 2206750. [CrossRef]
41. Momma, K.; Izumi, F. VESTA 3 for three-dimensional visualization of crystal, volumetric and morphology data. *J. Appl. Crystallogr.* **2011**, *44*, 1272–1276. [CrossRef]

Disclaimer/Publisher's Note: The statements, opinions and data contained in all publications are solely those of the individual author(s) and contributor(s) and not of MDPI and/or the editor(s). MDPI and/or the editor(s) disclaim responsibility for any injury to people or property resulting from any ideas, methods, instructions or products referred to in the content.

Article

On the Size of Superconducting Islands on the Density-Wave Background in Organic Metals

Vladislav D. Kochev [1], Seidali S. Seidov [1] and Pavel D. Grigoriev [1,2,*]

1. Department of Theoretical Physics and Quantum Technology, National University of Science and Technology "MISiS", 119049 Moscow, Russia; vd.kochev@gmail.com (V.D.K.); alikseidov@yandex.ru (S.S.S.)
2. L.D. Landau Institute for Theoretical Physics, 142432 Chernogolovka, Russia
* Correspondence: grigorev@itp.ac.ru

Abstract: Most high-T_c superconductors are spatially inhomogeneous. Usually, this heterogeneity originates from the interplay of various types of electronic ordering. It affects various superconducting properties, such as the transition temperature, the magnetic upper critical field, the critical current, etc. In this paper, we analyze the parameters of spatial phase segregation during the first-order transition between superconductivity (SC) and a charge- or spin-density wave state in quasi-one-dimensional metals with imperfect nesting, typical of organic superconductors. An external pressure or another driving parameter increases the transfer integrals in electron dispersion, which only slightly affects SC but violates the Fermi surface nesting and suppresses the density wave (DW). At a critical pressure P_c, the transition from a DW to SC occurs. We estimate the characteristic size of superconducting islands during this phase transition in organic metals in two ways. Using the Ginzburg–Landau expansion, we analytically obtain a lower bound for the size of SC domains. To estimate a more specific interval of the possible size of the superconducting islands in (TMTSF)$_2$PF$_6$ samples, we perform numerical calculations of the percolation probability via SC domains and compare the results with experimental resistivity data. This helps to develop a consistent microscopic description of SC spatial heterogeneity in various organic superconductors.

Keywords: superconductivity; CDW (charge-density waves); SDW (spin-density wave); phase diagram; organic superconductor

Citation: Kochev, V.D.; Seidov, S.S.; Grigoriev, P.D. On the Size of Superconducting Islands on the Density-Wave Background in Organic Metals. *Magnetochemistry* **2023**, 9, 173. https://doi.org/10.3390/magnetochemistry9070173

Academic Editor: Carlos J. Gómez García

Received: 23 May 2023
Revised: 15 June 2023
Accepted: 28 June 2023
Published: 4 July 2023

Copyright: © 2023 by the authors. Licensee MDPI, Basel, Switzerland. This article is an open access article distributed under the terms and conditions of the Creative Commons Attribution (CC BY) license (https://creativecommons.org/licenses/by/4.0/).

1. Introduction

Superconductivity (SC) often competes [1–3] with charge-density wave (CDW) or spin-density wave (SDW) electronic instabilities [3,4], as both create an energy gap on the Fermi level. In such materials, the density wave (DW) is suppressed by some external parameter, which deteriorates the nesting property of the Fermi surface (FS) and enables superconductivity. The driving parameters are usually the chemical composition (doping level) and pressure, as in cuprate- [5–10] or iron-based high-T_c superconductors [11,12], organic superconductors (OSs) [13–26], transition metal dichalcogenides [1–3], etc. The DW can also be suppressed [27] or enhanced [28,29] by disorder. The latter happens, e.g., in (TMTSF)$_2$ClO$_4$ organic superconductors [13,14,28–30], where the disorder is controlled by the cooling rate during the anion ordering transition. Anion ordering splits the electron spectrum, which deteriorates the FS nesting and dampens the SDW, enabling SC.

The SC–DW interplay is much more interesting than just a competition. Usually, the SC transition temperature, T_c, is the highest in the coexistence region near the quantum critical point where the DW disappears [1,2,20,21]. This is attributed to the enhancement of Cooper pairing by the critical DW fluctuations, similar to cuprate high-T_c superconductors [31]. This enhancement is also common for other types of quantum critical points, such as antiferromagnetic (AFM) points in cuprate [32,33] or heavy fermion [34] superconductors, ferromagnetic points [35], nematic phase transitions in Fe-based superconductors [36,37], etc. The enhancement of electron–electron (e–e) interactions in the Cooper channel already appears in the random-phase approximation, and the resulting strong momentum dependence of e–e coupling may lead to unconventional superconductivity [38]. The spin-dependent coupling to an SDW may additionally affect the SC in the case of their microscopic coexistence and even favor triplet SC pairing [39,40]. Generally, any antiferromagnetic background changes the spin structure of eigenstates and the electronic g-factor, as was studied both theoretically and experimentally in cuprate and organic superconductors [41]. The upper critical field H_{c2} is often several times higher in the coexistence region than in a pure SC phase [18,25], which may be useful for applications.

OSs are helpful for investigating the SC–DW interplay because they have rather weak electronic correlations and low DW and SC transition temperatures [13,14], which is convenient for their theoretical and experimental study. However, their phase diagram, layered crystal structure and many other features are very similar to those of high-T_c superconductors. Moreover, by changing the chemical composition or pressure in OSs, one can easily vary the electronic dispersion in a wide interval and even change the FS topology from quasi-1D (Q1D) to quasi-2D. Large and pure monocrystals of organic metals can be synthesized, so that their electronic structure can be experimentally studied by high-magnetic-field tools [42] and by other experimental techniques [13,14].

To understand the DW–SC interplay and the influence of DW on SC properties in OSs, one needs to know the microscopic structure of their coexistence. Each of these ground states creates an energy gap on the Fermi level and removes the FS instability. Hence, the DW and SC must be somehow separated in the momentum or coordinate space. The momentum space DW–SC separation assumes a spatially uniform structure, where the FS is only partially gapped by the DW, and the non-gapped parts of the FS maintain SC [3,40]. The resistivity hysteresis observed in (TMTSF)$_2$PF$_6$ [19] suggests the spatial DW/SC segregation in OSs. Microscopic SC domains of size d comparable to the DW coherence length ζ_{DW} may emerge due to the soliton DW structure [39,43–46]. However, such a small size of SC or metallic domains contradicts the angular magnetoresistance oscillations (AMROs) in the region of SC/DW coexistence, observed both in (TMTSF)$_2$ClO$_4$ [30] and in (TMTSF)$_2$PF$_6$ [21] and implying the domain width $d > 1$ μm [21,30].

The observed [18,25] enhancement of the SC upper critical field H_{c2} in OSs is possible in all of the above scenarios [40,45]. Spatial DW–SC segregation only requires a SC domain on the order of the penetration depth λ of the magnetic field into the superconductor [47]. In (TMTSF)$_2$ClO$_4$, the penetration depth within the TMTSF layers is [48] $\lambda_{ab}(T = 0) \approx 0.86$ μm, and increases with $T \to T_{cSC}$. Hence, the macroscopic spatial phase separation with a SC domain size $d > 1$ μm suggested by AMRO data [21,30] is consistent with the observed H_{c2} enhancement in the DW–SC coexistence phase.

Another interesting feature of SDW/SC coexistence in OSs is the anisotropic SC onset, opposite to a weak intrinsic interlayer Josephson coupling in high-T_c superconductors [47]; the SC transition and the zero resistance in OSs was first observed [20,21,28] only along the least-conducting interlayer z-direction, then along the two least-conducting directions, z and y, and only finally in all three directions. This anisotropic SC onset was explained recently [49] by assuming a spatial SC/DW separation and studying the percolation in finite-size samples with a thin elongated shape relevant to the experiments on (TMTSF)$_2$PF$_6$ [20,21] and (TMTSF)$_2$ClO$_4$ [28,29]. This additionally supports the scenario of spatial SC/DW segregation in the form of rather large domains of width $d > 1$ μm. However, the microscopic reason for such phase segregation remains unknown. Similar anisotropic SC onset and even T_c enhancement in FeSe mesa structures was observed and explained by heterogeneous SC inception [50]. The spatial segregation in FeSe and some other Fe-based high-T_c superconductors probably originates from the so-called nematic phase transition and domain structure, but similar electronic ordering is absent in OSs.

Recently, the DW–metal phase transition in OSs was shown to be of first order [51], which suggests that the spatial DW–SC segregation may be due to phase nucleation during this transition. In this paper, we estimate the typical size of superconducting islands in organic metals with two different methods. In Section 2, we formulate a model and the

Landau–Ginzburg functional for free energy in the DW state. In Section 3.1, we analytically obtain a lower bound for the size of the superconducting islands. In Section 3.2, we discuss the relationship between the DW coherence length and the SC nucleation size during the first-order phase transition. In Section 3.3, we perform numerical calculations of the percolation probability, from which we determine the interval of possible sizes of the superconducting islands in $(TMTSF)_2PF_6$. In Section 4, we discuss our results in connection with the experimental observations of $(TMTSF)_2PF_6$ and in other superconductors.

2. The Model

2.1. Q1D Electron Dispersion and the Driving Parameters of DW–Metal/SC Phase Transitions in OSs

In Q1D organic metals [13,14], the free electron dispersion near the Fermi level is approximately given by

$$\varepsilon(\mathbf{k}) = \hbar v_F(|k_x| - k_F) + t_\perp(\mathbf{k}_\perp), \tag{1}$$

where v_F and k_F are the Fermi velocity and Fermi momentum in the chain x-direction. The interchain electron dispersion $t_\perp(\mathbf{k}_\perp)$ is given by the tight-binding model:

$$t_\perp(\mathbf{k}_\perp) = 2t_b \cos(k_y b) + 2t'_b \cos(2k_y b), \tag{2}$$

where b is the lattice constant in the y-direction. The dispersion along the interlayer z-axis is usually significantly less than along the y-axis; thus, it is left out here. In $(TMTSF)_2PF_6$, the transfer integral t_b is ≈ 30 meV [52], and the "antinesting" parameter t'_b is ≈ 4.5 K [53] at ambient pressure.

As illustrated in Figure 1b, the FS of Q1D metals consists of two slightly warped sheets separated by $2k_F$ and roughly exhibits the nesting property.

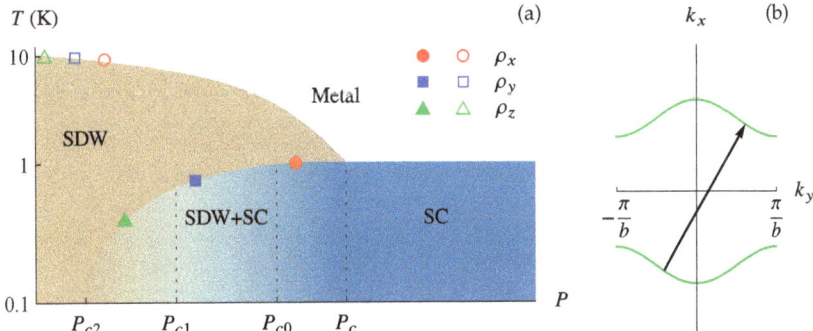

Figure 1. (a) Pressure–temperature phase diagram of $(TMTSF)_2PF_6$ recreated from resistivity data in ref. [20]; (b) schematic FS of $(TMTSF)_2PF_6$, obtained from the Q1D electron dispersion given by Equations (1) and (2). The nesting vector $\mathbf{Q}$ is indicated by the black arrow.

It leads to the Peierls instability and favors the formation of DWs at low temperatures $T < T_{cDW} \equiv T_c$, which competes with superconductivity. The quasiparticle dispersion in the DW state in the mean-field approximation is given by

$$E_\pm(\mathbf{k}) = \varepsilon_+(\mathbf{k}, \mathbf{k} - \mathbf{Q}) \pm \sqrt{|\Delta_Q|^2 + \varepsilon_-^2(\mathbf{k}, \mathbf{k} - \mathbf{Q})}, \tag{3}$$

where we have used the notations

$$\varepsilon_\pm(\mathbf{k}, \mathbf{k}') = \frac{\varepsilon(\mathbf{k}) \pm \varepsilon(\mathbf{k}')}{2}. \tag{4}$$

The FS has the property of perfect nesting at the wave vector Q if $\varepsilon_+(k, k - Q) = 0$. If $\varepsilon_+(k, k - Q) < |\Delta_Q|$ for the entire FS, all electron states are gapped at the Fermi level due to DW formation. Then, the DW converts to a semiconducting state at $T < T_{cDW}$ and SC does not emerge. If $\varepsilon_+(k, k - Q) > |\Delta_Q|$ in a finite interval of k at the Fermi level, the metallic state survives at $T < T_{cDW}$. Then, a uniform SC state may emerge, but its properties differ from those without DWs [39,40] because of the FS reconstruction and the change in electron dispersion by the DW. For $Q = Q_0 = (2k_F, \pi/b)$, only the second harmonic in the electron dispersion given by Equation (2) violates FS nesting: $\varepsilon_+(k, k - Q_0) = 2t'_b \cos(2k_y b)$. Hence, usually only $t'_b \sim t_b^2 / v_F k_F \ll t_b$ is important for the DW phase diagram.

With the increase in applied pressure P, the lattice constants decrease. This enhances the interchain electron tunneling and the transfer integrals. The increase in $t'_b(P)$ with pressure spoils the FS nesting and decreases the DW transition temperature $T_{cDW}(P)$. There is a critical pressure P_c and a corresponding critical value $t'^*_b = t'_b(P_c)$ at which $T_{cDW}(P_c) = 0$ and a quantum critical point (QCP) exists. The electronic properties at this DW QCP are additionally complicated by superconductivity emerging at $T < T_{cSC}$ at $P > P_c$. In organic metals, SC appears even earlier, at $P > P_{c1} < P_c$, and there is a finite region $P_{c1} < P < P_c$ of SC–DW coexistence [20,21,25]. This simple model qualitatively describes the phase diagram observed in (TMTSF)$_2$PF$_6$ [20,21,25], α-(BEDT-TTF)$_2$KHg(SCN)$_4$ [25], in various compounds of the (TMTTF)$_2$X family [26,54,55] and in many other OSs [13,14,16,17].

2.2. Mean Field Approach and the Landau–Ginzburg Expansion of DW Free Energy

Mean-field theory does not correctly describe strictly 1D conductors, where non-perturbative methods are helpful. However, in most DW materials, nonzero electron hopping between the conducting 1D chains and the 3D character of the electron–electron (e–e) interactions and lattice elasticity reduce the deviations from the mean-field solution and also make most of the methods and exactly solvable models developed for the strictly 1D case inapplicable. On the other hand, the interchain electron dispersion strongly dampens the fluctuations and validates the mean-field description [56,57]. The perpendicular-to-chain term $t_\perp(k_\perp)$ in Equations (1) and (2) is much greater than the energy scale of the DW transition temperature ($T_{c0} \approx 12.1$ K). Only the "imperfect nesting" term $\sim t'_b$ of $t_\perp(k_\perp)$ is on the order of T_{c0}. Hence, the criterion for the mean-field theory to be applicable [56,57], $t_\perp \gg T_{c0}$, is reliably satisfied in most Q1D organic metals.

For our analysis, we take the Landau–Ginzburg expansion of the free energy in the series of even powers of the DW order parameter $\Delta = \Delta_Q$:

$$F \simeq \frac{A(T, Q)}{2}|\Delta|^2 + \frac{B}{4}|\Delta|^4 + \frac{C}{6}|\Delta|^6 + \frac{D}{8}|\Delta|^8 + \dots \quad (5)$$

Usually, the minimum of the free energy corresponds to the uniform DW order parameter Δ when $Q = Q_0$. Since the coefficient $A(T_{cDW}, Q_0) = 0$, we keep its temperature and momentum dependence. The sign of the coefficient B determines the type of DW–metal phase transition. If $B_{DW} > 0$, the phase transition is of the second order, and only the first two coefficients A_{DW} and B_{DW} are sufficient for its description. If $B_{DW} < 0$, the phase transition may be of the first order and the coefficients C_{DW} and even D_{DW} if $C_{DW} \leq 0$ are required for its description. The self-consistency equation (SCE) for a DW is obtained by the variation in the free energy (5) with respect to Δ:

$$\Delta \left(A + B|\Delta|^2 + C|\Delta|^4 + D|\Delta|^6 + \dots \right) = 0. \quad (6)$$

The free energy (5) can also be calculated by integrating the SCE over Δ. In ref. [58], the SCE for the DW was derived in a magnetic field acting via Zeeman splitting and for two coupling constants of the e–e interaction, charge U_c and spin U_s (see Equation (17) in ref. [58]). Without a magnetic field, the charge U_c and spin U_s coupling constants do not couple, and the system chooses the largest one of them, corresponding to the highest transition temperature. We rewrite the SCE without a magnetic field and for only one charge or spin coupling constant U:

$$\Delta = -TU \sum_{k\omega} \frac{\Delta}{(\omega + i\varepsilon_+)^2 + \varepsilon_-^2 + |\Delta|^2}, \quad (7)$$

where $\varepsilon_\pm = \varepsilon_\pm(k, k-Q)$ are given by Equation (4), and ω takes the values $\pi T(2n+1)$, $n \in \mathbb{Z}$. In Appendix A, we briefly describe the derivation of Equation (7) and discuss the relation of coefficients in the Landau–Ginzburg expansion (5) with electronic susceptibility. The Landau–Ginzburg expansion coefficients in Equations (5) and (6) can be obtained by the expansion of Equation (7) in a power series of $|\Delta|^2$.

The sum over k in Equation (7) for a macroscopic sample is equivalent to the integral:

$$\sum_k = 2 \int \frac{dk_x}{2\pi} \int_{-\pi/b}^{\pi/b} \frac{dk_y}{2\pi}. \quad (8)$$

The factor of 2 appears because of two FS sheets are present at $k_x \approx \pm k_F$. Usually, for simplicity, the integration limits over k_x are taken to be infinite and the resulting logarithmic divergence of Equation (7) is regularized by the definition of the transition temperature T_{c0}. This procedure is briefly described in Appendix B of ref. [51]. When the Fermi energy $E_F \gg t_b$, for a linearized electron dispersion (1) near the Fermi level, one may integrate Equation (7) over k_x in infinite limits, which gives (cf. Equation (22) of ref. [58])

$$\Delta = \frac{\pi v_F |U| T}{2} \sum_\omega \left\langle \frac{\Delta}{\sqrt{(\omega + i\varepsilon_+)^2 + |\Delta|^2}} \right\rangle_{k_y}, \quad (9)$$

where the density of electron states at the Fermi level in the metallic phase per two spin components per unit length L_x of one chain is $\nu_F = 2/\pi \hbar v_F$. Averaging over k_y is denoted by triangular brackets, i.e., $\langle \cdot \rangle_{k_y} = b \int_{-\pi/b}^{\pi/b} dk_y/2\pi \cdot$. Equation (9) is similar to the self-consistency equation for superconductivity in a magnetic field, where the orbital effect of the magnetic field is neglected and the pair-breaking Zeeman splitting is replaced by $\varepsilon_+(k, k-Q)$.

3. Estimation of the Size of the SC Islands

3.1. Analytical Calculation of the Ginzburg–Landau Expansion Coefficients for $T \gg t_b'$

Expansion of Equation (9) over Δ yields

$$A = -\frac{4\pi T}{\hbar v_F} \sum_\omega \left\langle \frac{\text{sgn}\,\omega}{\omega + i\varepsilon_+} \right\rangle_{k_y} - \frac{1}{U} =$$
$$= -\frac{4}{\hbar v_F} \left[\ln \frac{T_{c0}}{T} + \psi\left(\frac{1}{2}\right) - \left\langle \text{Re}\,\psi\left(\frac{1}{2} + \frac{\varepsilon_+}{2\pi T}\right) \right\rangle_{k_y} \right], \quad (10)$$

where the logarithmic divergence $\pi T \sum_\omega |\omega|^{-1} \approx \ln(E_F/T)$ is contained in the definition of $T_{c0} = E_F \exp\{-1/(\nu_F|U|)\}$. In (TMTSF)$_2PF_6$, $T_{c0} \approx 12.1$ K [53].

The spatial modulation with the wave vector q of the DW order parameter Δ corresponds to the deviation of the DW wave vector Q from Q_0 by $\pm q$. Hence, the gradient term in the Ginzburg–Landau expansion of the DW free energy can be obtained by the expansion of $A(T, Q)$ given by Equation (10) in the powers of small deviation $q = Q - Q_0$. $A(Q, T)$ depends on Q via $\varepsilon_+ = \varepsilon_+(k, k-Q)$, given by Equation (4). For the quasi-1D electron dispersion in Equations (1) and (2), approximately describing (TMTSF)$_2$PF$_6$, we may use Equation (21) from ref. [58]:

$$\varepsilon_+ = \frac{\hbar v_F q_x}{2} + 2t_b \sin \frac{b q_y}{2} \sin\left(b\left[k_y - \frac{q_y}{2}\right]\right) - 2t_b' \cos(b q_y) \cos(b[2k_y - q_y]). \quad (11)$$

The general form of the Taylor series of $A(T, Q_0 + q)$, given by Equation (10), over the deviation $q = Q - Q_0$ of the DW wave vector Q from its optimal value Q_0 up to the second order is

$$A(q) \simeq -\frac{4}{\hbar v_F}\left[\ln\frac{T_{c0}}{T} + \int_{-\pi/b}^{\pi/b} dk_y \left(C_0 + c_x q_x + c_y q_y + c_{xy} q_x q_y + A_x q_x^2 + A_y q_y^2\right)\right]. \quad (12)$$

The linear terms and the cross term vanish when taking the integral $\int_{-\pi/b}^{\pi/b} dq_y$ (this is always the case if wave vector Q_0 is the optimal one). The constant and quadratic terms do not vanish, thus

$$A(q) \simeq A_0 + A_x q_x^2 + A_y q_y^2. \quad (13)$$

Expanding Equation (11) over the deviation $q = Q - Q_0$ up to the second order, substituting it in Equation (10) and expanding the digamma function over the same wave vector q, we obtain the coefficients A_i:

$$\begin{aligned}
A_0 &= -\frac{4}{\hbar v_F}\left[\frac{T_{c0}}{T} + \psi\left(\frac{1}{2}\right) - \left\langle \operatorname{Re}\psi\left(\frac{1}{2} - \frac{it_b' \cos(2bk_y)}{\pi T}\right)\right\rangle_{k_y}\right]; \\
A_x &= -\frac{4}{\hbar v_F}\frac{\hbar^2 v_F^2}{32\pi^2 T^2}\left\langle \operatorname{Re}\psi^{(2)}\left(\frac{1}{2} - \frac{it_b'\cos(2bk_y)}{\pi T}\right)\right\rangle_{k_y}; \\
A_y &= \frac{4}{\hbar v_F}\frac{b^2}{8\pi^2 T^2}\left\langle 2\pi T[t_b\cos(bk_y) - 4t_b'\cos(2bk_y)]\operatorname{Im}\psi^{(1)}\left(\frac{1}{2} - \frac{it_b'\cos(2bk_y)}{\pi T}\right) \right.\\
&\quad \left. - [t_b - 4t_b'\cos(bk_y)]^2 \sin^2(bk_y)\operatorname{Re}\psi^{(2)}\left(\frac{1}{2} - \frac{it_b'\cos(2bk_y)}{\pi T}\right)\right\rangle_{k_y}.
\end{aligned} \quad (14)$$

The integrals over k_y in A_y, A_x and A_0 can be calculated numerically and give the coherence lengthes ξ_x and ξ_y. From Equations (A10) and (A11), it follows that

$$\xi_i^2 = A_i/A_0. \quad (15)$$

Figure 2 shows ξ_x and ξ_y as functions of temperature T for two different values of t_b': $t_b' \approx 0.42\, t_b'^*$, corresponding to (TMTSF)$_2$PF$_6$ at ambient pressure [53] (solid orange and dashed green lines), and $t_b' = 0.95\, t_b'^*$ (solid red and dashed blue lines), i.e., close to the quantum critical point at $t_b' = t_b'^*$. These curves diverge at $T = T_c(t_b')$, where $A_0 = 0$. This divergence, being a general property of phase transitions, is well known in superconductors. When plotting Figure 2, we used Equations (14) and (15), and the parameters of (TMTSF)$_2$PF$_6$, i.e., $b = 0.767$ nm [59] and $v_F = 10^7$ cm/s [52].

At rather high temperatures ($2\pi T \gg \varepsilon_+ \sim t_b'$), we may expand the digamma function in Equation (10) to a Taylor series near $1/2$, which gives

$$\psi\left(\frac{1}{2}\right) - \left\langle \operatorname{Re}\psi\left(\frac{1}{2} + \frac{i\varepsilon_+(k, k-Q)}{2\pi T}\right)\right\rangle_{k_y} \simeq \frac{b}{2\pi}\int_{-\pi/b}^{\pi/b} dk_y \frac{\psi^{(2)}(1/2)\varepsilon_+^2(k, k-Q)}{8\pi^2 T^2}. \quad (16)$$

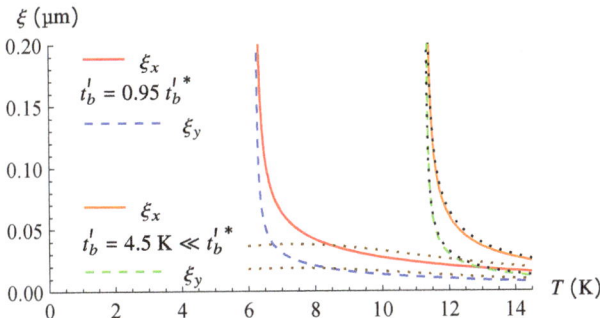

Figure 2. Temperature dependence of the DW coherence length ζ along two main axes at two different values of t'_b: $t'_b = 4.5$ K $= 0.42\, t'^{*}_b$, corresponding to (TMTSF)$_2$PF$_6$ at ambient pressure, and $t'_b = 0.95\, t'^{*}_b$. Solid and dashed lines correspond to the numerical solution of Equations (14), while the dotted lines correspond to the approximate analytical formulas in Equations (17) and (18).

Expanding Equation (11) over q up to the second order and substituting it into Equation (10) after using Equation (16), we obtain the coefficients A_i:

$$A_0 = -\frac{4}{\hbar v_F}\left[\ln\frac{T_{c0}}{T} + \frac{t'^{2}_b}{4\pi^2 T^2}\right];$$
$$A_x = -\frac{4}{\hbar v_F}\psi^{(2)}\left(\frac{1}{2}\right)\left[\frac{\hbar^2 v_F^2}{32\pi^2 T^2}\right]; \quad A_y = -\frac{4}{\hbar v_F}\psi^{(2)}\left(\frac{1}{2}\right)\left[\frac{b^2\left(t_b^2 - 4t'^{2}_b\right)}{16\pi^2 T^2}\right]. \quad (17)$$

Substituting them into Equation (15), we derive simple analytical formulas for the SDW coherence lengths, ζ_x and ζ_y, valid at $\pi T \gg t'_b$:

$$\zeta_x = \frac{\hbar v_F}{2\sqrt{2}}\sqrt{1/\left(4\pi^2 T^2 \ln(T_{c0}/T)/\psi^{(2)}(1/2) + t'^{2}_b\right)};$$
$$\zeta_y = \frac{b}{2}\sqrt{\frac{t_b^2 - 4t'^{2}_b}{4\pi^2 T^2 \ln(T_{c0}/T)/\psi^{(2)}(1/2) + t'^{2}_b}}. \quad (18)$$

At this limit of $\pi T \gg t'_b$, the ratio of coherent lengths along the y- and x-axes does not depend on temperature:

$$\frac{\zeta_y}{\zeta_x} = \frac{b}{\hbar v_F}\sqrt{2\left(t_b^2 - 4t'^{2}_b\right)} \approx 0.5. \quad (19)$$

The temperature dependence of the coherence lengthes ζ_x and ζ_y given by Equation (18) are shown in Figure 2 by dotted lines. The black dotted curves in Figure 2 are obtained from Equation (18) by setting $t'_b = 4.5$ K $= 0.42\, t'^{*}_b$, corresponding to (TMTSF)$_2$PF$_6$ at ambient pressure [53]. These curves coincide with the result of numerical integration in Equations (14), which confirms the applicability of Equations (17) and (18) with these parameters. From Equation (18), we obtain $\zeta_x \approx 0.06$ μm and $\zeta_y \approx 0.03$ μm at $T = T_{c0} = 12.1$ K, corresponding to $T/T_c - 1 \approx 0.075$. However, at $T/T_c - 1 \approx 0.01$, this gives $\zeta_x \approx 0.16$ μm and $\zeta_y \approx 0.08$ μm.

3.2. Relation between the Coherence Length and Nucleation Size during the First-Order Phase Transition

Despite an extensive study of the phase nucleation process during the first-order phase transition [60–63], its general quantitative description is still missing. The nucleation rate and size may strongly depend on minor factors relevant to a particular system. The DW–metal or DW–SC phase transitions also have peculiarities, such as a strong dependence on the details of electron dispersion. Nevertheless, one can roughly estimate the lower limit of the nucleus size using the Ginzburg–Landau expansion for DW free energy. The latter gives the energy of a phase nucleus Ω, described by the spatial variation $\Delta(r)$ of the DW order parameter during the first-order phase transition as

$$\Delta F_\Omega \approx \int_\Omega d^3 r \frac{1}{2} \left[A_0 \Delta^2 + \sum_i A_i (\partial_i \Delta)^2 \right] \approx \int_\Omega d^3 r \frac{A_0}{2} \left[\Delta^2 + \sum_i (\xi_i \partial_i \Delta)^2 \right]. \quad (20)$$

If the nucleus size d_i is $< 2\xi_i$, the second (always positive) gradient term exceeds the first term, which is energetically unfavorable. Hence, the minimal dimensions of phase nucleation during the first-order phase transition is given by the coherence lengths $d_i > 2\xi_i$. The latter diverges at the spinodal line $T_c(t'_b)$ of the phase transition where $A_0 = 0$, as illustrated in Figure 2 for our DW system. However, the first-order phase transition starts at a slightly different temperature T_{c1}, while the spinodal line $T_c(t'_b)$ corresponds to the instability of one phase. Hence, for the estimates of nucleus size d_i, one should take some finite interval $\Delta T = T_{c1} - T_c$, which is determined by the width of the first-order phase transition. Unfortunately, the latter is unknown and strongly depends on the physical system. In our case, this width ΔT depends on the details of electron dispersion, e.g., on the amplitude of higher harmonics in the electron dispersion given by Equation (2). If we take a reasonable estimate, i.e., $\Delta T = T_{c1} - T_c \approx 0.01\, T_c$, we obtain the SC domain size $d > 2\xi > 0.3$ μm.

3.3. Estimates of Superconducting Island Size from Transport Measurements and the Numerical Calculation of the Current Percolation Threshold

Another method of estimating the average SC island size is based on using the available transport measurements, especially the anisotropy of the SC transition temperature observed in various organic superconductors [20,21,28] and determined from the anisotropic zero-resistance onset in various samples. This anisotropy was explained both in organic superconductors [49] and in mesa structures of FeSe [50] by the direct calculation of the percolation threshold along different axes in samples of various spatial dimensions relevant to experiments. The qualitative idea behind this anisotropy is very simple. As the volume fraction ϕ of the SC phase grows, the isolated clusters of superconducting islands grow and become comparable to the sample size. When the percolation via superconducting islands between the opposite sample boundaries is established, zero resistance sets in. If the sample shape is flat or needle-like, as in organic metals, this percolation first establishes along the shortest sample dimension, when the SC cluster becomes comparable to the sample thickness (see Figure 4a in ref. [49] or Figure 4b in ref. [50] for illustration). With a further increase in the SC volume fraction ϕ, the zero resistance sets in along two axes, and only finally in all three directions, including the sample length.

In infinitely large samples, the percolation threshold is isotropic [64]. Hence, this anisotropy depends on the ratio of the average size d of superconducting islands to the sample size L. This dependence can be used for a qualitative estimate of SC island size d by analyzing the interval of d where the experimental data on conductivity anisotropy are consistent with theoretical calculations.

The algorithm and implementation details of percolation calculations are given in refs. [49,50]. Using this method, we calculated the probability of percolation of a random geometric configuration of superconducting islands in a sample of $(TMTSF)_2PF_6$ with typical experimental dimensions of $3 \times 0.2 \times 0.1$ mm^3 [19,20] for various island sizes. For simplicity, the geometry of the islands was taken as spherical.

Figure 3 shows the dependence of the percolation threshold ϕ_c of the SC phase on the geometric dimensions of the superconducting islands. By the percolation threshold, we mean the SC volume fraction ϕ at which the probability of percolation of a randomly chosen geometrical configuration of islands is 1/2. In order to take into account possible random fluctuations of this SC current percolation, in Figure 3, we also plot the interval of the SC volume fraction ϕ, corresponding to the large interval of percolation probability $p \in (0.1, 0.9)$ and denoted by the error bars. These error bars get bigger with the increase in size d of the spherical islands, because the larger the SC domain size d, the smaller the number N of SC domains required for percolation and hence, the stronger its relative fluctuations $\delta N/N \propto N^{-1/2}$. From Figure 3, we can see that for the sample dimensions used in the experiment [20], the percolation threshold via SC domains is considerably anisotropic, beyond the random fluctuations corresponding to a particular sample realization, if the domain size exceeds 2 µm. For smaller sizes of superconducting islands, the anisotropy is smaller than the "error bar" of ϕ_c, corresponding to the fluctuations in the percolation probability $p \in (0.1, 0.9)$. These error bars get bigger with the increase in the size d of the superconducting islands, because the larger the SC domain size, the smaller the number N of SC domains required for percolation and the stronger the fluctuations in this number $\delta N/N \propto N^{-1/2}$. For $d < 2$ µm, the percolation thresholds along all three axes converge to the known isotropic percolation threshold in infinite samples $\phi_{c\infty} \approx 0.2895$ (see page 253 of ref. [65]).

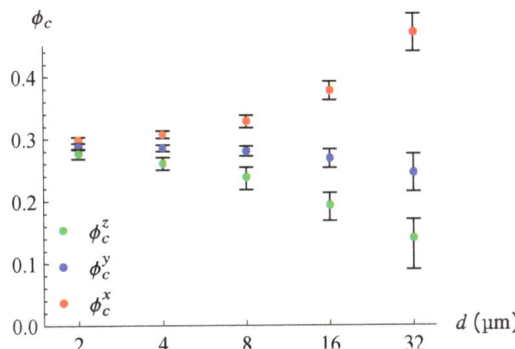

Figure 3. The dependence of the percolation threshold ϕ_c along different axes on the size of the spherical island d in (TMTSF)$_2$PF$_6$. The intervals of ϕ_c, corresponding to the percolation probability $p \in (0.5 \pm 0.4)$, are indicated by error bars.

4. Discussion and Conclusions

The observed strong anisotropy of the SC transition temperature T_{cSC} in (TMTSF)$_2$PF$_6$ [20,21] and (TMTSF)$_2$ClO$_4$ [28] samples of thicknesses of ~0.1 mm is consistent with our percolation calculations of the SC domain size of $d > 2$ µm. These estimates of the SC domain size d agree well with the result that $d_x > 1$ µm, implied by the clear observation of angular magnetoresistance oscillations and of a field-induced SDW in (TMTSF)$_2$PF$_6$ [21] and (TMTSF)$_2$ClO$_4$ [30]. The latter requires that the electron mean free path, l_τ, is $> l_B$, where $l_B = \hbar/eBb \sim 1$ µm is the so-called quasi-1D magnetic length [21,42]. Hence, all experimental observations agree and suggest an almost macroscopic spatial separation of SC and SDW phases in these organic superconductors.

The above SDW coherence length $\tilde{\xi}$ obtained from the Ginzburg–Landau expansion of the SDW free energy at the first-order SDW–SC phase transition in the organic superconductor (TMTSF)$_2$PF$_6$ gives the SC domain size $d > 2\tilde{\xi} > 0.3$ µm. This generally agrees with the experimental estimates of $d > 1$ µm, but gives a too weak limitation because of the following three possible reasons:

(1) The SC proximity effect [47]: The SC order parameter is nonzero not only in the SC domains themselves, but also in shells of width $\delta d \sim \xi_{SC}$ around these SC domains. The SC coherence length $\xi_{SC} \sim \hbar v_F/\pi \Delta_{SC}$ diverges near the SC transition temperature T_{cSC}, and even far from $T_{cSC} \approx 1$ K in organic superconduc-

tors $\delta d_x \sim \xi_{SC} \sim \hbar v_F/\pi T_{cSC} \approx 0.3$ µm. Hence, the resulting size of SC domains with this proximity effect shell is $d_x \gtrsim 2(\xi_{SC} + \xi) \approx 1$ µm, which well agrees with experimental data.

(2) The clusterization of superconducting islands with the formation of larger SC domains, glued by the Josephson junction: In current percolation and zero-frequency transport measurements, such a cluster is seen as a single SC domain. Since the SDW–SC transition is observed close to the SC percolation threshold (the SC volume fraction of $\phi_c > 0.1$), the formation of such SC clusters is very probable. Note that such clusterization may also explain the small difference between the estimates of the SC domain size from AMRO data and from current percolation.

(3) An oversimplified physical model: In our percolation calculations, we take all clusters of the same size, because the actual size distribution of superconducting islands is unknown. In addition, special types of disorder, such as local variations in (chemical) pressure, affect the SC–SDW balance.

The size of isolated SC domains allows an independent approximate measurement of the diamagnetic response. The diamagnetic response of small SC grains of size $d \lesssim \lambda$, where λ is the penetration depth of magnetic field into the superconductor, strongly depends on the d/λ [47] ratio. Note that this penetration depth in layered superconductors is anisotropic. Since the SC volume fraction ϕ is approximately known from the transport measurements and from the percolation threshold, by measuring the diamagnetic response at $\phi < \phi_c$ for three main orientations of a magnetic field B and comparing it with the susceptibility $\chi = -\phi/(4\pi)$ of large SC domains of volume fraction ϕ, one may roughly estimate the SC domain size along all three axes. A similar diamagnetic response in combination with transport measurements was used in FeSe to estimate the size and shape of superconducting islands above T_{cSC} [66,67]. A similar combined analysis of the diamagnetic response and transport measurements has also been used to obtain information about the SC domain size and shape above T_{cSC} in another organic superconductor, β-(BEDT-TTF)$_2$I$_3$ [68].

Spatial phase segregation may also happen near the quantum critical point of the Mott-AFM metal–insulator phase transition, e.g., as observed in the κ-(BEDT-TTF)$_2$X family of organic superconductors [69–72]. The observation of clear magnetic quantum oscillations [71,72] in the almost insulating phase of these materials indicates a rather large size d of metal/SC domains in the Mott insulator media, comparable to the electron cyclotron radius. Although the first of our methods, based on the Ginzburg–Landau SDW free energy expansion, is not applicable in this case, our second method [49,50], based on the calculation of percolation anisotropy in finite-sized samples, should work well and give valuable information about the shape and size of metal/SC domains.

The obtained, almost macroscopic spatial SDW–SC phase separation on a scale of $d \gtrsim 1$ µm implies a rather weak influence of the SDW quantum critical point on SC coupling. Indeed, while in cuprate high-T_c superconductors [5,31,73–76] and in transition metal dichalcogenides [1–3] the SC–DW coexistence is more "microscopic" and the corresponding T_{cSC} enhancement is several-fold, in organic superconductors, the T_{cSC} enhancement by quantum criticality is rather weak, at ∼10%. Note that in iron-based high-T_c superconductors [11,12], e.g., in FeSe, the T_{cSC} enhancement by quantum criticality is also rather weak, at ∼10%. A comparison of the observed [50,77] T_{cSC} anisotropy in thin FeSe mesa structures of various thicknesses with the numerical calculations of percolation anisotropy in finite-sized samples [50], similar to that in Section 3.3, suggests that the SC domain size in FeSe is also rather large, at $d \sim 0.1$ µm, close to the nematic domain width in this compound. Hence, similar to organic superconductors, in FeSe and other iron-based high-T_c superconductors, the large size of SC domains reduces the SC enhancement by critical fluctuations. This observation may give a hint about raising the transition temperature in high-T_c superconductors, which are always spatially inhomogeneous. The knowledge of the parameters of SC domains also helps to estimate and even propose possible methods to increase the upper critical field and critical current in such heterogeneous superconductors, considered as a network of SC nanoclusters linked by Josephson junctions [75,78].

To summarize, we have shown that the scenario in which the first-order phase transition results in the spatial phase separation of SC and SDW in organic superconductors is self-consistent and also agrees with the available experimental data. We estimated the size of SC domains d by two different methods. This estimate of $d > 1$ µm is consistent with various transport measurements, including the anisotropic zero resistance onset in thin samples [20,21,28] and with angular magnetoresistance oscillations and magnetic-field-induced spin-density waves [21,30]. We also discuss the relevance of our results, obtained for organic superconductors, to high-T_c superconductors, and why the knowledge of SC domain parameters is important for increasing the transition temperature, the critical magnetic field H_{c2} and the critical current density in various heterogeneous superconductors.

Author Contributions: Conceptualization, P.D.G.; methodology, P.D.G. and V.D.K.; software, V.D.K.; validation, V.D.K. and S.S.S.; formal analysis, V.D.K.; investigation, V.D.K. and P.D.G.; writing—original draft preparation, V.D.K.; writing—review and editing, P.D.G. and S.S.S.; supervision, P.D.G. All authors have read and agreed to the published version of the manuscript.

Funding: V.D.K. acknowledges the Foundation for the Advancement of Theoretical Physics and Mathematics "Basis" for grant # 22-1-1-24-1, and the RFBR grant # 21-52-12027. The work of S.S.S. was supported by the NUST "MISIS" grant no. K2-2022-025 in the framework of the federal academic leadership program Priority 2030. P.D.G. acknowledges the State assignment # 0033-2019-0001 and the RFBR grant # 21-52-12043.

Institutional Review Board Statement: Not applicable.

Informed Consent Statement: Not applicable.

Data Availability Statement: Data will be provided on request.

Conflicts of Interest: The authors declare no conflict of interest.

Appendix A. Mean-Field Theory for DW

The electronic Hamiltonian consists of the free-electron part H_0 and the interaction part H_{int}:

$$H = H_0 + H_{\text{int}},$$
$$H_0 = \sum_k \varepsilon(k) a_k^\dagger a_k, \quad (A1)$$
$$H_{\text{int}} = \frac{1}{2} \sum_{kk'Q} V_Q a_{k+Q}^\dagger a_k a_{k'-Q}^\dagger a_{k'}.$$

We consider the interactions at the wave vector Q close to the nesting vector $Q_0 = (\pm 2k_F, \pi/b)$. If the deviations from Q_0 are small, we can approximate the interaction function as $V(Q) \approx V(Q_0) = U$. In the case of CDW, U is the charge coupling constant, while for a SDW, U denotes the spin coupling constant. Next, in the mean-field approximation, we introduce the order parameter

$$\Delta_Q = 2U \sum_k g(k-Q, k, -0),$$
$$g(k, k', \tau - \tau') = \langle T_\tau a_{k'}^\dagger(\tau') a_k(\tau) \rangle. \quad (A2)$$

Then, the final mean-field Hamiltonian, which we will study further, is

$$H_{\text{int}} = \sum_{kQ} \Delta_Q a_{k+Q}^\dagger a_k + \text{H.c.} + \text{const}. \quad (A3)$$

The factor of 2 in Equations (A2) and (A7) comes from the summation over two spin components. The operators $a_{k+Q}^\dagger$ and a_k correspond to the same spin component for a CDW, and to different spin components for an SDW. From Equations (A1) and (A3), using the standard equation for the operator evolution, $i\hbar\, d\hat{A}/dt = [\hat{A}, \hat{H}]$, one obtains the equations of motion for the Fourier transform of the Green's function $g(k, k', \tau - \tau') = \int d\omega/(2\pi) e^{i\omega(\tau-\tau')} g(k', k, \omega)$:

$$[i\omega - \varepsilon(k)] g(k', k, \omega) - \sum_Q \Delta_Q g(k', k, \omega) = \delta_{k',k}. \tag{A4}$$

In the metallic phase, Δ_Q, given by Equation (A2), vanishes after the thermodynamic averaging denoted by triangular brackets in Equation (A2). If the DW at wave vector Q_0 is formed, the order parameter $\Delta_Q \neq 0$ for $Q = Q_0$, while for $Q \neq Q_0$, the average $\Delta_Q = 0$. The spatial variation in the order parameter $\Delta(r) = \int d^3q/(2\pi) \Delta(q) e^{iq\cdot r}$ is described by the deviation $q = Q - Q_0$ of the DW wave vector Q from its value Q_0, corresponding to the maximum susceptibility. If we now set $\Delta_{Q_1} = \Delta \delta_{Q_1, \pm Q}$, where $Q = Q_0 + q$ with $|q| \ll k_F$, the equations of motion (A4) can be solved, giving

$$g(k - Q, k, \omega) = -\frac{\Delta_Q}{(\omega + i\varepsilon_+)^2 + \varepsilon_-^2 + |\Delta_Q|^2}, \tag{A5}$$

where $\varepsilon_\pm = \varepsilon^\pm(k, k - Q)$ are given by Equation (4). From Equations (A2) and (A5), we obtain the self-consistency Equation (7), omitting the subscript Q in Δ_Q:

$$\Delta = -TU \sum_{k\omega} \frac{\Delta}{(\omega + i\varepsilon_+)^2 + \varepsilon_-^2 + |\Delta|^2}. \tag{A6}$$

The mean-field Hamiltonian given by Equations (A1) and (A3) decouples to a sum over k of 2×2 matrices. Their diagonalization gives the new quasiparticle dispersion given by Equation (3). Hence, the order parameter Δ_Q defined in Equation (A2) has the physical meaning of the DW energy gap for the case of perfect nesting. One could define the order parameter in a different way as

$$n_Q = \frac{\Delta_Q}{U} = 2 \sum_k \langle T_\tau a_{k'}^\dagger(\tau') a_k(\tau) \rangle, \tag{A7}$$

which has the physical meaning of electron density n_Q at wave vector Q. The latter couples to the external potential V_Q at the same wave vector in the Hamiltonian: $\delta H = \delta F = -\sum_Q n_Q V_Q$. The equilibrium value of the DW order parameter $\Delta_Q = U n_Q$ in the presence of an external field V_Q can be obtained from the minimization of the total free energy $F_{tot} = F + \delta F$, where the free energy F without an external field is given by Equation (5) at $\Delta_Q \to 0$:

$$\frac{\partial F_{tot}}{\partial n_Q} = -V_Q + U \frac{\partial F}{\partial \Delta_Q} = 0, \tag{A8}$$

or

$$-V_Q + U^2 n_Q \left[A(T, Q) + B |\Delta_Q|^2 + \ldots \right] = 0. \tag{A9}$$

Hence, the electronic susceptibility just above the DW phase transition temperature T_{cDW}, where $\Delta_Q = 0$, is related to the coefficient $A(T, Q) > 0$ of the Landau–Ginzburg expansion:

$$\chi(Q) = \frac{n_Q}{V_Q} = \frac{1}{A(T, Q) U^2}. \tag{A10}$$

At the DW transition temperature $T = T_{cDW}$, the coefficient $A(T, Q) = 0$ for some Q. Hence, the DW wave vector Q corresponds to the minimum of $A(T_{cDW}, Q)$ or to the maximum of susceptibility $\chi(Q)$ in Equation (A10). Near this extremum, one can expand

Equation (A10) over the deviation $q = Q - Q_0$ of the DW wave vector Q from its optimal value Q_0:

$$\chi(Q) = \frac{\chi(Q_0)}{1 + \xi^2 q^2}, \tag{A11}$$

which gives the estimate of the DW coherence length ξ.

Below the phase transition temperature T_{cDW}, Equation (A9) gives

$$\chi^{-1}(Q) = U^2 \left[A(T, Q) + B|\Delta|^2 + C|\Delta|^4 + .. \right] \to \infty, \tag{A12}$$

which corresponds to a finite Δ_Q at vanishing V_Q. Nevertheless, one can find the differential susceptibility

$$\chi^{-1}(Q) = \frac{dV_Q}{dn_Q} = \frac{\partial^2 F}{\partial n_Q^2} = U^2 \frac{\partial^2 F}{\partial \Delta_Q^2}, \tag{A13}$$

which generalizes Equation (A10).

References

1. Gabovich, A.M.; Voitenko, A.I.; Annett, J.F.; Ausloos, M. Charge- and spin-density-wave superconductors. *Supercond. Sci. Technol.* 2001, *14*, R1–R27. [CrossRef]
2. Gabovich, A.M.; Voitenko, A.I.; Ausloos, M. Charge- and spin-density waves in existing superconductors: Competition between Cooper pairing and Peierls or excitonic instabilities. *Phys. Rep.* 2002, *367*, 583–709. [CrossRef]
3. Monceau, P. Electronic crystals: An experimental overview. *Adv. Phys.* 2012, *61*, 325–581. [CrossRef]
4. Grüner, G. *Density Waves in Solids*; CRC Press: Boca Raton, FL, USA, 2000.
5. Chang, J.; Blackburn, E.; Holmes, A.; Christensen, N.; Larsen, J.; Mesot, J.; Liang, R.; Bonn, D.; Hardy, W.; Watenphul, A.; et al. Direct observation of competition between superconductivity and charge density wave order in YBa$_2$Cu$_3$O$_{6.67}$. *Nat. Phys.* 2012, *8*, 871–876. [CrossRef]
6. Blanco-Canosa, S.; Frano, A.; Loew, T.; Lu, Y.; Porras, J.; Ghiringhelli, G.; Minola, M.; Mazzoli, C.; Braicovich, L.; Schierle, E.; et al. Momentum-Dependent Charge Correlations in YBa$_2$Cu$_3$O$_{6+\delta}$ Superconductors Probed by Resonant X-ray Scattering: Evidence for Three Competing Phases. *Phys. Rev. Lett.* 2013, *110*, 187001. [CrossRef]
7. Tabis, W.; Yu, B.; Bialo, I.; Bluschke, M.; Kolodziej, T.; Kozlowski, A.; Blackburn, E.; Sen, K.; Forgan, E.; Zimmermann, M.; et al. Synchrotron x-ray scattering study of charge-density-wave order in HgBa$_2$CuO$_{4+\delta}$. *Phys. Rev. B* 2017, *96*, 134510. [CrossRef]
8. Tabis, W.; Li, Y.; Tacon, M.L.; Braicovich, L.; Kreyssig, A.; Minola, M.; Dellea, G.; Weschke, E.; Veit, M.J.; Ramazanoglu, M.; et al. Charge order and its connection with Fermi-liquid charge transport in a pristine high-Tc cuprate. *Nat. Commun.* 2014, *5*, 5875. [CrossRef]
9. da Silva Neto, E.H.; Comin, R.; He, F.; Sutarto, R.; Jiang, Y.; Greene, R.L.; Sawatzky, G.A.; Damascelli, A. Charge ordering in the electron-doped superconductor Nd$_{2-x}$Ce$_x$CuO$_4$. *Science* 2015, *347*, 282–285.
10. Wen, J.J.; Huang, H.; Lee, S.J.; Jang, H.; Knight, J.; Lee, Y.S.; Fujita, M.; Suzuki, K.M.; Asano, S.; Kivelson, S.A.; et al. Observation of two types of charge-density-wave orders in superconducting La$_{2-x}$Sr$_x$CuO$_4$. *Nat. Commun.* 2019, *10*, 3269. [CrossRef]
11. Si, Q.; Yu, R.; Abrahams, E. High-temperature superconductivity in iron pnictides and chalcogenides. *Nat. Rev. Mater.* 2016, *1*, 16017. [CrossRef]
12. Liu, X.; Zhao, L.; He, S.; He, J.; Liu, D.; Mou, D.; Shen, B.; Hu, Y.; Huang, J.; Zhou, X.J. Electronic structure and superconductivity of FeSe-related superconductors. *J. Phys. Condens. Matter* 2015, *27*, 183201. [CrossRef] [PubMed]
13. Ishiguro, T.; Yamaji, K.; Saito, G. *Organic Superconductors*; Springer: Berlin/Heidelberg, Germany, 1998. [CrossRef]
14. Lebed, A. (Ed.) *The Physics of Organic Superconductors and Conductors*; Springer: Berlin/Heidelberg, Germany, 2008. [CrossRef]
15. Naito, T. Modern History of Organic Conductors: An Overview. *Crystals* 2021, *11*, 838. [CrossRef]
16. Yasuzuka, S.; Murata, K. Recent progress in high-pressure studies on organic conductors. *Sci. Technol. Adv. Mater.* 2009, *10*, 024307.
17. Clay, R.; Mazumdar, S. From charge- and spin-ordering to superconductivity in the organic charge-transfer solids. *Phys. Rep.* 2019, *788*, 1–89.
18. Lee, I.J.; Chaikin, P.M.; Naughton, M.J. Critical Field Enhancement near a Superconductor-Insulator Transition. *Phys. Rev. Lett.* 2002, *88*, 207002. [CrossRef]
19. Vuletić, T.; Auban-Senzier, P.; Pasquier, C.; Tomić, S.; Jérome, D.; Héritier, M.; Bechgaard, K. Coexistence of superconductivity and spin density wave orderings in the organic superconductor (TMTSF)$_2$PF$_6$. *Eur. Phys. J. B* 2002, *25*, 319–331. [CrossRef]
20. Kang, N.; Salameh, B.; Auban-Senzier, P.; Jerome, D.; Pasquier, C.R.; Brazovskii, S. Domain walls at the spin-density-wave endpoint of the organic superconductor (TMTSF)$_2$PF$_6$ under pressure. *Phys. Rev. B* 2010, *81*, 100509(R). [CrossRef]
21. Narayanan, A.; Kiswandhi, A.; Graf, D.; Brooks, J.; Chaikin, P. Coexistence of Spin Density Waves and Superconductivity in (TMTSF)$_2$PF$_6$. *Phys. Rev. Lett.* 2014, *112*, 146402. [CrossRef]

22. Lee, I.J.; Brown, S.E.; Yu, W.; Naughton, M.J.; Chaikin, P.M. Coexistence of Superconductivity and Antiferromagnetism Probed by Simultaneous Nuclear Magnetic Resonance and Electrical Transport in $(TMTSF)_2PF_6$ System. *Phys. Rev. Lett.* **2005**, *94*, 197001. [CrossRef]
23. Lee, I.J.; Naughton, M.J.; Danner, G.M.; Chaikin, P.M. Anisotropy of the Upper Critical Field in $(TMTSF)_2PF_6$. *Phys. Rev. Lett.* **1997**, *78*, 3555–3558. [CrossRef]
24. Lee, I.J.; Brown, S.E.; Clark, W.G.; Strouse, M.J.; Naughton, M.J.; Kang, W.; Chaikin, P.M. Triplet Superconductivity in an Organic Superconductor Probed by NMR Knight Shift. *Phys. Rev. Lett.* **2001**, *88*, 017004. [CrossRef] [PubMed]
25. Andres, D.; Kartsovnik, M.V.; Biberacher, W.; Neumaier, K.; Schuberth, E.; Muller, H. Superconductivity in the charge-density-wave state of the organic metal $\alpha-(BEDT-TTF)_2KHg(SCN)_4$. *Phys. Rev. B* **2005**, *72*, 174513. [CrossRef]
26. Itoi, M.; Nakamura, T.; Uwatoko, Y. Pressure-Induced Superconductivity of the Quasi-One-Dimensional Organic Conductor $(TMTTF)_2TaF_6$. *Materials* **2022**, *15*, 4638. [CrossRef] [PubMed]
27. Cho, K.; Kończykowski, M.; Teknowijoyo, S.; Tanatar, M.; Guss, J.; Gartin, P.; Wilde, J.; Kreyssig, A.; McQueeney, R.; Goldman, A.; et al. Using controlled disorder to probe the interplay between charge order and superconductivity in $NbSe_2$. *Nat. Commun.* **2018**, *9*, 2796. [CrossRef]
28. Gerasimenko, Y.A.; Sanduleanu, S.V.; Prudkoglyad, V.A.; Kornilov, A.V.; Yamada, J.; Qualls, J.S.; Pudalov, V.M. Coexistence of superconductivity and spin-density wave in $(TMTSF)_2ClO_4$: Spatial structure of the two-phase state. *Phys. Rev. B* **2014**, *89*, 054518. [CrossRef]
29. Yonezawa, S.; Marrache-Kikuchi, C.A.; Bechgaard, K.; Jerome, D. Crossover from impurity-controlled to granular superconductivity in $(TMTSF)_2ClO_4$. *Phys. Rev. B* **2018**, *97*, 014521. [CrossRef]
30. Gerasimenko, Y.A.; Prudkoglyad, V.A.; Kornilov, A.V.; Sanduleanu, S.V.; Qualls, J.S.; Pudalov, V.M. Role of anion ordering in the coexistence of spin-density-wave and superconductivity in $(TMTSF)_2ClO_4$. *JETP Lett.* **2013**, *97*, 419–424. [CrossRef]
31. Wang, Y.; Chubukov, A.V. Enhancement of superconductivity at the onset of charge-density-wave order in a metal. *Phys. Rev. B* **2015**, *92*, 125108. [CrossRef]
32. Armitage, N.P.; Fournier, P.; Greene, R.L. Progress and perspectives on electron-doped cuprates. *Rev. Mod. Phys.* **2010**, *82*, 2421–2487. [CrossRef]
33. Helm, T.; Kartsovnik, M.V.; Proust, C.; Vignolle, B.; Putzke, C.; Kampert, E.; Sheikin, I.; Choi, E.S.; Brooks, J.S.; Bittner, N.; et al. Correlation between Fermi surface transformations and superconductivity in the electron-doped high-T_c superconductor $Nd_{2-x}Ce_xCuO_4$. *Phys. Rev. B* **2015**, *92*, 094501. [CrossRef]
34. Mukuda, H.; Fujii, T.; Ohara, T.; Harada, A.; Yashima, M.; Kitaoka, Y.; Okuda, Y.; Settai, R.; Onuki, Y. Enhancement of Superconducting Transition Temperature due to the Strong Antiferromagnetic Spin Fluctuations in the Noncentrosymmetric Heavy–Fermion Superconductor $CeIrSi_3$: A ^{29}Si NMR Study under Pressure. *Phys. Rev. Lett.* **2008**, *100*, 107003. [CrossRef] [PubMed]
35. Manago, M.; Kitagawa, S.; Ishida, K.; Deguchi, K.; Sato, N.K.; Yamamura, T. Enhancement of superconductivity by pressure-induced critical ferromagnetic fluctuations in UCoGe. *Phys. Rev. B* **2019**, *99*, 020506. [CrossRef]
36. Eckberg, C.; Campbell, D.J.; Metz, T.; Collini, J.; Hodovanets, H.; Drye, T.; Zavalij, P.; Christensen, M.H.; Fernandes, R.M.; Lee, S.; et al. Sixfold enhancement of superconductivity in a tunable electronic nematic system. *Nat. Phys.* **2020**, *16*, 346–350. [CrossRef] [PubMed]
37. Mukasa, K.; Ishida, K.; Imajo, S.; Qiu, M.; Saito, M.; Matsuura, K.; Sugimura, Y.; Liu, S.; Uezono, Y.; Otsuka, T.; et al. Enhanced Superconducting Pairing Strength near a Pure Nematic Quantum Critical Point. *Phys. Rev. X* **2023**, *13*, 011032. [CrossRef]
38. Tanaka, Y.; Kuroki, K. Microscopic theory of spin-triplet f-wave pairing in quasi-one-dimensional organic superconductors. *Phys. Rev. B* **2004**, *70*, 060502. [CrossRef]
39. Gor'kov, L.P.; Grigoriev, P.D. Nature of the superconducting state in the new phase in $(TMTSF)_2PF_6$ under pressure. *Phys. Rev. B* **2007**, *75*, 020507(R). [CrossRef]
40. Grigoriev, P.D. Properties of superconductivity on a density wave background with small ungapped Fermi surface parts. *Phys. Rev. B* **2008**, *77*, 224508. [CrossRef]
41. Ramazashvili, R.; Grigoriev, P.D.; Helm, T.; Kollmannsberger, F.; Kunz, M.; Biberacher, W.; Kampert, E.; Fujiwara, H.; Erb, A.; Wosnitza, J.; et al. Experimental evidence for Zeeman spin–orbit coupling in layered antiferromagnetic conductors. *Npj Quantum Mater.* **2021**, *6*, 11. [CrossRef]
42. Kartsovnik, M.V. High Magnetic Fields: A Tool for Studying Electronic Properties of Layered Organic Metals. *Chem. Rev.* **2004**, *104*, 5737–5782. [CrossRef]
43. Brazovskii, S.; Kirova, N. Electron selflocalization and superstructures in quasi one-dimensional dielectrics. *Sov. Sci. Rev. A* **1984**, *5*, 99–166.
44. Su, W.P.; Kivelson, S.; Schrieffer, J.R. Theory of Polymers Having Broken Symmetry Ground States. In *Physics in One Dimension*; Bernascony, J., Schneider, T., Eds.; Springer Series in Solid-State Sciences; Springer: Berlin/Heidelberg, Germany, 1981; pp. 201–211. [CrossRef]
45. Grigoriev, P.D. Superconductivity on the density-wave background with soliton-wall structure. *Phys. B* **2009**, *404*, 513–516. [CrossRef]
46. Gor'kov, L.P.; Grigoriev, P.D. Soliton phase near antiferromagnetic quantum critical point in Q1D conductors. *Europhys. Lett.* **2005**, *71*, 425–430. [CrossRef]

Equation (A10) over the deviation $q = Q - Q_0$ of the DW wave vector Q from its optimal value Q_0:

$$\chi(Q) = \frac{\chi(Q_0)}{1 + \xi^2 q^2}, \quad (A11)$$

which gives the estimate of the DW coherence length ξ.

Below the phase transition temperature T_{cDW}, Equation (A9) gives

$$\chi^{-1}(Q) = U^2 \left[A(T, Q) + B|\Delta|^2 + C|\Delta|^4 + .. \right] \to \infty, \quad (A12)$$

which corresponds to a finite Δ_Q at vanishing V_Q. Nevertheless, one can find the differential susceptibility

$$\chi^{-1}(Q) = \frac{dV_Q}{dn_Q} = \frac{\partial^2 F}{\partial n_Q^2} = U^2 \frac{\partial^2 F}{\partial \Delta_Q^2}, \quad (A13)$$

which generalizes Equation (A10).

References

1. Gabovich, A.M.; Voitenko, A.I.; Annett, J.F.; Ausloos, M. Charge- and spin-density-wave superconductors. *Supercond. Sci. Technol.* **2001**, *14*, R1–R27. [CrossRef]
2. Gabovich, A.M.; Voitenko, A.I.; Ausloos, M. Charge- and spin-density waves in existing superconductors: Competition between Cooper pairing and Peierls or excitonic instabilities. *Phys. Rep.* **2002**, *367*, 583–709. [CrossRef]
3. Monceau, P. Electronic crystals: An experimental overview. *Adv. Phys.* **2012**, *61*, 325–581. [CrossRef]
4. Grüner, G. *Density Waves in Solids*; CRC Press: Boca Raton, FL, USA, 2000.
5. Chang, J.; Blackburn, E.; Holmes, A.; Christensen, N.; Larsen, J.; Mesot, J.; Liang, R.; Bonn, D.; Hardy, W.; Watenphul, A.; et al. Direct observation of competition between superconductivity and charge density wave order in YBa$_2$Cu$_3$O$_{6.67}$. *Nat. Phys.* **2012**, *8*, 871–876. [CrossRef]
6. Blanco-Canosa, S.; Frano, A.; Loew, T.; Lu, Y.; Porras, J.; Ghiringhelli, G.; Minola, M.; Mazzoli, C.; Braicovich, L.; Schierle, E.; et al. Momentum-Dependent Charge Correlations in YBa$_2$Cu$_3$O$_{6+\delta}$ Superconductors Probed by Resonant X-ray Scattering: Evidence for Three Competing Phases. *Phys. Rev. Lett.* **2013**, *110*, 187001. [CrossRef]
7. Tabis, W.; Yu, B.; Bialo, I.; Bluschke, M.; Kolodziej, T.; Kozlowski, A.; Blackburn, E.; Sen, K.; Forgan, E.; Zimmermann, M.; et al. Synchrotron x-ray scattering study of charge-density-wave order in HgBa$_2$CuO$_{4+\delta}$. *Phys. Rev. B* **2017**, *96*, 134510. [CrossRef]
8. Tabis, W.; Li, Y.; Tacon, M.L.; Braicovich, L.; Kreyssig, A.; Minola, M.; Dellea, G.; Weschke, E.; Veit, M.J.; Ramazanoglu, M.; et al. Charge order and its connection with Fermi-liquid charge transport in a pristine high-Tc cuprate. *Nat. Commun.* **2014**, *5*, 5875. [CrossRef]
9. da Silva Neto, E.H.; Comin, R.; He, F.; Sutarto, R.; Jiang, Y.; Greene, R.L.; Sawatzky, G.A.; Damascelli, A. Charge ordering in the electron-doped superconductor Nd$_{2-x}$Ce$_x$CuO$_4$. *Science* **2015**, *347*, 282–285.
10. Wen, J.J.; Huang, H.; Lee, S.J.; Jang, H.; Knight, J.; Lee, Y.S.; Fujita, M.; Suzuki, K.M.; Asano, S.; Kivelson, S.A.; et al. Observation of two types of charge-density-wave orders in superconducting La$_{2-x}$Sr$_x$CuO$_4$. *Nat. Commun.* **2019**, *10*, 3269. [CrossRef]
11. Si, Q.; Yu, R.; Abrahams, E. High-temperature superconductivity in iron pnictides and chalcogenides. *Nat. Rev. Mater.* **2016**, *1*, 16017. [CrossRef]
12. Liu, X.; Zhao, L.; He, S.; He, J.; Liu, D.; Mou, D.; Shen, B.; Hu, Y.; Huang, J.; Zhou, X.J. Electronic structure and superconductivity of FeSe-related superconductors. *J. Phys. Condens. Matter* **2015**, *27*, 183201. [CrossRef] [PubMed]
13. Ishiguro, T.; Yamaji, K.; Saito, G. *Organic Superconductors*; Springer: Berlin/Heidelberg, Germany, 1998. [CrossRef]
14. Lebed, A. (Ed.) *The Physics of Organic Superconductors and Conductors*; Springer: Berlin/Heidelberg, Germany, 2008. [CrossRef]
15. Naito, T. Modern History of Organic Conductors: An Overview. *Crystals* **2021**, *11*, 838. [CrossRef]
16. Yasuzuka, S.; Murata, K. Recent progress in high-pressure studies on organic conductors. *Sci. Technol. Adv. Mater.* **2009**, *10*, 024307.
17. Clay, R.; Mazumdar, S. From charge- and spin-ordering to superconductivity in the organic charge-transfer solids. *Phys. Rep.* **2019**, *788*, 1–89. [CrossRef]
18. Lee, I.J.; Chaikin, P.M.; Naughton, M.J. Critical Field Enhancement near a Superconductor-Insulator Transition. *Phys. Rev. Lett.* **2002**, *88*, 207002. [CrossRef]
19. Vuletić, T.; Auban-Senzier, P.; Pasquier, C.; Tomić, S.; Jérome, D.; Héritier, M.; Bechgaard, K. Coexistence of superconductivity and spin density wave orderings in the organic superconductor (TMTSF)$_2$PF$_6$. *Eur. Phys. J. B* **2002**, *25*, 319–331. [CrossRef]
20. Kang, N.; Salameh, B.; Auban-Senzier, P.; Jerome, D.; Pasquier, C.R.; Brazovskii, S. Domain walls at the spin-density-wave endpoint of the organic superconductor (TMTSF)$_2$PF$_6$ under pressure. *Phys. Rev. B* **2010**, *81*, 100509(R). [CrossRef]
21. Narayanan, A.; Kiswandhi, A.; Graf, D.; Brooks, J.; Chaikin, P. Coexistence of Spin Density Waves and Superconductivity in (TMTSF)$_2$PF$_6$. *Phys. Rev. Lett.* **2014**, *112*, 146402. [CrossRef]

22. Lee, I.J.; Brown, S.E.; Yu, W.; Naughton, M.J.; Chaikin, P.M. Coexistence of Superconductivity and Antiferromagnetism Probed by Simultaneous Nuclear Magnetic Resonance and Electrical Transport in (TMTSF)$_2$PF$_6$ System. *Phys. Rev. Lett.* **2005**, *94*, 197001. [CrossRef]
23. Lee, I.J.; Naughton, M.J.; Danner, G.M.; Chaikin, P.M. Anisotropy of the Upper Critical Field in (TMTSF)$_2$PF$_6$. *Phys. Rev. Lett.* **1997**, *78*, 3555–3558. [CrossRef]
24. Lee, I.J.; Brown, S.E.; Clark, W.G.; Strouse, M.J.; Naughton, M.J.; Kang, W.; Chaikin, P.M. Triplet Superconductivity in an Organic Superconductor Probed by NMR Knight Shift. *Phys. Rev. Lett.* **2001**, *88*, 017004. [CrossRef] [PubMed]
25. Andres, D.; Kartsovnik, M.V.; Biberacher, W.; Neumaier, K.; Schuberth, E.; Muller, H. Superconductivity in the charge-density-wave state of the organic metal α−(BEDT−TTF)$_2$KHg(SCN)$_4$. *Phys. Rev. B* **2005**, *72*, 174513. [CrossRef]
26. Itoi, M.; Nakamura, T.; Uwatoko, Y. Pressure-Induced Superconductivity of the Quasi-One-Dimensional Organic Conductor (TMTTF)$_2$TaF$_6$. *Materials* **2022**, *15*, 4638. [CrossRef] [PubMed]
27. Cho, K.; Kończykowski, M.; Teknowijoyo, S.; Tanatar, M.; Guss, J.; Gartin, P.; Wilde, J.; Kreyssig, A.; McQueeney, R.; Goldman, A.; et al. Using controlled disorder to probe the interplay between charge order and superconductivity in NbSe$_2$. *Nat. Commun.* **2018**, *9*, 2796. [CrossRef]
28. Gerasimenko, Y.A.; Sanduleanu, S.V.; Prudkoglyad, V.A.; Kornilov, A.V.; Yamada, J.; Qualls, J.S.; Pudalov, V.M. Coexistence of superconductivity and spin-density wave in(TMTSF)$_2$ClO$_4$: Spatial structure of the two-phase state. *Phys. Rev. B* **2014**, *89*, 054518. [CrossRef]
29. Yonezawa, S.; Marrache-Kikuchi, C.A.; Bechgaard, K.; Jerome, D. Crossover from impurity-controlled to granular superconductivity in (TMTSF)$_2$ClO$_4$. *Phys. Rev. B* **2018**, *97*, 014521. [CrossRef]
30. Gerasimenko, Y.A.; Prudkoglyad, V.A.; Kornilov, A.V.; Sanduleanu, S.V.; Qualls, J.S.; Pudalov, V.M. Role of anion ordering in the coexistence of spin-density-wave and superconductivity in (TMTSF)$_2$ClO$_4$. *JETP Lett.* **2013**, *97*, 419–424. [CrossRef]
31. Wang, Y.; Chubukov, A.V. Enhancement of superconductivity at the onset of charge-density-wave order in a metal. *Phys. Rev. B* **2015**, *92*, 125108. [CrossRef]
32. Armitage, N.P.; Fournier, P.; Greene, R.L. Progress and perspectives on electron-doped cuprates. *Rev. Mod. Phys.* **2010**, *82*, 2421–2487. [CrossRef]
33. Helm, T.; Kartsovnik, M.V.; Proust, C.; Vignolle, B.; Putzke, C.; Kampert, E.; Sheikin, I.; Choi, E.S.; Brooks, J.S.; Bittner, N.; et al. Correlation between Fermi surface transformations and superconductivity in the electron-doped high-T_c superconductor Nd$_{2-x}$Ce$_x$CuO$_4$. *Phys. Rev. B* **2015**, *92*, 094501. [CrossRef]
34. Mukuda, H.; Fujii, T.; Ohara, T.; Harada, A.; Yashima, M.; Kitaoka, Y.; Okuda, Y.; Settai, R.; Onuki, Y. Enhancement of Superconducting Transition Temperature due to the Strong Antiferromagnetic Spin Fluctuations in the Noncentrosymmetric Heavy–Fermion Superconductor CeIrSi$_3$: A ^{29}Si NMR Study under Pressure. *Phys. Rev. Lett.* **2008**, *100*, 107003. [CrossRef] [PubMed]
35. Manago, M.; Kitagawa, S.; Ishida, K.; Deguchi, K.; Sato, N.K.; Yamamura, T. Enhancement of superconductivity by pressure-induced critical ferromagnetic fluctuations in UCoGe. *Phys. Rev. B* **2019**, *99*, 020506. [CrossRef]
36. Eckberg, C.; Campbell, D.J.; Metz, T.; Collini, J.; Hodovanets, H.; Drye, T.; Zavalij, P.; Christensen, M.H.; Fernandes, R.M.; Lee, S.; et al. Sixfold enhancement of superconductivity in a tunable electronic nematic system. *Nat. Phys.* **2020**, *16*, 346–350. [CrossRef] [PubMed]
37. Mukasa, K.; Ishida, K.; Imajo, S.; Qiu, M.; Saito, M.; Matsuura, K.; Sugimura, Y.; Liu, S.; Uezono, Y.; Otsuka, T.; et al. Enhanced Superconducting Pairing Strength near a Pure Nematic Quantum Critical Point. *Phys. Rev. X* **2023**, *13*, 011032. [CrossRef]
38. Tanaka, Y.; Kuroki, K. Microscopic theory of spin-triplet f-wave pairing in quasi-one-dimensional organic superconductors. *Phys. Rev. B* **2004**, *70*, 060502. [CrossRef]
39. Gor'kov, L.P.; Grigoriev, P.D. Nature of the superconducting state in the new phase in (TMTSF)$_2$PF$_6$ under pressure. *Phys. Rev. B* **2007**, *75*, 020507(R). [CrossRef]
40. Grigoriev, P.D. Properties of superconductivity on a density wave background with small ungapped Fermi surface parts. *Phys. Rev. B* **2008**, *77*, 224508. [CrossRef]
41. Ramazashvili, R.; Grigoriev, P.D.; Helm, T.; Kollmannsberger, F.; Kunz, M.; Biberacher, W.; Kampert, E.; Fujiwara, H.; Erb, A.; Wosnitza, J.; et al. Experimental evidence for Zeeman spin–orbit coupling in layered antiferromagnetic conductors. *Npj Quantum Mater.* **2021**, *6*, 11. [CrossRef]
42. Kartsovnik, M.V. High Magnetic Fields: A Tool for Studying Electronic Properties of Layered Organic Metals. *Chem. Rev.* **2004**, *104*, 5737–5782. [CrossRef]
43. Brazovskii, S.; Kirova, N. Electron selflocalization and superstructures in quasi one-dimensional dielectrics. *Sov. Sci. Rev. A* **1984**, *5*, 99–166.
44. Su, W.P.; Kivelson, S.; Schrieffer, J.R. Theory of Polymers Having Broken Symmetry Ground States. In *Physics in One Dimension*; Bernascony, J., Schneider, T., Eds.; Springer Series in Solid-State Sciences; Springer: Berlin/Heidelberg, Germany, 1981; pp. 201–211. [CrossRef]
45. Grigoriev, P.D. Superconductivity on the density-wave background with soliton-wall structure. *Phys. B* **2009**, *404*, 513–516. [CrossRef]
46. Gor'kov, L.P.; Grigoriev, P.D. Soliton phase near antiferromagnetic quantum critical point in Q1D conductors. *Europhys. Lett.* **2005**, *71*, 425–430. [CrossRef]

47. Tinkham, M. *Introduction to Superconductivity*, 2nd ed.; International Series in Pure and Applied Physics; McGraw-Hill, Inc.: New York, NY, USA, 1996.
48. Pratt, F.L.; Lancaster, T.; Blundell, S.J.; Baines, C. Low-Field Superconducting Phase of $(TMTSF)_2ClO_4$. *Phys. Rev. Lett.* **2013**, *110*, 107005. [CrossRef] [PubMed]
49. Kochev, V.D.; Kesharpu, K.K.; Grigoriev, P.D. Anisotropic zero-resistance onset in organic superconductors. *Phys. Rev. B* **2021**, *103*, 014519. [CrossRef]
50. Grigoriev, P.D.; Kochev, V.D.; Orlov, A.P.; Frolov, A.V.; Sinchenko, A.A. Inhomogeneous Superconductivity Onset in FeSe Studied by Transport Properties. *Materials* **2023**, *16*, 1840. [CrossRef]
51. Seidov, S.S.; Kochev, V.D.; Grigoriev, P.D. First-order phase transition between superconductivity and charge/spin-density wave as the reason of their coexistence in organic metals. *arXiv* **2023**, arXiv:2305.06957. [CrossRef]
52. Valfells, S.; Brooks, J.S.; Wang, Z.; Takasaki, S.; Yamada, J.; Anzai, H.; Tokumoto, M. Quantum Hall transitions in $(TMTSF)_2PF_6$. *Phys. Rev. B* **1996**, *54*, 16413–16416. [CrossRef] [PubMed]
53. Danner, G.M.; Chaikin, P.M.; Hannahs, S.T. Critical imperfect nesting in $(TMTSF)_2PF_6$. *Phys. Rev. B* **1996**, *53*, 2727–2731. [CrossRef] [PubMed]
54. Araki, C.; Itoi, M.; Hedo, M.; Uwatoko, Y.; Mori, H. Electrical Resistivity of $(TMTSF)_2PF_6$ under High Pressure. *J. Phys. Soc. Jpn.* **2007**, *76*, 198–199.
55. Auban-Senzier, P.; Pasquier, C.; Jérome, D.; Carcel, C.; Fabre, J. From Mott insulator to superconductivity in $(TMTTF)_2BF_4$: High pressure transport measurements. *Synth. Met.* **2003**, *133–134*, 11–14.
56. Horovitz, B.; Gutfreund, H.; Weger, M. Interchain coupling and the Peierls transition in linear-chain systems. *Phys. Rev. B* **1975**, *12*, 3174–3185. [CrossRef]
57. McKenzie, R.H. Microscopic theory of the pseudogap and Peierls transition in quasi-one-dimensional materials. *Phys. Rev. B* **1995**, *52*, 16428–16442. [CrossRef] [PubMed]
58. Grigoriev, P.D.; Lyubshin, D.S. Phase diagram and structure of the charge-density-wave state in a high magnetic field in quasi-one-dimensional materials: A mean-field approach. *Phys. Rev. B* **2005**, *72*, 195106. [CrossRef]
59. Kim, J.; Yun, M.; Jeong, D.W.; Kim, J.J.; Lee, I. Structural and Electrical Properties of the Single-crystal Organic Semiconductor Tetramethyltetraselenafulvalene (TMTSF). *J. Korean Phys. Soc.* **2009**, *55*, 212–216. [CrossRef]
60. Oxtoby, D.W. Nucleation of First-Order Phase Transitions. *Acc. Chem. Res.* **1998**, *31*, 91–97. [CrossRef]
61. Umantsev, A. *Field Theoretic Method in Phase Transformations*; Lecture Notes in Physics; Springer: New York, NY, USA, 2012. [CrossRef]
62. Kalikmanov, V. *Nucleation Theory*; Lecture Notes in Physics; Springer: Dordrecht, The Netherlands, 2012. [CrossRef]
63. Karthika, S.; Radhakrishnan, T.K.; Kalaichelvi, P. A Review of Classical and Nonclassical Nucleation Theories. *Cryst. Growth Des.* **2016**, *16*, 6663–6681. [CrossRef]
64. Efros, A.L. *Physics and Geometry of Disorder: Percolation Theory*; Science for Everyone; Mir Publishers: Moscow, Russia, 1987.
65. Torquato, S. *Random Heterogeneous Materials*; Springer: New York, NY, USA, 2002. [CrossRef]
66. Sinchenko, A.A.; Grigoriev, P.D.; Orlov, A.P.; Frolov, A.V.; Shakin, A.; Chareev, D.A.; Volkova, O.S.; Vasiliev, A.N. Gossamer high-temperature bulk superconductivity in FeSe. *Phys. Rev. B* **2017**, *95*, 165120. [CrossRef]
67. Grigoriev, P.D.; Sinchenko, A.A.; Kesharpu, K.K.; Shakin, A.; Mogilyuk, T.I.; Orlov, A.P.; Frolov, A.V.; Lyubshin, D.S.; Chareev, D.A.; Volkova, O.S.; et al. Anisotropic effect of appearing superconductivity on the electron transport in FeSe. *JETP Lett.* **2017**, *105*, 786–791. [CrossRef]
68. Seidov, S.S.; Kesharpu, K.K.; Karpov, P.I.; Grigoriev, P.D. Conductivity of anisotropic inhomogeneous superconductors above the critical temperature. *Phys. Rev. B* **2018**, *98*, 014515. [CrossRef]
69. Miyagawa, K.; Kawamoto, A.; Kanoda, K. Proximity of Pseudogapped Superconductor and Commensurate Antiferromagnet in a Quasi-Two-Dimensional Organic System. *Phys. Rev. Lett.* **2002**, *89*, 017003. [CrossRef]
70. Sasaki, T.; Yoneyama, N. Spatial mapping of electronic states in κ-(BEDT-TTF)2X using infrared reflectivity. *Sci. Technol. Adv. Mater.* **2009**, *10*, 024306. [CrossRef]
71. Zverev, V.N.; Biberacher, W.; Oberbauer, S.; Sheikin, I.; Alemany, P.; Canadell, E.; Kartsovnik, M.V. Fermi surface properties of the bifunctional organic metal $\kappa-(BETS)_2Mn[N(CN)_2]_3$ near the metal-insulator transition. *Phys. Rev. B* **2019**, *99*, 125136. [CrossRef]
72. Oberbauer, S.; Erkenov, S.; Biberacher, W.; Kushch, N.D.; Gross, R.; Kartsovnik, M.V. Coherent heavy charge carriers in an organic conductor near the bandwidth-controlled Mott transition. *Phys. Rev. B* **2023**, *107*, 075139. [CrossRef]
73. Lang, K.M.; Madhavan, V.; Hoffman, J.E.; Hudson, E.W.; Eisaki, H.; Uchida, S.; Davis, J.C. Imaging the granular structure of high-Tc superconductivity in underdoped $Bi_2Sr_2CaCu_2O_{8+\delta}$. *Nature* **2002**, *415*, 412–416. [CrossRef] [PubMed]
74. Wise, W.D.; Chatterjee, K.; Boyer, M.C.; Kondo, T.; Takeuchi, T.; Ikuta, H.; Xu, Z.; Wen, J.; Gu, G.D.; Wang, Y.; et al. Imaging nanoscale Fermi-surface variations in an inhomogeneous superconductor. *Nat. Phys* **2009**, *5*, 213–216. [CrossRef]
75. Kresin, V.; Ovchinnikov, Y.; Wolf, S. Inhomogeneous superconductivity and the "pseudogap" state of novel superconductors. *Phys. Rep.* **2006**, *431*, 231–259. [CrossRef]
76. Campi, G.; Bianconi, A.; Poccia, N.; Bianconi, G.; Barba, L.; Arrighetti, G.; Innocenti, D.; Karpinski, J.; Zhigadlo, N.D.; Kazakov, S.M.; et al. Inhomogeneity of charge-density-wave order and quenched disorder in a high-Tc superconductor. *Nature* **2015**, *525*, 359–362. [CrossRef]

77. Mogilyuk, T.I.; Grigoriev, P.D.; Kesharpu, K.K.; Kolesnikov, I.A.; Sinchenko, A.A.; Frolov, A.V.; Orlov, A.P. Excess Conductivity of Anisotropic Inhomogeneous Superconductors Above the Critical Temperature. *Phys. Solid State* **2019**, *61*, 1549–1552. [CrossRef]
78. Kresin, V.Z.; Ovchinnikov, Y.N. Nano-based Josephson Tunneling Networks and High Temperature Superconductivity. *J. Supercond. Nov. Magn.* **2021**, *34*, 1705–1708. [CrossRef]

Disclaimer/Publisher's Note: The statements, opinions and data contained in all publications are solely those of the individual author(s) and contributor(s) and not of MDPI and/or the editor(s). MDPI and/or the editor(s) disclaim responsibility for any injury to people or property resulting from any ideas, methods, instructions or products referred to in the content.

Article

Band Structure Evolution during Reversible Interconversion between Dirac and Standard Fermions in Organic Charge-Transfer Salts

Ryuhei Oka [1], Keishi Ohara [1,2], Kensuke Konishi [1,2], Ichiro Yamane [3], Toshihiro Shimada [3] and Toshio Naito [1,2,4,*]

1. Graduate School of Science and Engineering, Ehime University, Matsuyama 790-8577, Ehime, Japan; okari910@gmail.com (R.O.)
2. Research Unit for Materials Development for Efficient Utilization and Storage of Energy, Ehime University, Matsuyama 790-8577, Ehime, Japan
3. Graduate School of Engineering, Hokkaido University, Kita 13, Nishi 8, Kita-ku, Sapporo 060-8628, Hokkaido, Japan
4. Geodynamics Research Center (GRC), Ehime University, Matsuyama 790-8577, Ehime, Japan
* Correspondence: tnaito@ehime-u.ac.jp

Abstract: Materials containing Dirac fermions (DFs) have been actively researched because they often alter electrical and magnetic properties in an unprecedented manner. Although many studies have suggested the transformation between standard fermions (SFs) and DFs, the non-availability of appropriate samples has prevented the observation of the transformation process. We observed the interconversion process of DFs and SFs using organic charge-transfer (CT) salts. The samples are unique in that the constituents (the donor D and acceptor A species) are particularly close to each other in energy, leading to the temperature- and D-A-combination-sensitive CT interactions in the solid states. The three-dimensional weak D–A CT interactions in low-symmetry crystals induced the continuous reshaping of flat-bottomed bands into Dirac cones with decreasing temperature; this is a characteristic shape of bands that converts the behavior of SFs into that of DFs. Based on the first-principles band structures supported by the observed electronic properties, round-apex-Dirac-cone-like features appear and disappear with temperature variation. These band-structure snapshots are expected to add further detailed understanding to the related research fields.

Keywords: organic Dirac electron systems; band structure reshaping; coexistence of standard and Dirac fermions; first-principles band calculation; donor-anion interactions; self-doping

Citation: Oka, R.; Ohara, K.; Konishi, K.; Yamane, I., Shimada, T.; Naito, T. Band Structure Evolution during Reversible Interconversion between Dirac and Standard Fermions in Organic Charge-Transfer Salts. *Magnetochemistry* **2023**, *9*, 153. https://doi.org/10.3390/magnetochemistry9060153

Academic Editor: Laura C. J. Pereira

Received: 16 May 2023
Revised: 2 June 2023
Accepted: 5 June 2023
Published: 9 June 2023

Copyright: © 2023 by the authors. Licensee MDPI, Basel, Switzerland. This article is an open access article distributed under the terms and conditions of the Creative Commons Attribution (CC BY) license (https://creativecommons.org/licenses/by/4.0/).

1. Introduction

The physical and chemical properties of common materials are governed by electrons or holes with finite effective masses, i.e., massive fermions [1]. Recently, a series of materials that contain unique particles called Dirac fermions (DFs) have gained considerable scientific attention [1–6]. These particles possess unusually small effective masses and unusually high mobilities because of their Fermi velocities. Owing to these characteristics, the electrons in these materials behave as relativistic massless fermions. DFs regulate the physical properties when located near the Fermi level; these fermions exhibit unique electrical properties such as temperature (T)-independent resistivity (ρ), dρ/dT ~0. Such behaviors belong to neither metals (dρ/dT > 0) nor nonmetals (dρ/dT < 0). These intriguing physical properties can be attributed to a unique feature that DF systems share in their band structures around the Fermi levels: the band dispersion linearly depends on the wave vectors in the reciprocal space, which leads to cone-shaped bands called Dirac cones [7–32] (Figure 1a). Unlike the unique feature of the band structures of the DF system, the band dispersions in common materials are described by parabolic or cosine bands [1].

A single electron cannot be a DF (massless fermion) and a standard fermion (SF; massive fermion) at the same time. Therefore, the transformation between them, driven by band reshaping, has been focused on in terms of how and why this conversion should occur and how the physical properties should change. Most studies on such transformations are based on theoretical physics [33–44], and the actual transformation, where chemistry also plays an important role, remains elusive.

The organic charge-transfer (CT) salts α-D_2I_3 (D = $C_{10}H_8S_8$ (ET), $C_{10}H_8S_6Se_2$ (STF), and $C_{10}H_8S_4Se_4$ (BETS); Figure 1b) under ambient pressure are extensively studied, as well as their aspects regarding organic DF systems [19–36,45–62]. However, most studies on them are focused on whether they belong to the zero-gap semiconductors (ZGSs) [19–31,63–66]. It has not been examined experimentally how the reversible conversion of massive and massless fermions occurs in conjunction with the formation of a Dirac-cone-type band. Since the discovery of DFs in graphene and related inorganic substances, many studies have focused on stable two-dimensional (2D) compounds with simple chemical formulae and high structural symmetries [1–16]. Such samples are often more easily available than the organic DF systems and favor clear discussion based on precise theoretical analyses such as first-principles band calculations. Stable inorganic compounds are often obtained as large robust single crystals in large-scale syntheses, whereas few of these conditions are generally observed in organic compounds. However, simple electronic systems rarely exhibit the interconversion between DFs and SFs, as they are thermodynamically too stable and possess lesser degrees of freedom than complicated electronic systems. In this regard, the bulk DF systems in α-D_2I_3 exhibited gradual and reversible transformation between DFs and SFs with variation in the temperature T, owing to their low-symmetry crystal structures and various weak but non-negligible competing interactions. We demonstrate the validities of the calculated band structures by comparing them with the observed physical properties. Further, we discuss the mechanism of the appearance and disappearance of DFs based on the corresponding band structures. Our finding revealed that the chemical equilibrium in the D-I_3 redox reaction governs the "physical equilibrium" that determines whether the fermions behave as DFs or SFs. Most previous studies on organic DFs have discussed their findings in terms of physics rather than chemistry and focused on the local band structure around the Dirac cones. By providing the whole band structures connected with D-I_3 redox reaction, this study sheds new light on the research on the production mechanism and stability of DFs.

2. Materials and Methods

2.1. Materials

The single crystals of α-D_2I_3 (D = ET, STF, and BETS) and the donor molecules (STF and BETS) were synthesized by following the reported procedures [45–47]. The details of the syntheses are described in Supporting Information (Scheme S1). The neutral ET was purchased from Tokyo Chemical Industry Co., Ltd. (Tokyo, Japan), and used as received. Prior to the physical property measurements, all single crystals were checked by X-ray oscillation photography in terms of their crystal quality and the orientation of crystallographic axes.

2.2. Electrical Resistivity Measurements

The electrical resistivities of α-D_2I_3 (D = ET, STF, and BETS) were measured by a standard four-probe method using single crystals, gold wires (15 µm in diameter, Nilaco), carbon paste, and a physical property measurement system (PPMS-9; 10 mW—5 MW) with EverCool II (Quantum Design). A direct current of 100–200 µA was applied depending on the resistivity values. For the resistivity measurement of the insulating phase of α-ET_2I_3, different equipment with a wider dynamic range of resistivity measurement was used (the maximum output = 200 V, resolution = 1 pA) with the same four-probe configuration. The equipment was a home-made cryostat consisting of a digital voltmeter (Keithley Nanovoltmeter 2182A, Tektronics, Tokyo, Japan), a current/voltage source (Keithley SourceMeter 2400, Tokyo, Japan, Tektronics), a digital temperature controller (LakeShore,

Model 331, Lake Shore Cryotronics, Inc., Westerville, OH, USA), and a diffusion/rotary pumping system (DIAVAC, DS-A412N, Diavac Limited, Yachiyo, Japan). To minimize the joule heating of the sample and a resultant thermoelectric power, a constant current ($\leq$2 mA) was applied for 20 ms, and the voltage drop was measured in a synchronized way with the current source. Immediately (~20 ms) after the first measurement, the voltage was reversed, and the same measurement was carried out. Then, the average was taken to cancel out the thermoelectric power. At every data acquisition during the resistivity measurement of the insulating phase of α-ET$_2$I$_3$, we confirmed that 98–100% values of the set current flew in the sample. In both types of equipment, gold wires (25 μmφ) were attached along the b-axis on the ab planes of the crystals for the electrical contacts. The linearity between current and voltage was checked at every beginning of the measurement at ~300 K.

2.3. Magnetic Susceptibility Measurements

The magnetic susceptibility was measured using the polycrystalline samples and DC mode of a SQUID susceptometer (Quantum Design MPMS-XL with an EverCool) in both field cooling (FC) and zero-field cooling (ZFC) processes at 2–300 K. The sample was contained in a gelatin capsule with ventilation holes and was set in the middle of a polystyrene straw (Quantum Design); a magnetic field of 1 T was applied. Background signals from the capsule and the straw were independently measured, which were then subtracted from the total susceptibility. The diamagnetic susceptibilities (emu mol^{-1}G^{-1}) of the ET (-2.18×10^{-4}), STF (-1.55×10^{-4}), BETS (-2.12×10^{-4}), and I$_3^-$ (-3.61×10^{-5}) species were cited from previous studies [26,67,68].

2.4. ESR Spectra Measurements

The ESR spectra of the X band (9.3 GHz) were measured using the single crystals of α-D$_2$I$_3$ in the temperature range of 120–300 K using JEOL JES-FA100 (JEOL Ltd., Tokyo, Japan). The single crystal was mounted on a Teflon piece settled with a minimal amount of Apiezon N grease and sealed in a 5-mm-diameter quartz sample tube in a low-pressure (~20 mm Hg) helium atmosphere. The background signals were measured prior to the sample measurement under conditions identical to those of the samples, and the resultant spectra were subtracted from the raw sample spectra. The Q-values, time constant, sweep time, modulation, and its amplitude were 4700–8200, 0.03 s, 1 min, 100 kHz, and 2 mT, respectively. Here, the Q-values are the factors indicating the resonance specification of the cavity and are defined as the ratio between the energy stored in the cavity and that consumed inside the cavity. The root dependencies of the ESR intensities on the microwave power were checked (1–10 mW), and the power was selected to be 4 mW or 9 mW depending on the signal intensity. The magnetic field was corrected by a Gauss meter (JEOL NMR Field Meter ES-FC5, Akishima, Tokyo, Japan) at the end of every measurement. The temperature was controlled using a continuous flow-type liquid N$_2$ cryostat with a digital temperature controller (JEOL). The temperature variation did not exceed $\pm$0.5 K during the field sweep. The cooling rate was $-$10 K/min. No hysteretic behavior was observed in the ESR spectra between the heating and cooling processes.

2.5. X-ray Structural Analyses

The single-crystal X-ray structural analyses of α-ET$_2$I$_3$ (296 K and 100 K) and α-BETS$_2$I$_3$ (296 K and 100 K) were performed using RIGAKU VariMax SaturnCCD724/α (graphite monochromated Mo-K$_\alpha$ radiation; λ = 0.71073 Å) (Rigaku Co., Ltd., Tokyo, Japan) equipped with a nitrogen-gas–flow temperature controller (Cobra, Oxford Cryosystems, Long Hanborough, UK). Regarding the low-temperature measurements, the cooling rate was $-$1 K/min. The fluctuation in temperature during the data collections was ~$\pm$1 K. The collected data were processed using CrysAlisPro ver. 171.38.46 or ver. 171.41_64.93a (Rigaku Oxford Diffraction, Tokyo, Japan) prior to the structure determination and refinement using CrystalStructure 4.3.2 (Rigaku Oxford Diffraction, Tokyo, Japan) or Olex 2-1.5 [69]. The hydrogen atoms on the ethylene groups of the ET and BETS molecules were located at the calculated positions. Further details of the data collection and analyses are described in the .cif files deposited to the Cambridge Crystallographic Data Centre (CCDC). The CCDC deposit numbers of the .cif files are as follows: α-BETS$_2$I$_3$ 2217842 (100 K) and 2217843 (296 K); α-ET$_2$I$_3$ 2217847 (100 K) and 2217848 (296 K).

2.6. Band Structure Calculation

The electronic band structures of the materials were calculated using the Vienna *Ab initio* Simulation Package (VASP) [70]. PBE functional [71] was used along with the augmented plane wave method. Crystal structures derived from the X-ray diffraction experiments were used, and no structure optimization was performed. The atomic parameters used in the calculation were submitted as cif files. The structural data were partly (80 K and 30 K for α-BETS$_2$I$_3$ [25] and 150 K and 30 K for α-ET$_2$I$_3$ [49]) provided by Prof. H. Sawa at Nagoya University in Japan; the remaining structural data (296 K and 100 K for both α-BETS$_2$I$_3$ and α-ET$_2$I$_3$) were obtained in this study as described above. The details of structural analyses were included in the cif files. The energy convergence with the cut-off energy and K-point density was carefully examined and the final cut-off of 1000 eV and $8 \times 8 \times 4$ K-points were found sufficient to achieve the convergence of the total energy $< 10^{-3}$ eV/atom.

3. Results and Discussion

Below, we briefly summarize the structural and physical properties of α-D$_2$I$_3$. These properties are known [19–26,45–50], but are necessary for proving the validity of our calculated band structures. As the observed electronic properties and calculated band structures in previous papers do not always provide necessary information for comparison with our calculated band structures in detail, we re-examined the physical properties shown below. The disorder at the inner-chalcogen (Se and S) atom sites prevented us from evaluating the band structure of α-STF$_2$I$_3$ using the same calculation as that employed for the other two salts [21,26,63]. Thus, we will not discuss the band structure of α-STF$_2$I$_3$ and will only discuss the physical properties of α-STF$_2$I$_3$ for comparison. There are papers showing that the STF salt belongs to the ZGSs at ambient pressure despite the disorder at the STF sites [21,26,36,63]. Thus, the physical properties of α-STF$_2$I$_3$ augment the current understanding of the behavior of the DFs in α-D$_2$I$_3$.

3.1. Crystal Structures of α-D$_2$I$_3$

Except for the change in the space group in the phase transition at 135 K for α-ET$_2$I$_3$, the crystal structures of α-D$_2$I$_3$ (D = ET, STF, and BETS; Figure 1c,d) are qualitatively identical and practically independent of the temperature T, which is consistent with that reported in the previous studies [19,45–49]. An important feature is that there are weak and T-dependent charge-transfer (CT) interactions between D and I$_3$ in addition to the strong D–D CT interactions that yield D$^{(0.5-\delta/2)+}$ and I$_3^{(1-\delta)-}$ ($0 < \delta << 1$) [72,73]; this indicates that both D and I$_3$ are radical species. The band formation is dominated by D–D interactions, whereas band reshaping is driven by D–I$_3$ interactions [72,73]. The D cations aggregate via S---S short (<van der Waals distance of 3.70 Å) contacts, and they form 2D conduction sheets in the *ab* planes. In addition, the I$_3$ anions are sandwiched between these sheets, and they develop H bonds with the ethylene groups of the D cations (C-H---I). The D–I$_3$ interactions occur via different pathways based on the C-H---I contacts (H---I distances of *ca.* 3.0–3.5 Å). The amount of CT between D and I$_3$ depends on T. This structural feature produces rather complicated hyperfine structures in the electron spin resonance (ESR) spectra, as discussed hereinafter, which provides evidence for the D–I$_3$ interactions.

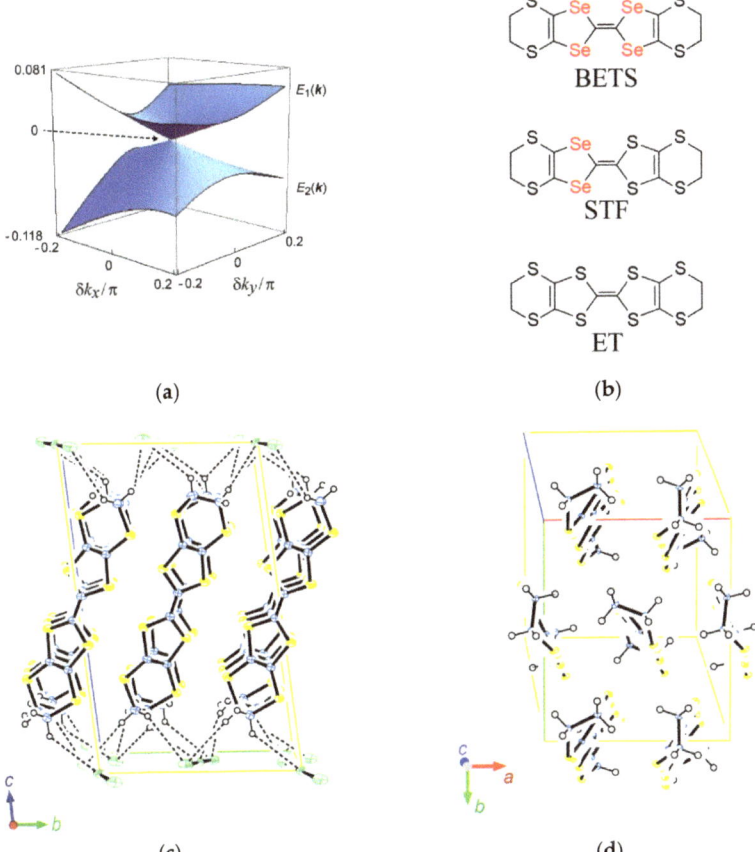

Figure 1. Organic Dirac fermion systems: (**a**) A characteristic band structure of Dirac fermion (DF) systems in the zero-gap semiconductors around a Dirac point, which is the apex of the cone indicated by a broken arrow. In this example, a pair of bands ($E_1(k)$ and $E_2(k)$) form a cone-type band structure in the $k_x k_y$ space. © 2020 The Physical Society of Japan (J. Phys. Soc. Jpn. 89, 023701) [21]; (**b**) Chemical structures of ET = bis(ethylenedithio)tetrathiafulvalene, STF = bis(ethylenedithio)diselenadithiafulvalene, and BETS = bis(ethylenedithio)tetraselenafulvalene; (**c**) Crystal structures (view along the a axis) of α-ET$_2$I$_3$. The broken lines represent (C-)H---I and (C-)H---S contacts that are shorter than the van der Waals distances, i.e., 3.35 and 3.05 Å, respectively. These contacts demonstrate the ET–I$_3$ interactions, which drive band reshaping and transformation between SFs and DFs; (**d**) Crystal structures (view along the c axis) of α-ET$_2$I$_3$. The I atoms are omitted in Figure 1d for clarity.

3.2. Physical Properties of α-D$_2$I$_3$

Figure 2 shows the T dependencies of the electrical resistivity (ρ: Figure 2a,b) and paramagnetic susceptibility (χ_p: Figure 2c,d) of α-D$_2$I$_3$. Below ~120 K, the resistivity of α-ET$_2$I$_3$ exhibited a very small temperature dependence (activation energy ~5.2 meV at 78–105 K). This activation energy well agreed with the present band calculation (4.23 meV at 100 K) as discussed below. Additionally, our observation of the temperature-dependent activation energy, which clearly decreased at ~70 K, agrees quantitatively with the previous study [74]. The paramagnetic susceptibility χ_p manifests the net contribution of the carriers to the magnetic susceptibility. The high values of χ_p below 135 K appears to contradict the fact that this salt is in the insulating phase below 135 K. However, a substantial number of residual spins were observed in the insulating phase of α-ET$_2$I$_3$.

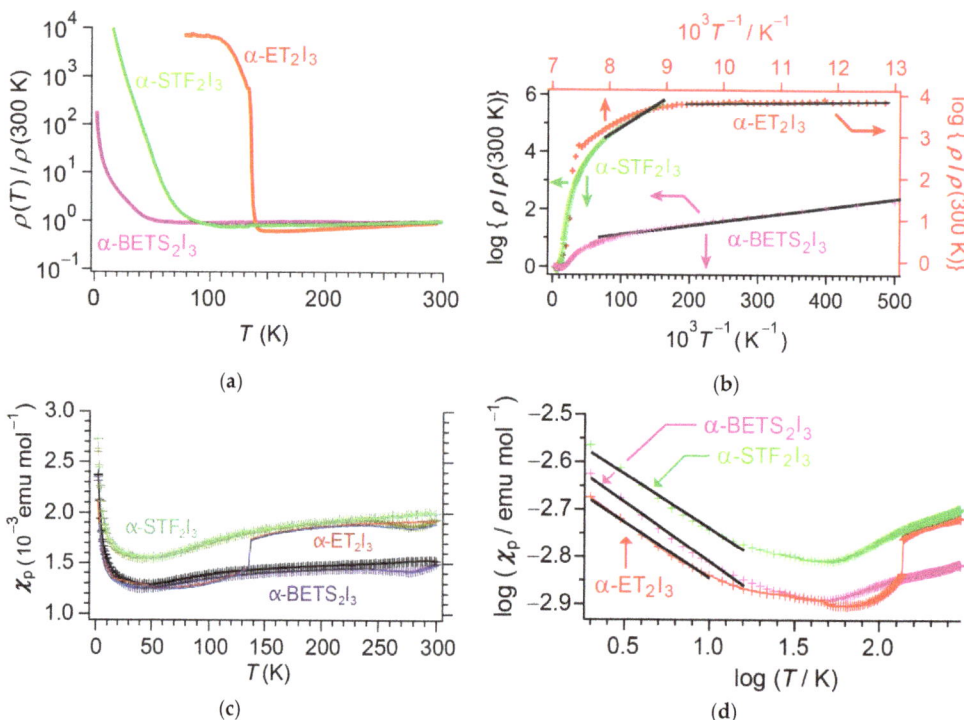

Figure 2. Temperature (T) dependencies of the physical properties (ρ) of α-D$_2$I$_3$: (**a**) Electrical resistivity (ρ). D = ET (red), STF (green), and BETS (violet). The vertical axis for ρ is normalized by ρ at 300 K. The absolute values of ρ at 300 K for the three materials are 47 (D = ET), 2.2 (D = STF), and 2.9 (D = BETS) in the unit of 10^{-2} Ω cm; (**b**) Enlarged view of ρ at low T, which shows thermally activated T-dependence near the ground states. Energy gaps estimated from the best-fit lines are 5.2 (ET, 78–105 K), 6.7 (STF, $T \leq 16$ K), and 1.4 (BETS, $T \leq 16$ K) meV; (**c**) Paramagnetic susceptibility (χ_p). D = ET (red = field-cooling (FC) and blue = zero-field cooling (ZFC)), STF (green = FC and pale brown = ZFC), and BETS (black = FC and violet = ZFC). The absolute values of χ_p at 300 K for the three materials are 1.9 (D = ET), 2.0 (D = STF), and 1.5 (D = BETS) in the unit of 10^{-3} emu G^{-1} mol^{-1}; (**d**) Enlarged view of χ_p at low T, which shows a unique power-law T-dependence near the ground states. The T-dependencies of χ_p near the ground states are different from those of ρ in each α-D$_2$I$_3$. D = ET (red), STF (green), and BETS (violet). Black lines represent the best-fit lines in the indicated T ranges: $\chi_p \propto T^{-0.24}$ ($T < 10$ K), $\chi_p \propto T^{-0.23}$ ($T < 16$ K), and $\chi_p \propto T^{-0.25}$ ($T < 16$ K) for D = ET, STF, and BETS, respectively.

For example, Venturini et al. [53] and Sugano et al. [54] measured the ESR of the single crystals of α-ET$_2$I$_3$ down to 76 K and 5 K, respectively. Uji et al. measured the resistances in the ab plane at low-bias voltages of α-ET$_2$I$_3$ down to ~30 K [55]. These studies indicate the existence of (thermally excited) unpaired electrons/holes well below 135 K. Both ρ and χ_p exhibit a metal–insulator (M–I) transition at 135 K for α-ET$_2$I$_3$ (Figure 2a,c); the metallic carriers (SFs) disappear below this temperature [48,50]. The values of χ_p and ρ at $T < 135$ K are considerably high and low, respectively, for the insulating solids, which suggests that there is a different type of carriers (fermions) that dominates χ_p and ρ at ~$100 \leq T < 135$ K. The calculated band structure (Figure 3) indicates that the carriers contain thermally excited holes and electrons across the band gaps of a few millielectronvolts at the Dirac points.

Depending on the chemical potential varying with T, these carriers should behave as DFs when they are thermally excited and are located at the linear band curvatures of the Dirac cones. At $T \leq 30$ K, the thermally excited carriers are massive (SFs) because there are no more Dirac cones in the bands. The complicated T dependencies of the band structures (Figure 3, Supplementary Figure S1) create a striking contrast with the monotonical T dependencies of the electrical and magnetic properties of α-D$_2$I$_3$ (Figure 2) except for the M–I transition of α-ET$_2$I$_3$. This is because SFs and DFs coexist to contribute to the physical properties in a wide T range; the ratio (SFs/DFs) varies sensitively with T. The coexistence of SFs and DFs in some materials was previously suggested for α-ET$_2$I$_3$ [28,64,65] and was reviewed for other materials [75]. Therefore, we cannot explain all physical properties from room temperature to the ground states based exclusively on either SFs or DFs in α-D$_2$I$_3$. Similar cases are reported in different types of materials: the physical properties of some DF systems can be understood by assuming that they contain different types of fermions [26,76–78]. The importance of electronic correlation effects on the physical properties of α-ET$_2$I$_3$ is also discussed [19,28,29].

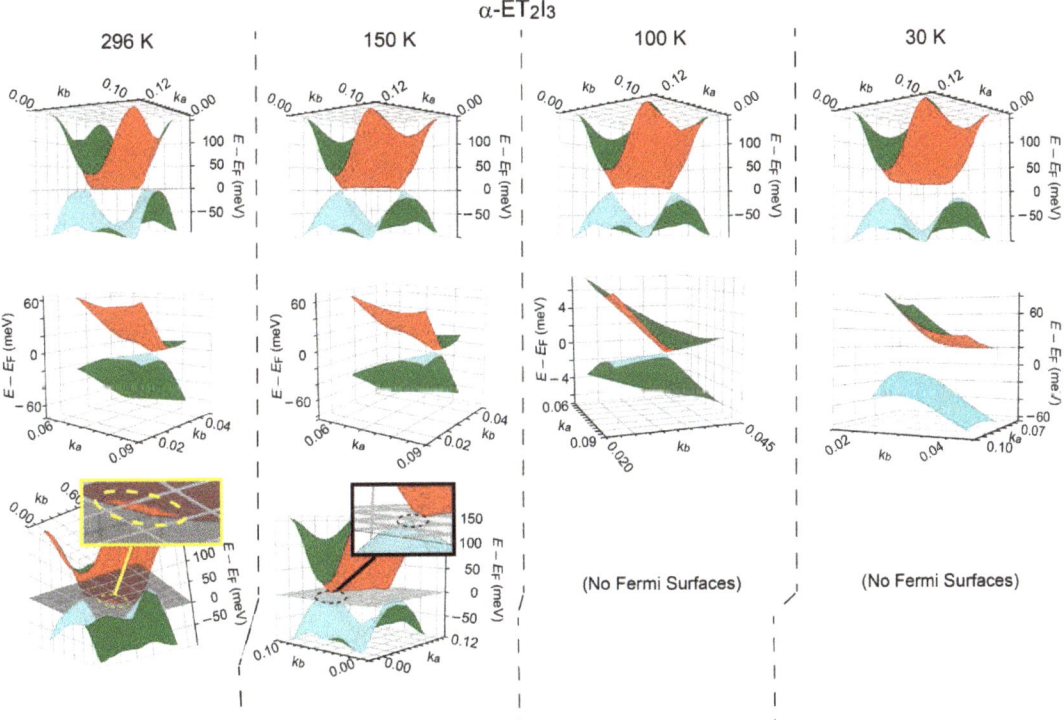

Figure 3. T-dependent band structures for α-ET$_2$I$_3$. The figures in the middle row from the left to the right show the gradual growth of Dirac cones from the flat-bottomed bands with decreasing T followed by the disappearance of the Dirac cones at 30 K. The Fermi levels (E_F) are indicated by the gray plane in the bottom figures (296 and 150 K). (Top) The outline, (middle) close-up views of the calculated band structures at the Dirac points, and (bottom) close-up views of the calculated band structures on the Fermi surfaces. The broken ovals in the bottom figures indicate the locations where E_F intersects the bottom of the top (red) band (296 K) or the top of the bottom (blue) band (150 K). The band gap between the Dirac cones at 100 K in the figure appears significantly smaller than the actual band gap (4.23 meV) because the viewing angle is highly inclined against the vertical (energy) axis.

3.3. Calculated Band Structures of α-ET$_2$I$_3$

The accuracy of the present band calculation is demonstrated to be ~0.04 meV by a series of closely related calculations of different groups [64,79,80]. Initially, the acquired band structures of α-ET$_2$I$_3$ (Figure 3) are discussed by comparing them with the observed physical properties of α-ET$_2$I$_3$ (Figure 2) to confirm the validity of the band calculation. The closely related band structures of α-BETS$_2$I$_3$ (Figure 4) and physical properties of the remaining α-D$_2$I$_3$ (D = STF and BETS) are described in this section to demonstrate the general aspects of band reshaping associated with the physical properties. Figure 3 depicts the band structures of α-ET$_2$I$_3$ calculated using the Vienna *ab initio* simulation package based on the crystal structures at 296, 150, 100, and 30 K under ambient pressure. We initially discuss the band structures obtained in the metallic phase (296 and 150 K) followed by those acquired in the insulating phase (100 and 30 K) because α-ET$_2$I$_3$ demonstrates an M–I transition at 135 K, where drastic changes in the band structures and physical properties are expected [48–50].

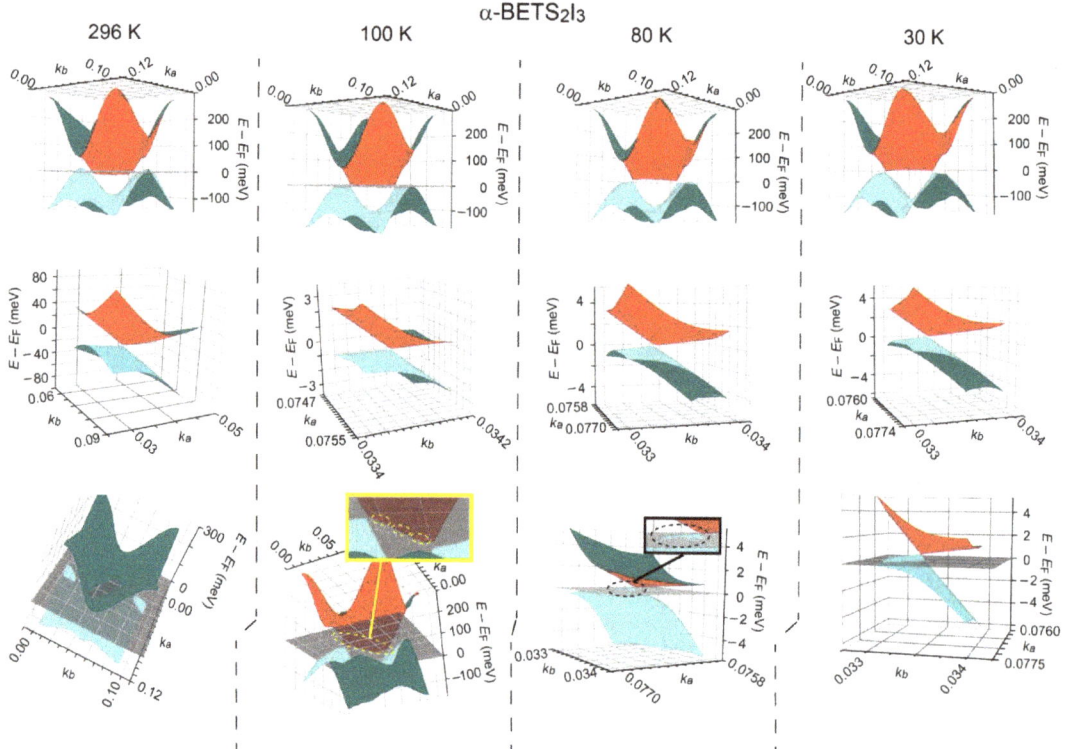

Figure 4. *T*-dependent band structures for α-BETS$_2$I$_3$. The figures in the middle row from the left to the right show the gradual changes in band filling and band shape with decreasing *T*. The Fermi levels (E_F) are indicated by the gray plane in the bottom figures. (Top) The outline, (middle) close-up views of the calculated band structures at the Dirac points, and (bottom) close-up views of the calculated band structures on the Fermi surfaces. The yellow and black broken ovals in the bottom figures represent the Fermi surfaces at 100 and 80 K, respectively.

α-ET$_2$I$_3$ exhibits the characteristic 2D ET arrangement and isotropic metallic conduction in the *ab* plane at 296 K. The obtained band structure at 296 K indicates significantly isotropic metallic dispersion in the $k_a k_b$ plane, and it is qualitatively consistent with the crystal structure and conducting properties. Furthermore, the acquired band structures have two important features in terms of band reshaping, which are driven by the ET–I$_3$ CT interactions.

First, the Fermi level is located at the top bands (red) of the four highest occupied molecular orbitals (HOMOs) of ET and the second band (blue) is fully filled. This implies that ET cation radicals are not quarter-filled (ET$^{+0.5}$ and I$_3$$^-$) but are electron doped by the I$_3$ anions (ET$^{(0.5-d/2)+}$ and I$_3$$^{(1-d)-}$). This contradicts the early-stage band calculation finding of a semi-metallic band structure with almost-equal areas of electron and hole pockets [51], where they did not consider ET–I$_3$ interactions. In most of the previous band calculations, some did not consider the CT interaction between I$_3$ and D [25,51,60,61,64,66], while some included the I$_3$ anions but obtained band structures with a negligible effect of the I$_3$-D CT interactions [24,62]. As a result, many of them suffered from some inconsistency with the observed physical properties or adjusted some parameters such as spin-orbit coupling constants to agree with the observations. The present band structure contains both electrons and holes as thermally excited Dirac electrons and holes, in addition to the normal electrons at the Fermi surface. Additionally, the calculated band gap between the Dirac cones is less than a few meV (~5–50 K) depending on the temperature T, implying that there are thermally excited Dirac electrons and holes as a part of the carriers down to the lowest T of the measurements in this work (~2 K). Their contribution to physical properties depends not only on their numbers but also on their mobilities and anisotropies. Thus, we cannot tell whether such a type of material would behave as a hole or electron conductor solely from the calculated band structures. The T-sensitive band structures make the major carriers' sign more unpredictable. This feature of band structure is common for α-D$_2$I$_3$ (D = ET (Figure 3), BETS (Figure 4)). The validity of these bands should be confirmed by the data of Hall coefficients and thermoelectric power for α-ET$_2$I$_3$ at 1 bar and $T \geq 30$ K; yet, they are not reported. For α-BETS$_2$I$_3$ at 1 bar, only the Hall coefficients are reported, which are small and noisy at $T \geq 50$ K [81]. The observed change in the sign of Hall coefficients of α-BETS$_2$I$_3$ at ~150 K can be partly explained by the present band calculation indicating that either holes or electrons are doped from I$_3$ to BETS depending on T (Figure 4). Therefore, our calculation does not contradict with previous experimental results. The bottom of the top band is apparently flat, and it produces a small Fermi surface as a shallow electron pocket at approximately $(k_a, k_b) \sim (0.05, 0.05)$. The nonlinear band curvature around the Fermi surface renders the doped electrons to behave as SFs. Dirac cones can barely be recognized because of the flattening of the bottom band.

The second feature is a small indirect bandgap at the Dirac point $E_g(296\ K) = 2.91$ meV. The band structure at 296 K is metallic with a Fermi surface at approximately $(k_a, k_b) = (0.05, 0.05)$; however, an energy gap exists between two Dirac points at approximately $(k_a, k_b) = (0.025, 0.06)$ and $(0.08, 0.035)$. The developing Dirac points are situated at substantially different positions from the Fermi surfaces, which supports the theory that massless (Dirac) and massive (standard) fermions can coexist. Below 150 K, a further decrease in T causes the curvatures of the band structures to gradually transform around the Fermi level. The nearly flat-bottomed top (red) band along with the second (blue) band at 296 K starts to develop a Dirac-cone-like shape at 150–100 K. Indirect band gaps between "primitive Dirac points", i.e., the lowest and highest energy points in the top and second bands, respectively, were 3.71 meV at 150 K and 4.23 meV at 100 K. This allows the coexistence of DFs and SFs, and is quantitatively consistent with the observed electrical, magnetic, and optical properties.

At 100 K, the small Fermi surface located at the bottom of the top band at 296 K disappears, which results in an insulating band structure consistent with the observation of the electrical and magnetic properties (Figure 2). The band gap ($E_g(100\ K) = 4.23$ meV) is indirect and located around the Dirac points. Although the Dirac cones are unclear, band dispersion around both ends of the top bands show linear curvatures, which render the electrons (and holes) massless (DFs). Similarly, thermal excitation across the band gap between "primitive Dirac points" produces Dirac-like electrons and holes, which account for the observed high electrical conductivity in the insulating phase ($T < 135$ K). The activation energy at 78–105 K (5.2 meV), derived from the T-dependent ρ (Figure 2b), quantitatively agreed with our calculated band gap (4.23 meV at 100 K). There is a theoretical paper suggesting the essential change in Dirac cones at ~100 K, in addition to the coexistence of the standard and Dirac fermions depending on the temperature [64]. Despite the sharp M–I transition at 135 K, the band structure, which includes the band gap, almost retained its shape between metallic (150 K) and insulating (100 K) phases. The curvature of the top band slightly changed from convex (296 K) to concave (100 K). The electrons accommodated in the top band at 296 K completely disappeared at 100 K, which led to the disappearance of the Fermi surface. Unlike common M–I transitions, the transition in α-ET$_2$I$_3$ does not originate from band splitting with the loss of all unpaired electrons or holes; instead, it originates from the band reshaping associated with the ET–I$_3$ CT interactions. As a CT interaction is a kind of redox reaction, the degree of CT interactions depends on T, and affects details of the band structures and band fillings (the number of carriers) simultaneously. This scenario of band reshaping across the M–I transition is consistent with that discussed in previous studies [49,52–55]. Such a T-sensitive band structure results in the coexistence and reversible interconversion between DFs and SFs. A previous optical study on α-ET$_2$I$_3$ at 17 K revealed that the energy gap in the insulating phase was 75 meV [52]. This optically estimated value should be compared with the corresponding energy difference in our calculation, namely, ~85 meV, because this value corresponds to a direct gap around the location of the Fermi surface in the metallic band structure, which agrees with the optically observed value (75 meV).

At 30 K, the bottom of the top band became nearly flat again, and the band gap (~26.8 meV) became evident at the Fermi level, which completed the M–I transition. This band reshaping can be referred to as merging; that is, a pair of symmetry-related Dirac cones merge into a flat-bottomed band, which simultaneously lose Dirac points, Dirac cones, and DFs.

The calculated band structures of α-BETS$_2$I$_3$ at 296 K, 100 K, 80 K, and 30 K are shown in Figure 4. At 296 K, both top (red) and second (blue) BETS HOMO bands intersect with the Fermi level, indicating that α-BETS$_2$I$_3$ at 296 K belongs to a semimetal. This is consistent with the observed behavior of electrical resistivity and magnetic susceptibility (Figure 2). Notably, the bottom of the top (red) band remains rather flat at all the temperatures, suggestive of merging between two overtilted Dirac cones [23,25,33–35,81]. Both top and second HOMO bands in the BETS salt possess about twice widths of those of corresponding bands in the ET salt, reflecting stronger D–D interactions in the BETS salt than those in the ET salt. There are band gaps between primitive Dirac points, namely 2.60 meV (296 K), 0.014 meV (100 K), 1.28 meV (80 K), and -0.04 meV (30 K). Herein, the negative sign of the band gap at 30 K implies that the energy of the bottom of the top-band Dirac cone is higher than that of the top of the second-band Dirac cone. The calculated band gaps strongly agree with previously reported values [25,81]. All these band-structure features indicate the BETS salt is close to overtilting and the merging of a pair of Dirac cones, as well as close to a ZGS at ambient pressure in the ground state as previously suggested by calculations [24,25]. Regarding the band reshaping, Figures 3 and 4 show that α-D$_2$I$_3$ (D = ET and BETS) share a qualitative T-dependence. In our calculation, α-BETS$_2$I$_3$ retained the metallic band structure down to 30 K. However, the electrical resistivity of α-BETS$_2$I$_3$ smoothly commenced to increase at ~50 K (Figure 2a), while the magnetic susceptibility took the broad minimum at ~30 K (Figure 2c). These observed electronic behaviors suggest the significant effects of fluctuation and/or strong electron correlation at low T in α-BETS$_2$I$_3$ [22]. The calculated band structures would be inconsistent with the observed electronic properties owing to fluctuation and/or strong electron correlation, because they were not considered in the calculation.

In previous work, the reflectance spectra clearly exhibited Drude-type dispersions at 300–25 K for α-BETS$_2$I$_3$ and α-STF$_2$I$_3$, indicating that both salts should be metallic without M–I transitions ($T \geq 25$ K) [26,46]. Meanwhile, for both salts, the electrical resistivities and magnetic susceptibilities began to increase at T_0 ~50 K (BETS) and T_0 ~80 K (STF) with decreasing temperature toward the ground states (Figure 2), suggesting that their ground states should be non-metallic. In both STF and BETS salts, if we assume an M–I transition from a semimetal with a flat-bottomed band at E_F and almost-vanishing Fermi surfaces to a band insulator with a band gap of the order of 1 meV = 8 cm^{-1}, the reflectance spectra (450 cm^{-1}–25,000 cm^{-1} at 25–300 K) cannot be distinguished from those of metallic materials. Accordingly, the M-I transition and the metallic reflectance spectra can be reconciled with each other based on the present band calculation.

The obscure M-I transitions in the ρ and χ in the BETS and STF salts are explained by continuous band and Fermi surface deformations with temperature variation. With decreasing temperature across the transition, the effects of continuous shrinking (and finally disappearance) of Fermi surfaces make the M-I transition obscure, as observed in the electrical and magnetic properties of the BETS and STF salts. In fact, their unit cell volumes did not exhibit any discontinuous change at 30–300 K [25,26]. Accordingly, the M–I transitions should be of second or higher order for both STF and BETS salts. Furthermore, the M-I phase transition model proposed here accounts for the observed non-Arrhenius behavior immediately below T_0 (STF and BETS) or T_{MI} (ET) in all α-D$_2$I$_3$ regarding both of ρ vs. T and χ vs. T (Figure 2).

A unique chemistry underlies the abovementioned reshaping of band structures in α-ET$_2$I$_3$ and α-BETS$_2$I$_3$. We will discuss this problem more in detail in Section 3.5. The D-I$_3$ interactions has three kinds of effects. Namely, they produce a small number of carriers in the D-main bands, they make the conduction system weakly three-dimensional, and they weaken the D-D interactions in exchange for adding the D-I$_3$ interactions. To lower the energy of such a small carrier system, shallow convex curvature in the band around the Fermi level would be effective to accommodate the electrons at the bottoms of the downward convex curvature, whether such curvature is a shallow hollow at the bottom of the band at the Fermi level (SFs) or a pair of shallow round-apex cones (DFs). In α-D$_2$I$_3$ (D = ET, BETS), such slight band structure reshaping would be realized only by changes in a part of the electron distribution without the cost of lattice deformation. This is one of the possible mechanisms to explain the T-sensitive band structure. If they had many carriers at the Fermi level in a wide band with a higher anisotropy, and if the D-I$_3$ interactions were negligible, the metal–insulator transition would involve band splitting with an evident crystal structure change in the D arrangement.

3.4. Spectroscopic Evidence for the Occurrence of DFs: ESR

Figure 5 shows the T dependencies of the ESR spectra of the ET salt under ambient pressure. The direction of the magnetic field B was parallel (0°, B // a-axes) or perpendicular (90°) to the main conduction sheets (ab planes). Even during the M-I transition, the T dependence was hardly observed in either the 0° or the 90° spectra. Instead, a significant angular dependence was noticed between the 0° and 90° spectra for the peaks centered at ~328 mT (g = 2.006); they were assigned to the carriers based on the following discussion. First, we discuss the T dependencies of these peaks in both spectra. If these peaks originate from metallic carriers, i.e., massive fermions, they should disappear below 135 K, because the salt is an insulator below 135 K. However, this is not the case here: we still observed ESR signals below 135 K. As stated above, a substantial number of spins were observed below 135 K in different physical properties of α-ET$_2$I$_3$ in previous studies [53–55]. The ESR signal intensity is not simply proportional to the number of spins when the signal originates from different kinds of spin systems with different relaxation times. The signal intensities from the different spin systems may exhibit different anisotropies, which would also deviate the T-dependence of the intensity from what is anticipated based on the number ratio between each type of spin. In this case, (Figure 5), the signal (line shape) changed from a Dysonian type (asymmetric, characteristic to metals) to a Lorentzian type (symmetric, characteristic to insulators) at the phase transition as shown in Figure 5c, which is consistent with previous work [53,54,82]. Therefore, the ESR results reconcile with the metal–insulator transition. The observation of spins in the insulating phase is explained by assuming that there are two types of carriers, i.e., metallic carriers (massive fermions) and thermally excited Dirac (massless) fermions (Figure 6), and that only the metallic carriers disappear below 135 K.

Yet, a substantial number of the carriers remain by fluctuation of metal–insulator transition and thermal excitation. This assumption is consistent with the substantially high χ_P and low ρ of the "insulating" phase (Figure 2).

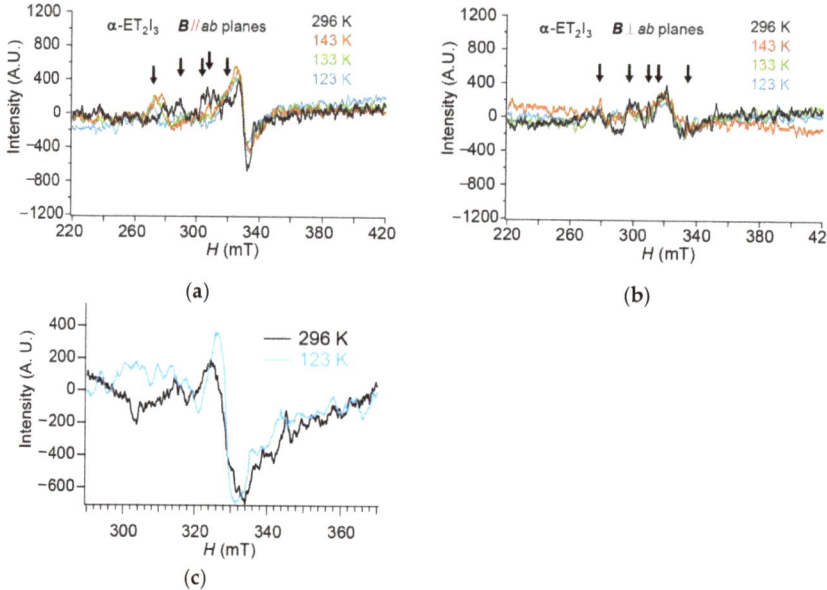

Figure 5. T-dependent ESR spectra of α-ET$_2$I$_3$: (**a**) 0° spectra; (**b**) 90° spectra; and (**c**) enlarged view for comparison of the line shapes between metallic (296 K) and insulating (123 K) phases. The 123 K spectrum is vertically offset for easy comparison of the height/depth of the peaks/valleys. The spectra in a and b were obtained from the same single crystal. The spectra in c were obtained from the different sample from those in a and b to show the reproducibility of the Dysonian (296 K) and Lorentzian (123 K) line shapes. The black arrows indicate possible hyperfine (nuclear spin–electron spin) or superhyperfine (nuclear spin–nuclear spin) structures.

The evidently weaker intensities of the peaks at all measured T in the 90° spectra compared to those in the case of the 0° spectra are noteworthy because they indicate the occurrence of DFs. The difference is prominent only around the peak at ~328 mT. This observation is contrary to the general observation in the ESR spectra of 2D organic conductors that contain only SFs [82]. The application of B in the 90° direction generally yields a stronger peak than that when B is applied in the 0° direction [82]. However, the thermally excited DFs in α-ET$_2$I$_3$ at 1 bar contribute significantly to electrical conduction in the ab plane, and they generate only weak ESR signals because of their short relaxation times originating from their unusually large Fermi velocities.

Based on the calculated band structures, this finding can be explained by the peak at ~328 mT being associated with both SFs and DFs (Figure 5). Both fermions are unpaired electrons and holes, and some are produced by thermal excitation depending on the temperature. The contributions of these carriers to the ESR signal continuously crossover with a variation in T. Thermally excited DFs are important below the M–I transition T (135 K). Accordingly, at T = ~100–135 K, the weak signal at ~328 mT in the 90° spectra is associated with the dominant roles of thermally excited DF in the electrical and magnetic properties of α-ET$_2$I$_3$ in the ab planes.

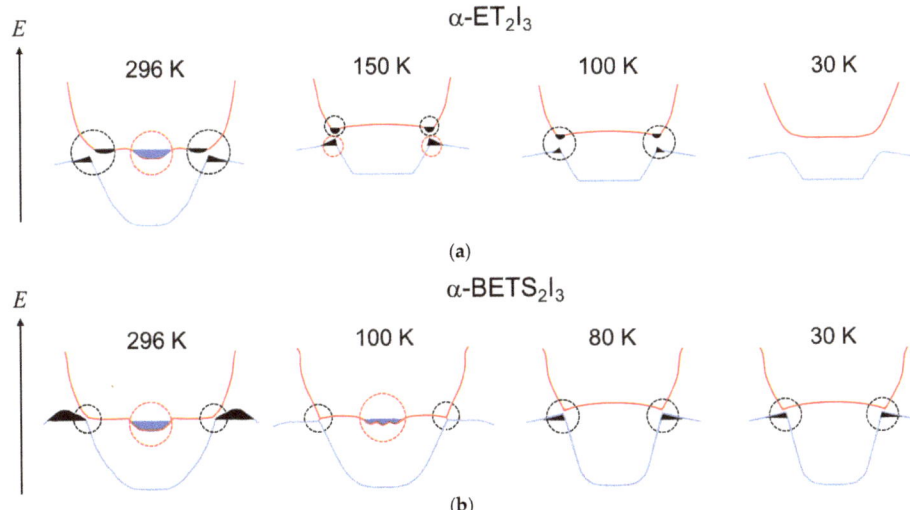

Figure 6. Schematic T-driven band reshaping of: (**a**) α-ET$_2$I$_3$; and (**b**) α-BETS$_2$I$_3$. From left to right, the figures show small but qualitative changes in the band structure and band filling of the electrons and holes with decreasing T. Red and blue curves represent the top and second HOMO bands, respectively. For the ET salt, at 296, 150, and 100 K, the shaded areas in the black broken circles at the growing Dirac points indicate thermally excited electrons (in the top band) and holes (in the second band), both of which are Dirac fermions (massless). For the ET salt, at 296 and 150 K, the shaded areas in the red broken circles represent the metallic carriers on the Fermi surface (massive fermions at 296 K and massless fermions at 150 K). Similarly, for the BETS salt, the shaded areas in the black (at 296, 80, and 30 K) and red (at 296 and 100 K) broken circles at the Dirac points indicate metallic carriers (black = filled with holes, blue = filled with electrons) on the Fermi surfaces. For the BETS salts, thermally excited electrons/holes at the Dirac points are not drawn, for they may be confused with the metallic carriers at the Dirac points. In the case of the BETS salt at 100 K, the Dirac points are slightly lower than the Fermi level in energy.

Complicated structures are noticed at ~220–420 mT in both 0° and 90° spectra at all T. As the spins on the I atoms ($I = 5/2$) in the I$_3$ anions exhibited hyperfine or superhyperfine structures, some structures were intrinsic. In fact, they partially demonstrated consistent and reproducible responses to different microwave powers (Supplementary Figure S1) and different samples/independent measurements. The observed microwave power dependencies of the peaks are quantitatively explained by hypothesizing three types of spin systems with different saturation behaviors (for details, see the note in the caption of Supplementary Figure S1). In the absence of CT interactions between the ET and I$_3$ species, all carriers (spins) should be confined within the conducting ET (C$_{10}$H$_8$S$_8$) sheets, and no hyperfine structures are expected. Owing to the H bonding shown in Figure 1c, all four terminal I atoms in the two crystallographically independent I$_3^-$ species are inequivalent in the hyperfine interactions. Consequently, a single peak is divided into many fine peaks.

3.5. Chemistry Driving the Massless–Massive Fermion Transformation

The occurrence and disappearance of Dirac (massless) fermions in the standard (massive) fermion systems discussed above revealed the importance of a chemical equilibrium in the D–I$_3$ interaction. This is a spontaneous solid-state redox reaction described by the following formula,

$$D^{(0.5-\varepsilon)+} + I_3^{(1-\varepsilon)-} \rightleftarrows D^{(0.5-\delta)+} + I_3^{(1-\delta)-} \quad (\varepsilon \neq \delta, \quad D = \text{ET, STF, and BETS}) \quad (1)$$

$$K = \frac{\left[D^{(0.5-\delta)+}\right]\left[I_3^{(1-\delta)-}\right]}{\left[D^{(0.5-\varepsilon)+}\right]\left[I_3^{(1-\varepsilon)-}\right]} \quad (\varepsilon \neq \delta, \quad D = ET, STF, \text{ and BETS}) \quad (2)$$

As the equilibrium constants (K in Equation (2)) generally depend on the temperature, the redox reaction (Equation (1)) proceeds depending on the temperature. Similarly, the chemical potentials generally depend on both thermodynamic conditions and chemical environments when intermolecular interactions are important, such as those in solid states. Still, the reaction described by Equations (1) and (2) in α-D$_2$I$_3$ is unusual in that ε and δ can take any non-integer number, respectively, resulting in continuous change in the (metastable) oxidation states of all the chemical species involved in the reaction (Supplementary Figure S2). This is only possible when both D and I$_3$ species form energy bands based on close D–D and D–I$_3$ interactions. When the temperature changes, a slight change in the electron density distribution in the unit cell of α-D$_2$I$_3$ caused by the reaction (Equation (1)) drives band deformation as shown in Figures 3 and 4. With decreasing temperature (296 K → 150 K in the ET salt and 296 K → 80 K in the BETS salt), more and more electrons or holes (massive fermions) travel to the Dirac points, because the energy at the developing Dirac points continuously becomes lower and lower until it is finally lowest in the respective bands. Thus, in a wide temperature range, a part of the doped electrons/holes from the I$_3$ anions (massive fermions) collaborate with the thermally activated electrons/holes at the Dirac points (massless fermions) in the electronic properties. Accordingly, the transformation between massless and massive fermions can occur in a continuous manner. In summary, the abovementioned unusual chemical situation drives a continuous but qualitative change in the electrical conductivity, magnetic susceptibility, dimensionality in the electronic system, and the direction of electron transfer (electron- or hole-doping from the I$_3$ species), in addition to the effective mass of the fermions by temperature variation.

4. Conclusions

The band reshaping of α-D$_2$I$_3$ (D = ET, BETS) under ambient pressure with respect to T demonstrates the lifecycle of DF, i.e., the entire process of their occurrence, development, and disappearance. Massless and massive fermions coexist and crossover each other via band reshaping. The D-I$_3$ CT interactions play key roles in T-sensitive band reshaping. As the organic DF systems share many features with other types of DF systems, these findings shed new light on the entirety of DF systems and accelerate the progression of related studies. However, the electronic properties specific to DFs, such as T dependencies of the electrical resistivity and magnetic susceptibility and wavenumber dependence of the optical conductivity, are often featureless when compared with those of SFs [1–36]. Such monotonical behaviors of the electronic properties could be explained in various ways. As the analyses of the calculation and experimental results in this study were based on assumption and approximation as in other studies, further research from different perspectives and methods for crosschecking are required for a deeper insight into the nature of DFs.

Supplementary Materials: The following supporting information can be downloaded at: https://www.mdpi.com/article/10.3390/magnetochemistry9060153/s1, Scheme S1: Syntheses of neutral BETS and STF; Figure S1: Microwave-power (P) dependencies of ESR spectra of α-ET$_2$I$_3$ (single crystal, B // a-axis) at 296 K; Figure S2: Temperature dependence of the molecular (Bader) charges on the crystallographically independent D (A, A', B, and C) and I$_3$ (1 and 2) species in (a) α-ET$_2$I$_3$ and (b) α-BETS$_2$I$_3$.

Author Contributions: R.O. measured the resistivity and ESR spectra; R.O. and T.N. analyzed the resistivity, magnetic susceptibility, and ESR data; I.Y. and T.S. measured the magnetic susceptibility and calculated the band structures; K.O. and K.K. assisted R.O. in measuring the ESR spectra and resistivity, respectively; T.N., I.Y. and T.S. drafted the manuscript. All authors have read and agreed to the published version of the manuscript.

Funding: This work was partially funded by a Grant-in-Aid for Challenging Exploratory Research (18K19061) and a Grant-in-Aid for Scientific Research (B) (22H02034) of JSPS, the Iketani Science and Technology Foundation (ISTF; 0331005-A), the Research Grant Program of the Futaba Foundation, the CASIO Science Promotion Foundation, an Ehime University Grant for Project for the Promotion of Industry/University Cooperation, and the Canon Foundation (Science and Technology that Achieve a Good Future).

Institutional Review Board Statement: Not applicable.

Informed Consent Statement: Not applicable.

Data Availability Statement: The relevant data supporting our key findings are available within the article and the Supplementary Information file. All raw data generated during our current study are available from the corresponding authors upon reasonable request, which should be addressed to T. Naito.

Acknowledgments: The authors are grateful to H. Sawa (Nagoya University) for providing a part of the atomic coordinates of α-D_2I_3. We would like to acknowledge the assistance provided by S. Mori and R. Konishi (ADRES, Ehime University) in the X-ray structural analyses. We would also like to thank R. Nakamura for partly conducting the resistivity measurements with the help of T. Yamamoto (Ehime University). The computational resources were provided by ISSP, University of Tokyo supercomputer (Projects: 2022-Ba-0061 and 2021-Ba-0028) and the Research Center for Computational Science, Okazaki, Japan (Projects: 21-IMS-C088 and 22-IMS-C088).

Conflicts of Interest: The authors declare no conflict of interest.

References

1. Castro Neto, A.H.; Guinea, F.; Peres, N.M.R.; Novoselov, K.S.; Geim, A.K. The electronic properties of graphene. *Rev. Mod. Phys.* **2009**, *81*, 109–162. [CrossRef]
2. Park, S.H.; Sammon, M.; Mele, E.; Low, T. Plasmonic gain in current biased tilted Dirac nodes. *Nat. Commun.* **2022**, *13*, 7667. [CrossRef] [PubMed]
3. Geng, D.; Zhou, H.; Yue, S.; Sun, Z.; Cheng, P.; Chen, L.; Meng, S.; Wu, K.; Feng, B. Observation of gapped Dirac cones in a two-dimensional Su-Schrieffer-Heeger lattice. *Nat. Commun.* **2022**, *13*, 7000. [CrossRef] [PubMed]
4. Lu, Q.; Cook, J.; Zhang, X.; Chen, K.Y.; Snyder, M.; Nguyen, D.T.; Sreenivasa Reddy, P.V.; Qin, B.; Zhan, S.; Zhao, L.-D.; et al. Realization of unpinned two-dimensional dirac states in antimony atomic layers. *Nat. Commun.* **2022**, *13*, 4603. [CrossRef]
5. Bauernfeind, M.; Erhardt, J.; Eck, P.; Thakur, P.K.; Gabel, J.; Lee, T.-L.; Schäfer, J.; Moser, S.; Di Sante, D.; Claessen, R.; et al. Design and realization of topological Dirac fermions on a triangular lattice. *Nat. Commun.* **2021**, *12*, 5396. [CrossRef] [PubMed]
6. Guo, C.; Alexandradinata, A.; Putzke, C.; Estry, A.; Tu, T.; Kumar, N.; Fan, F.-R.; Zhang, S.; Wu, Q.; Yazyev, O.V.; et al. Temperature dependence of quantum oscillations from non-parabolic dispersions. *Nat. Commun.* **2021**, *12*, 6213. [CrossRef]
7. Park, H.; Gao, W.; Zhang, X.; Oh, S.S. Nodal lines in momentum space: Topological invariants and recent realizations in photonic and other systems. *Nanophotonics* **2022**, *11*, 2779–2801. [CrossRef]
8. Bai, Y.; Li, N.; Li, R.; Liu, P. Ultrafast dynamics of helical Dirac fermions in the topological insulators. *Adv. Physics: X* **2022**, *7*, 2013134. [CrossRef]
9. Jana, S.; Bandyopadhyay, A.; Datta, S.; Bhattacharya, D.; Jana, D. Emerging properties of carbon based 2D material beyond graphene. *J. Phys.: Cond. Mat.* **2022**, *34*, 053001. [CrossRef]
10. Yilmatz, T.; Tong, X.; Dai, Z.; Sadowski, J.T.; Schwier, E.F.; Shimada, K.; Hwang, S.; Kisslinger, K.; Kaznatcheev, K.; Vescovo, E.; et al. Emergent flat band electronic structure in a VSe_2/Bi_2Se_3 heterostructure. *Commun. Mater.* **2021**, *2*, 11. [CrossRef]
11. Khoury, J.F.; Schoop, L.M. Chemical bonds in topological materials. *Trends Chem.* **2021**, *3*, 700–715. [CrossRef]
12. Kumar, N.; Guin, S.N.; Manna, K.; Shekhar, C.; Felser, C. Topological quantum materials from the viewpoint of chemistry. *Chem. Rev.* **2021**, *121*, 2780–2815. [CrossRef] [PubMed]
13. Bhattacharyya, S.; Akhgar, G.; Gebert, M.; Karel, J.; Edmonds, M.T.; Fuhrer, M.S. Recent progress in proximity coupling of magnetism to topological insulators. *Adv. Mater.* **2021**, *33*, 2007795. [CrossRef] [PubMed]
14. Mandal, I.; Saha, K. Thermopower in an anisotropic two-dimensional Weyl semimetal. *Phys. Rev. B* **2020**, *101*, 045101. [CrossRef]
15. Neubauer, D.; Yaresko, A.; Li, W.; Löhle, A.; Hübner, R.; Schilling, M.B.; Shekhar, C.; Felser, C.; Dressel, M.; Pronin, A.V. Optical conductivity of the Wyle semimetal NbP. *Phys. Rev. B* **2018**, *98*, 195203. [CrossRef]
16. Chen, Z.-G.; Shi, Z.; Yang, W.; Lu, X.; Lai, Y.; Yan, H.; Wang, F.; Zhang, G.; Li, Z. Observation of an intrinsic bandgap and Landau level renormalization in graphene/boron-nitride heterostructures. *Nat. Commun.* **2014**, *5*, 4461. [CrossRef]
17. Timusk, T.; Carbotte, J.P.; Homes, C.C.; Basov, D.N.; Sharapov, S.G. Three-dimensional Dirac fermions in quasicrystals as seen via optical conductivity. *Phys. Rev. B* **2013**, *87*, 235121. [CrossRef]

18. Pop, F.; Mézière, C.; Allain, M.; Auban-Senzier, P.; Tajima, N.; Hirobe, D.; Yamamoto, H.M.; Canadell, E.; Avarvari, N. Unusual stoichiometry, band structure and band filling in conducting enantiopure radical cation salts of TM-BEDT-TTF showing helical packing of the donors. *J. Mater. Chem. C* **2021**, *9*, 10777–10786. [CrossRef]
19. Hirata, M.; Kobayashi, A.; Berthier, C.; Kanoda, K. Interacting chiral electrons at the 2D Dirac points: A review. *Rep. Prog. Phys.* **2021**, *84*, 036502. [CrossRef]
20. Kajita, K.; Nishio, Y.; Tajima, N.; Suzumura, Y.; Kobayashi, A. Molecular Dirac fermion systems –Theoretical and experimental approaches –. *J. Phys. Soc. Jpn.* **2014**, *83*, 072002. [CrossRef]
21. Naito, T.; Doi, R.; Suzumura, Y. Exotic Dirac cones on the band structure of α-STF$_2$I$_3$ at ambient temperature and pressure. *J. Phys. Soc. Jpn.* **2020**, *89*, 023701. [CrossRef]
22. Nomoto, T.; Imajo, S.; Akutsu, H.; Nakazawa, Y.; Kohama, Y. Correlation-driven organic 3D topological insulator with relativistic fermions. *Nat. Commun.* **2023**, *14*, 2130. [CrossRef] [PubMed]
23. Kawasugi, Y.; Masuda, H.; Uebe, M.; Yamamoto, H.M.; Kato, R.; Nishio, Y.; Tajima, N. Pressure-induced phase switching of Shubnikov-de Haas oscillations in the molecular Dirac fermion system a-(BETS)$_2$I$_3$. *Phys. Rev. B* **2021**, *103*, 205140. [CrossRef]
24. Tsumuraya, T.; Suzumura, Y. First-principles study of the effective Hamiltonian for Dirac fermions with spin-orbit coupling in two-dimensional molecular conductor α-(BETS)$_2$I$_3$. *Eur. Phys. J. B* **2021**, *94*, 17. [CrossRef]
25. Kitou, S.; Tsumuraya, T.; Sawahata, H.; Ishii, F.; Hiraki, K.-I.; Nakamura, T.; Katayama, N.; Sawa, H. Ambient-pressure Dirac electron system in the quasi-two-dimensional molecular conductor α-(BETS)$_2$I$_3$. *Phys. Rev. B* **2021**, *103*, 035135. [CrossRef]
26. Naito, T.; Doi, R. Band structure and physical properties of α-STF$_2$I$_3$: Dirac electrons in disordered conduction sheets. *Crystals* **2020**, *10*, 270. [CrossRef]
27. Ohki, D.; Yoshimi, K.; Kobayashi, A. Transport properties of the organic Dirac electron system α-(BEDT-TSeF)$_2$I$_3$. *Phys. Rev. B* **2020**, *102*, 235116. [CrossRef]
28. Uykur, E.; Li, W.; Kuntscher, C.A.; Dressel, M. Optical signatures of energy gap in correlated Dirac fermions. *npj Quantum Mater.* **2019**, *4*, 19. [CrossRef]
29. Hirata, M.; Ishikawa, K.; Matsuno, G.; Kobayashi, A.; Miyagawa, K.; Tamura, M.; Berthier, C.; Kanoda, K. Anomalous spin correlations and excitonic instability of interacting 2D Weyl fermions. *Science* **2017**, *358*, 1403–1406. [CrossRef]
30. Beyer, R.; Dengl, A.; Peterseim, T.; Wackerow, S.; Ivek, T.; Pronin, A.V.; Schweitzer, D.; Dressel, M. Pressure-dependent optical investigations of α−(BEDT-TTF)$_2$I$_3$: Tuning charge order and narrow gap towards a Dirac semimetal. *Phys. Rev. B* **2016**, *93*, 195116. [CrossRef]
31. Hirata, M.; Ishikawa, K.; Miyagawa, K.; Tamura, M.; Berthier, C.; Basko, D.; Kobayashi, A.; Matsuno, G.; Kanoda, K. Observation of an anisotropic Dirac cone reshaping and ferrimagnetic spin polarization in an organic conductor. *Nat. Commun.* **2016**, *7*, 12666. [CrossRef] [PubMed]
32. Liu, D.; Ishikawa, K.; Takehara, R.; Miyagawa, K.; Tamura, M.; Kanoda, K. Insulating nature of strongly correlated massless Dirac fermions in an organic crystal. *Phys. Rev. Lett.* **2016**, *116*, 226401. [CrossRef] [PubMed]
33. Ogata, M.; Ozaki, S.; Matsuura, H. Anomalous Spin Transport Properties of Gapped Dirac Electrons with Tilting. *J. Phys. Soc. Jpn.* **2022**, *91*, 023708. [CrossRef]
34. Suzumura, Y.; Tsumuraya, T. Electric and magnetic responses of two-dimensional Dirac electrons in organic conductor a-(BETS)$_2$I$_3$. *J. Phys. Soc. Jpn* **2021**, *90*, 124707. [CrossRef]
35. Morinari, T.; Suzumura, Y. On the possible zero-gap state in organic conductor a-(BEDT-TSF)$_2$I$_3$ under pressure. *J. Phys. Soc. Jpn* **2014**, *83*, 094701. [CrossRef]
36. Suzumura, Y.; Naito, T. Conductivity of two-dimensional Dirac electrons close to merging in organic conductor a-STF$_2$I$_3$ at ambient pressure. *J. Phys. Soc. Jpn.* **2022**, *91*, 064701. [CrossRef]
37. Oroszlány, L.; Dóra, B.; Cserti, J.; Cortijo, A. Topological and trivial magnetic oscillations in nodal loop semimetals. *Phys. Rev. B* **2018**, *97*, 205107. [CrossRef]
38. Feilhauer, J.; Apel, W.; Schweitzer, L. Merging of the Dirac points in electronic artificial graphene. *Phys. Rev. B* **2015**, *92*, 245424. [CrossRef]
39. Delplace, P.; Gómez-León, A.; Platero, G. Merging of Dirac points and Floquet topological transitions in ac-driven graphene. *Phys. Rev B* **2013**, *88*, 245422. [CrossRef]
40. Dóra, B.; Herbut, I.F.; Moessner, R. Coupling, merging, and splitting Dirac points by electron-electron interaction. *Phys. Rev. B* **2013**, *88*, 075126. [CrossRef]
41. Wang, L.; Fu, L. Interaction-induced merging of Dirac points in non-Abelian optical lattices. *Phys. Rev. A* **2013**, *87*, 053612. [CrossRef]
42. Tarruell, L.; Greif, D.; Uehlinger, T.; Jotzu, G.; Esslinger, T. Creating, moving and merging Dirac points with a Fermi gas in a tunable honeycomb lattice. *Nature* **2012**, *483*, 302–305. [CrossRef] [PubMed]
43. Montambaux, G.; Piéchon, F.; Fuchs, J.-N.; Goerbig, M.O. A universal Hamiltonian for motion and merging of Dirac points in a two-dimensional crystal. *Eur. Phys. J. B* **2009**, *72*, 509–520. [CrossRef]
44. Pereira, V.M.; Castro Neto, A.H.; Peres, N.M.R. Tight-binding approach to uniaxial strain in graphene. *Phys. Rev. B* **2009**, *80*, 045401. [CrossRef]

45. Naito, T.; Kobayashi, H.; Kobayashi, A. The electrical behavior of charge-transfer salts based on an unsymmetrical donor bis(ethylenedithio)diselenadithiafulvalene (STF): Disorder effect on the transport properties. *Bull. Chem. Soc. Jpn.* **1997**, *70*, 107–114. [CrossRef]
46. Inokuchi, M.; Tajima, H.; Kobayashi, A.; Ohta, T.; Kuroda, H.; Kato, R.; Naito, T.; Kobayashi, H. Electrical and optical properties of α-(BETS)$_2$I$_3$ and α-(BEDT-STF)$_2$I$_3$. *Bull. Chem. Soc. Jpn.* **1995**, *68*, 547–553. [CrossRef]
47. Kato, R.; Kobayashi, H.; Kobayashi, A. Synthesis and properties of bis(ethylenedithio)tetraselenafulvalene (BEDT-TSeF) compounds. *Synth. Met.* **1991**, *42*, 2093–2096. [CrossRef]
48. Bender, K.; Hennig, I.; Schweitzer, D.; Dietz, K.; Endres, H.; Keller, H.J. Synthesis, structure and physical properties of a two-dimensional organic metal, di[bis(ethylenedithiolo)tetrathiofulvalene]triiodide, (BEDT-TTF)$_2$$^+I_3$$^-$. *Mol. Cryst. Liq. Cryst.* **1984**, *108*, 359–371. [CrossRef]
49. Kakiuchi, T.; Wakabayashi, Y.; Sawa, H.; Takahashi, T.; Nakamura, T. Charge ordering in a-(BEDT-TTF)$_2$I$_3$ by synchrotron X-ray diffraction. *J. Phys. Soc. Jpn.* **2007**, *76*, 113702. [CrossRef]
50. Rothaemel, B.; Forró, L.; Cooper, J.R.; Schilling, J.S.; Weger, M.; Bele, P.; Brunner, H.; Schweitzer, D.; Keller, H.J. Magnetic susceptibility of α and β phases of di[bis(ethylenedithiolo)tetrathiofulvalene]tri-iodide [(BEDT-TTF)$_2$I$_3$] under pressure. *Phys. Rev. B* **1986**, *34*, 704–712. [CrossRef]
51. Tajima, N.; Tamura, M.; Nishio, Y.; Kajita, K.; Iye, Y. Transport property of an organic conductor a-(BEDT-TTF)$_2$I$_3$ under high pressure—Discovery of a novel type of conductor. *J. Phys. Soc. Jpn.* **2000**, *69*, 543–551. [CrossRef]
52. Ivek, T.; Korin-Hamzić, B.; Milat, O.; Tomić, S.; Clauss, C.; Drichko, N.; Schweitzer, D.; Dressel, M. Electrodynamic response of the charge ordering phase: Dielectric and optical studies of α-(BEDT-TTF)$_2$I$_3$. *Phys. Rev. B* **2011**, *83*, 165128. [CrossRef]
53. Venturini, E.L.; Azevedo, L.J.; Schirber, J.E.; Williams, J.M.; Wang, H.H. ESR study of two phases of di[bis(ethylenedithio)tetrathiafulvalene]triiodide [(BEDT-TTF)$_2$I$_3$]. *Phys. Rev. B* **1985**, *32*, 2819–2823. [CrossRef] [PubMed]
54. Sugano, T.; Saito, G.; Kinoshita, M. Conduction-electron-spin resonance in organic conductors: α and β phases di[bis(ethylenedithio)tetrathiafulvalene]triiodide [(BEDT-TTF)$_2$I$_3$]. *Phys. Rev. B* **1986**, *34*, 117–125. [CrossRef] [PubMed]
55. Uji, S.; Kodama, K.; Sugii, K.; Takahide, Y.; Terashima, T.; Kurita, N.; Tsuchiya, S.; Kohno, M.; Kimata, M.; Yamamoto, K.; et al. Kosterlitz-Thouless-type transition in a charge ordered state of the layered organic conductor a-(BEDT-TTF)$_2$I$_3$. *Phys. Rev. Lett.* **2013**, *110*, 196602. [CrossRef] [PubMed]
56. Seo, H. Charge ordering in organic ET compounds. *J. Phys. Soc. Jpn.* **2000**, *69*, 805–820. [CrossRef]
57. Seo, H.; Merino, J.; Yoshioka, H.; Ogata, M. Theoretical aspects of charge ordering in molecular conductors. *J. Phys. Soc. Jpn.* **2006**, *75*, 051009. [CrossRef]
58. Takano, Y.; Hiraki, K.; Yamamoto, H.M.; Nakamura, T.; Takahashi, T. Charge disproportionation in the organic conductor, α-(BEDT-TTF)$_2$I$_3$. *J. Phys. Chem. Solids* **2001**, *62*, 393–395. [CrossRef]
59. Yamamoto, K.; Kowalska, A.A.; Yakushi, K. Direct observation of ferroelectric domains created by Wigner crystallization of electrons in α-[Bis (ethylenedithio) tetrathiafulvalene]$_2$I$_3$. *Appl. Phys. Lett.* **2010**, *96*, 122901. [CrossRef]
60. Kondo, R.; Kagoshima, S. Crystal and electronic structures of a-(BEDT-TTF)$_2$I$_3$ under uniaxial strains. *J. Phys. IV France* **2004**, *114*, 523–525. [CrossRef]
61. Kondo, R.; Kagoshima, S.; Harada, J. Crystal structure analysis under uniaxial strain at low temperature using a unique design of four-axis x-ray diffractometer with a fixed sample. *Rev. Sci. Instrum.* **2005**, *76*, 093902. [CrossRef]
62. Ishibashi, S.; Tamura, T.; Kohyama, M.; Terakura, K. *Ab initio* electronic-structure calculations for a-(BEDT-TTF)$_2$I$_3$. *J. Phys. Soc. Jpn.* **2006**, *75*, 015005. [CrossRef]
63. Naito, T.; Suzumura, Y. Theoretical model for novel electronic state in a Dirac electron system close to merging: An imaginary element between sulfur and selenium. *Crystals* **2022**, *12*, 346. [CrossRef]
64. Kino, H.; Miyazaki, T. First-Principles Study of Electronic Structure in a-(BEDT-TTF)$_2$I$_3$ at Ambient Pressure and with Uniaxial Strain. *J. Phys. Soc. Jpn.* **2006**, *75*, 034704. [CrossRef]
65. Monteverde, M.; Goerbig, M.O.; Auban-Senzier, P.; Navarin, F.; Henck, H.; Pasquier, C.R.; Mézière, C.; Batail, P. Coexistence of Diarc and massive carriers in a-(BEDT-TTF)$_2$I$_3$ under hydrostatic pressure. *Phys. Rev. B* **2013**, *87*, 245110. [CrossRef]
66. Katayama, S.; Kobayashi, A.; Suzumura, Y. Pressure-induced zero-gap semiconducting state in organic conductor a-(BEDT-TTF)$_2$I$_3$ salt. *J. Phys. Soc. Jpn.* **2006**, *75*, 054705. [CrossRef]
67. Naito, T.; Inabe, T. Molecular hexagonal perovskite: A new type of organic–inorganic hybrid conductor. *J. Solid State Chem.* **2003**, *176*, 243–249. [CrossRef]
68. Naito, T.; Matsuo, S.; Inabe, T.; Toda, Y. Carrier dynamics in a series of organic magnetic superconductors. *J. Phys. Chem. C* **2012**, *116*, 2588–2593. [CrossRef]
69. Dolomanov, O.V.; Bourhis, L.J.; Gildea, R.J.; Howard, J.A.K.; Puschmann, H. OLEX2: A complete structure solution, refinement and analysis program. *J. Appl. Cryst.* **2009**, *42*, 339–341. [CrossRef]
70. Kresse, G.; Hafner, J. *Ab initio* molecular dynamics for liquid metals. *Phys. Rev. B* **1993**, *47*, 558–561. [CrossRef]
71. Mochalin, V.N.; Shenderova, O.; Ho, D.; Gogotsi, Y. The properties and applications of nanodiamonds. *Nat. Nanotechnol.* **2012**, *7*, 11–23. [CrossRef] [PubMed]
72. Pouget, J.-P.; Alemany, P.; Canadell, E. Donor-anion interactions in quarter-filled low-dimensional organic conductors. *Mater. Horiz.* **2018**, *5*, 590–640. [CrossRef]

73. Alemany, P.; Pouget, J.-P.; Canadell, E. Essential role of anions in the charge ordering transition of a-(BEDT-TTF)$_2$I$_3$. *Phys. Rev. B* **2012**, *85*, 195118. [CrossRef]
74. Ivek, T.; Čulo, M.; Kuveždić, M.; Tutiš, E.; Basletić, M.; Mihaljević, B.; Tafra, E.; Tomić, S.; Löhle, A.; Dressel, M.; et al. Semimetallic and charge-ordered a-(BEDT-TTF)$_2$I$_3$: On the role of disorder in dc transport and dielectric properties. *Phys. Rev. B* **2017**, *96*, 075141. [CrossRef]
75. Kim, A.S.; Walter, A.L.; Moreschini, L.; Seyller, T.; Horn, K.; Rotenberg, E.; Bostwick, A. Coexisting massive and massless Dirac fermions in symmetry-broken bilayer graphene. *Nat. Mater.* **2013**, *12*, 887–892. [CrossRef]
76. Zhou, B.; Dong, S.; Wang, X.; Zhang, K. Prediction of two-dimensional *d*-block elemental materials with normal honeycomb, triangular-dodecagonal, and square-octagonal structures from first principles. *Appl. Surf. Sci.* **2017**, *419*, 484–496. [CrossRef]
77. Yuan, X.; Yan, Z.; Song, C.; Zhang, M.; Li, Z.; Zhang, C.; Liu, Y.; Wang, W.; Zhao, M.; Lin, Z.; et al. Chiral Landau levels in Wyle semimetal NbAs with multiple topological carriers. *Nat. Commun.* **2018**, *9*, 1854. [CrossRef]
78. Luyang, W.; Dao-Xin, Y. Coexistence of spin-1 fermion and Dirac fermion on the triangular kagome lattice. *Phys. Rev. B* **2018**, *98*, 161403. [CrossRef]
79. Miyazaki, T.; Kino, H. Atomic and electronic structures of the high-pressure superconductor *b*'-(BEDT-TTF)$_2$ICl$_2$: A first-principles study of the pressure effects. *Phys. Rev. B* **2003**, *68*, 220511. [CrossRef]
80. Miyazaki, T.; Kino, H. First-principles study of the pressure effects on *b*'-(BEDT-TTF)$_2$AuCl$_2$. *Phys. Rev. B* **2006**, *73*, 035107. [CrossRef]
81. Fujiyama, S.; Maebashi, H.; Tajima, N.; Tsumuraya, T.; Cui, H.-B.; Ogata, M.; Kato, R. Large diamagnetism and electromagnetic duality in two-dimensional Dirac electron system. *Phys. Rev. Lett.* **2022**, *128*, 027201. [CrossRef] [PubMed]
82. Coulon, C.; Clérac, R. Electron spin resonance: A major probe for molecular conductors. *Chem. Rev.* **2004**, *104*, 5655–5687. [CrossRef] [PubMed]

Disclaimer/Publisher's Note: The statements, opinions and data contained in all publications are solely those of the individual author(s) and contributor(s) and not of MDPI and/or the editor(s). MDPI and/or the editor(s) disclaim responsibility for any injury to people or property resulting from any ideas, methods, instructions or products referred to in the content.

Article

Driving a Molecular Spin-Peierls System into a Short Range Ordered State through Chemical Substitution

Adam Berlie [1,*], Ian Terry [2] and Marek Szablewski [2]

1. ISIS Neutron and Muon Source, Rutherford Appleton Laboratory, Science and Technology Facilities Council, Didcot OX11 0QX, UK
2. Physics Department, Durham University, South Road, Durham DH1 3LE, UK; ian.terry@durham.ac.uk (I.T.); marek.szablewski@durham.ac.uk (M.S.)
* Correspondence: adam.berlie@stfc.ac.uk

Abstract: Chemically altering molecules can have dramatic effects on the physical properties of a series of very similar molecular compounds. A good example of this is within the quasi-1D spin-Peierls system potassium TCNQ (TCNQ = 7,7,8,8-tetracyanoqunidimethane), where substitution of TCNQF$_4$ for TCNQ has a dramatic effect on the 1D interactions, resulting in a drop in the corresponding spin-Peierls transition temperature. Within this work, we extend the investigation to potassium TCNQBr$_2$, where only two protons of TCNQ can be substituted with bromine atoms due to steric constraints. The new system exhibits evidence for a residual component of the magnetism when probed via magnetic susceptibility measurements and muon spin spectroscopy. The observations suggest that the system is dominated by short range, and potentially disordered, correlations within the bulk phase.

Keywords: molecular magentism; muon spin relaxation; spin-Peierls; tetracyanoquinodimethane

Citation: Berlie, A.; Terry, I.; Szablewski, M. Driving a Molecular Spin-Peierls System into a Short Range Ordered State through Chemical Substitution. *Magnetochemistry* 2023, 9, 150. https://doi.org/10.3390/magnetochemistry9060150

Academic Editor: Laura C. J. Pereira

Received: 26 April 2023
Revised: 31 May 2023
Accepted: 6 June 2023
Published: 8 June 2023

Copyright: © 2023 by the authors. Licensee MDPI, Basel, Switzerland. This article is an open access article distributed under the terms and conditions of the Creative Commons Attribution (CC BY) license (https://creativecommons.org/licenses/by/4.0/).

1. Introduction

There is a large interest in how to tune and drive molecular-based systems into different ground states to create a novel series of functional materials [1–3]. One way to achieve this is to use the application of pressure [4]; however, another option within molecular materials is to use chemical substitution [2], inducing different properties [5] or where ground states can be altered by using different ligands [6–9].

A simple, well known class of materials to use as a test bed for tuning magnetic interactions is spin-Peierls (SP) systems. The 1D nature of the chemical magnetic structure relies on anisotropic interactions, where the coupling of the structural and magnetic properties leads to a dimerisation of magnetic atoms, ions or molecules, where, in the case of $S = 1/2$ magnetic materials, below the SP transition the system is dominated by singlet-triplet excitations.

A well-known molecular SP system is potassium TCNQ (KTCNQ, where TCNQ = 7,7,8,8-tetracyanoquinodimethane), where the TCNQ molecules form quasi-1D chains of $S = 1/2$ anions that are dominated by π–π interactions [10–12], with TCNQ stacking along the a-axis of the crystal structure. At high temperatures, the TCNQ anions are evenly spaced within the crystal structure. Below the SP transition temperature (T_{SP}), at 396 K and identified by using magnetic susceptibility data, the TCNQ anions dimerise, with a strong intra-dimer exchange, J', calculated to be approximately -1800 K [10]. This transition is mediated by electron–phonon coupling and one observes a change in the vibrational modes on going through T_{SP} [13].

Modern interest in these systems has arisen from the ability to be able to chemically tune their properties. Recent work by us [14] has shown that when replacing TCNQ with TCNQF$_4$ (2,3,5,6-Tetrafluoro-7,7,8,8-tetracyanoquinodimethane), there is a dramatic shift in T_{SP}, from 396 K to 160 K, as indicated by the magnetic susceptibility, which is a result of the change in interaction strength between the magnetic dimers. Using muon spin relaxation (μSR) [15,16], we were able to study the dynamics of the singlet-triplet excitations within the SP state, and the corresponding singlet–triplet gap was also shown to decrease from 0.20 eV to 0.11 eV on substitution of the fluorine atoms. Although the SP state is expected to be quasi-1D, the magnetic fluctuations follow the critical power-law expected for a 3D Heisenberg system; therefore, these types of systems exhibit strongly correlated 3D behaviour. Additionally, further work also illustrated that the highly concentrated defect states can separately order, where within KTCNQF$_4$, these states were suggestive of 2D ordering [17].

Within this work, we go a step further and replace TCNQ with TCNQBr$_2$ (2,5-dibromo-7,7,8,8-tetracyanoquinodimethane), where the slightly bent structure and steric hindrance imposed by the larger bromine atoms on the TCNQ molecule present an interesting comparison. We show that the system, KTCNQBr$_2$, is a highly disordered, where the bulk behaviour of the magnetic susceptibility is dominated by paramagnetic defects. However, there is also a broad peak in the temperature-dependent magnetic susceptibility, which suggests the presence of quasi-1D chains with antiferromagnetic interactions that exist between $S = 1/2$ spins residing on TCNQBr$_2$ anions. Importantly, there is no convincing evidence that KTCNQBr$_2$ undergoes an SP transition in the temperature range investigated in this work.

2. Materials and Methods

The TCNQBr$_2$, was synthesised using the route described by Wheland and Martin [18]. The KTCNQBr$_2$ was then synthesised through a metathesis reaction of potassium iodide and TCNQBr$_2$ in anhydrous acetonitrile, as described by Melby et al. [19]. As previously reported with KTCNQF$_4$, the crystallisation is extremely fast, and this creates an intrinsic level of disorder within the crystallites. UV-vis spectroscopy confirmed that the resulting blue/purple powder contained the reduced form of TCNQBr$_2$.

The magnetic susceptibility measurements were taken using a Quantum Design MPMS. μSR data were collected on the ARGUS spectrometer at the RIKEN-RAL Muon Facility within the ISIS Neutron and Muon Source. Due to the small amount of sample, the measurements were performed using a mini cold-finger, helium flow cryostat in flypast, where a small sample holder is used that allows all muons that do not stop within the sample to "fly past" and not be detected. Further information on the technique can be found elsewhere [15,16].

3. Results and Discussions

The magnetic susceptibility was measured in both Zero-Field Cooled (ZFC) and Field Cooled (FC) within an applied field of 1 kG and is dominated by a Curie tail, indicative of the presence of paramagnetic defects. Figure 1A shows the magnetic susceptibility between base temperature and 130 K, where despite the presence of a Curie tail, there is a divergence of the ZFC and FC data sets at approximately 120 K that is outside of the error within the measurement. This behaviour suggests that there may be a hysteric effect in addition to the bulk paramagnetism. Plotting the FC data as χT vs. T (inset of Figure 1A) shows a gradual increase as the temperature is increased, with the dotted line at 120 K indicating a point of inflection. This provides evidence that there is a change in the type of magnetic interactions within this region, with the downturn in χT vs. T suggesting that these interactions may be antiferromagnetic.

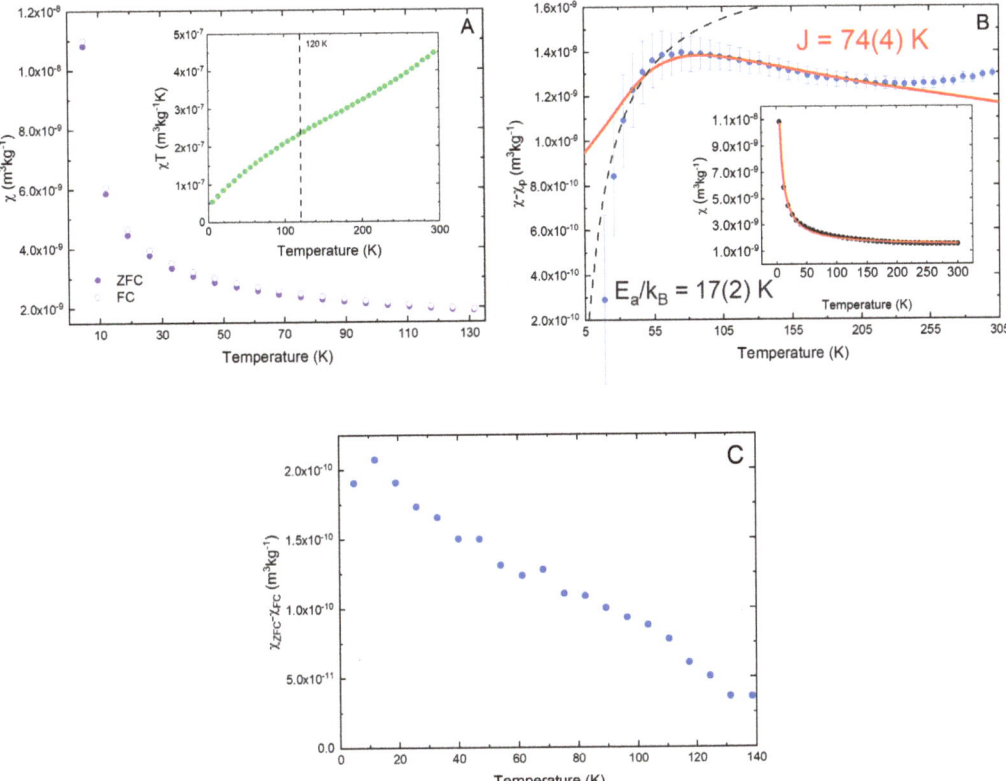

Figure 1. Magnetic susceptibility data on KTCNQBr$_2$: (**A**) The zero-field cooled (ZFC) and field cooled (FC) magnetic susceptibility in 1 kG, where one can see the splitting of the curves at approximately 120 K, which is not accounted for by the error within the measurement (the error bars are smaller then the data points). The inset shows the FC data plotted as χT vs. T. (**B**) The residual FC magnetic susceptibility once the paramagnetic, Curie-tail has been subtracted, with the fit to Curie's law shown in the inset. The solid line in the main figure is a fit to the Bonner–Fisher law and the dashed line a fit to Arrhenius behaviour. (**C**) Subtraction of the ZFC and FC magnetic susceptibilities highlighting the divergence of the two data sets.

In order to extract more information from these results, it is prudent to attempt to remove the Curie tail from the data. To do so, a fit was performed on the FC data set using the fitting function:

$$\chi = C/(T-\theta) + \chi_{BG}, \qquad (1)$$

where C is the Curie constant and θ is the Weiss parameter. The parameter χ_{BG} is an offset accounting for a temperature-independent contribution to the susceptibility, which was needed to obtain the best description of the data at high temperature. The general fit is shown in the inset of Figure 1B, with $\theta = -2.4(2)$ K, $C = 7.0(2) \times 10^{-8}$ m^3 kg^{-1} K^{-1} and $\chi_{BG} = 1.32(2) \times 10^{-9}$ m^3 kg^{-1}. On subtraction of the Curie tail from the data, one can see (Figure 1B) that there is a slight peak in the residual susceptibility at about 65 K. This is characteristic of a low-dimensional contribution to the bulk magnetic behaviour and is likely to be quasi-one-dimensional, being related to stacks of the TCNQBr$_2$ anions forming a system where short-range magnetic correlations exist. To further extract information on this quasi-1D state, the Bonner–Fisher equation [20] with a constant background was fit to the data that describe the temperature dependence of antiferromagnetic interactions along an $S = 1/2$ chain. The fit, shown in Figure 1B, produces a exchange constant, $|J| = 74(4)$ K, with $N = 1.5(2) \times 10^{16}$ kg^{-1} and a background value of the magnetic susceptibility of $9.2(4) \times 10^{-10}$ kg m^{-3}. The value of $|J|$, is consistent with the broad maxima within the data; however, intriguingly, the low value of N indicates that the 1D state does not account for the bulk of the sample. Additionally, in order to determine the size of a spin gap associated with a potential singlet–triplet excited state, the low temperature data were fitted to an Arrhenius function (see Figure 1B, dashed line) where an activation energy of 17(2) K (1.5 meV) was extracted. This result suggests that there may possibly be an SP transition occurring at the lowest temperatures investigated in this work, but one must be cautious with this interpretation; the Curie–Weiss law describing the defect paramagnetism may hold at higher temperatures but could break down at lower temperatures, within the region of interest, as other exchange pathways that are not governed by mean–field interactions may become dominant. In fact, the complexity of the defect paramagnetic response is highlighted by the underlying difference between the ZFC and FC curves, as shown in Figure 1C. There is a clear increase in the difference between the ZFC and FC magnetic susceptibilities as temperature is lowered, which is suggestive of a magnetically ordered component with remanence; however, this ordered state is not well defined and appears to have a slow onset as one cools the sample. This perhaps highlights the local and disordered nature of the underlying magnetism within the sample. It is also noteworthy that the number densities of each magnetic component derived from the data analyses are smaller then expected, and this may point to the predominance of itinerant electrons in the material, whilst the magnetic susceptibility measurements are dominated by the localised moments. Given the strong π–π interactions, it may not be surprising that electrons are able to delocalise across molecules or 1D stacks, and more work is needed to study this further.

In order to probe the evidence for a magnetic transition further, muon spin relaxation (μSR) was used. μSR is an exceptionally sensitive probe of weak magnetic order and dynamics on a local scale. The interaction between the muon, in this case μ^+, and the surrounding electronic and nuclear moments can provide information on both the static and dynamic parts of the magnetic susceptibility. The muon will respond to static magnetic fields that are transverse to its initial polarisation by precessing, where the precessional frequency can be related to the internal field through $\nu = \gamma_\mu B$, $\gamma_\mu/2\pi = 135.5$ MHz/T. Due to the small amount of sample, the measurements were performed using the fly-past set-up and this has to be optimised to collect the best quality data possible.

The μSR data were collected in zero-field, and one measures the time dependence of the asymmetry in the decay of muon polarisation when the muon is implanted within the sample. All spectra were fit using an exponential function:

$$G(t) = A_r \exp(-\lambda t) + A_B, \qquad (2)$$

where A_r is the relaxing asymmetry, λ is the muon spin relaxation rate and A_B is the baseline. For the whole data set, the baseline was fixed at 16.4%. The temperature dependencies of other fitting parameters can be seen within Figure 2. The relaxation rate is sensitive to both static and dynamic magnetic behaviour, where in the fast fluctuating limit, $\lambda \propto \Delta^2/\tau$, where τ is the correlation time and Δ is the static field distribution. The gradual increase in λ within Figure 2 is a hallmark of a slowing down of dynamic electronic fluctuations, and this has been observed within other KTCNQ salts [14] as well as other TCNQ-based salts [21–23]. Key to this is the gradual increase in λ, where the slowing down of electronic fluctuations appears to be very broad and broader then that observed for other TCNQ-based SP systems [14,23,24], with the onset of this increase in λ being well above the drop observed in the magnetic susceptibility below 65 K, as shown in Figure 1B. The $\lambda(T)$ data are more indicative of a slowly evolving magnetic system, where there are a range of interactions, likely due to short-range order, that create local areas of magnetic order which drive the system into a quasi-static state. Such a magnetic state may be equivalent to that suggested by the magnetic susceptibility difference shown in Figure 1C. Accompanying this change in $\lambda(T)$ is a drop in the relaxing asymmetry, which can be indicative of the formation of static order with an internal field that pushes the precessional frequency beyond the experimental time resolution and is consistent with the rise in λ. An effective way to model the onset of a transition is to use an error function, a function describing the first derivative of a Gaussian distribution of transition temperatures, which has also been used elsewhere [25]. Using an error function to model the data within the inset of Figure 2, the midpoint of the curve is 167(9) K. Therefore, it is likely that this is not an SP transition, and instead, there is a build up of 1D correlations, where the electronic moments move into the experimental time scale and then become quasi-static. Below 50 K, the relaxing asymmetry increases, and this may be caused by a reduction in the internal field due to a rearrangement of the spin creating a weaker internal field, with the saturation of λ indicating that persistent dynamics are present. However, one cannot discount that the flattening out of λ could also be due to a quasi-static, disordered magnetic state, like a spin-glass [16]. Interestingly, this is the also the point at which the magnetic susceptibility shown within the Figure 1B turns over and so the two behaviours may be correlated.

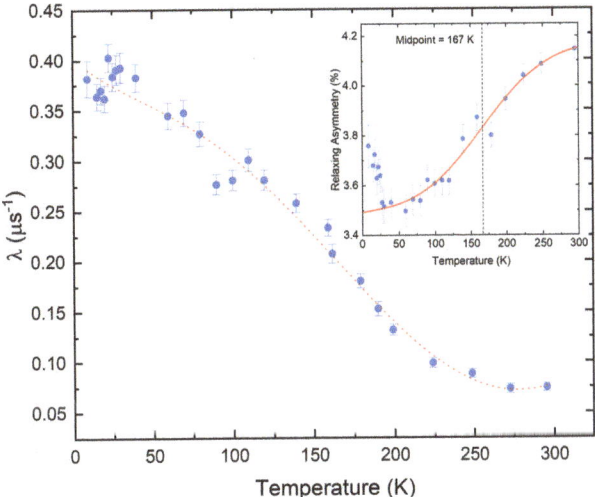

Figure 2. The temperature dependence of the muon spin relaxation rate, λ in zero-field. The dotted line is a guide to the eye. The inset shows the temperature dependence of the relaxing asymmetry; the solid line is a fit to the data using an error function, with a midpoint of 167 K.

4. Conclusions

The substitution of TCNQ for TCNQBr$_2$ within the simple spin-Peierls system KTCNQ has created a system where there is little evidence for a spin-Peierls transition as seen within similar molecular systems. The magnetic susceptibility data provide some evidence that there is a residual component that could be treated as showing 1D behaviour; however, it is worth nothing that the fits are merely suggestive and not conclusive on their own. Therefore, whilst one cannot discount that there is a change in behaviour below 65 K, there is evidence that the residual component undergoes a different transition or is dominated by a different type of interaction at higher temperatures. This is further highlighted by the difference between the ZFC and FC data sets. The μSR data present complementary information, but the data are not consistent with an spin-Peierls system that would align with that from the magnetic susceptibility shown in Figure 1B. Instead, the μSR measurements point to a system that has been driven into a state where short-range magnetic correlations dominate. The two different regions of behaviour within the μSR certainly warrant more exploration, but the current data suggest that, on cooling, there is a build-up of magnetic correlations and these may order or enter a magnetic state where persistent dynamics are present. Given the planar nature of the TCNQ molecules, it makes sense that this system is also dominated by 1D $\pi - \pi$ interactions, but the bulky bromine atoms likely prevent the structural change needed for the system to dimerise and fall into the spin-Peierls ground state.

Author Contributions: Conceptualization, A.B.; Methodology, A.B. and I.T.; Formal analysis, A.B. and I.T.; Investigation, A.B., I.T. and M.S.; Data curation, A.B.; Writing—original draft, A.B.; Writing—review & editing, I.T. and M.S. All authors have read and agreed to the published version of the manuscript.

Funding: This research received no external funding.

Institutional Review Board Statement: Not applicable.

Informed Consent Statement: Not applicable.

Data Availability Statement: Not applicable.

Acknowledgments: The authors thank the ISIS Neutron and Muon Source and RIKEN for access to the ARGUS spectrometer.

Conflicts of Interest: The authors declare no conflict of interest.

References

1. Wang, C.; Liu, D.; Lin, W. Metal–Organic Frameworks as A Tunable Platform for Designing Functional Molecular Materials. *J. Am. Chem. Soc.* **2013**, *135*, 13222–13234. [CrossRef] [PubMed]
2. Coronado, E. Molecular magnetism: From chemical design to spin control in molecules, materials and devices. *Nat. Rev. Mater.* **2020**, *5*, 87–104. [CrossRef]
3. Reczyński, M.; Nakabayashi, K.; Ohkoshi, S.-I. Tuning the Optical Properties of Magnetic Materials. *Eur. J. Inorg. Chem.* **2020**, *28*, 2669–2678. [CrossRef]
4. McMillan, P.F. New materials from high-pressure experiments. *Nat. Mater.* **2002**, *1*, 19–25. [CrossRef] [PubMed]
5. Coronado, E.; Martí-Gastaldo, C.; Navarro-Moratalla, E.; Ribera, A.; Blundell, S.J.; Baker, P.J. Coexistence of superconductivity and magnetism by chemical design. *Nat. Chem.* **2010**, *2*, 1031–1036. [CrossRef] [PubMed]
6. Zhang, K.; Montigaud, V.; Cador, O.; Li, G.-P.; Guennic, B.L.; Tang, J.-K.; Wang, Y.-Y. Tuning the Magnetic Interactions in Dy(III)$_4$ Single-Molecule Magnets. *Inorg. Chem.* **2018**, *57*, 8550–8557. [CrossRef]
7. Yin, Z.; Zhou, Y.-L.; Zeng, M.-H.; Kurmoo, M. The concept of mixed organic ligands in metal–organic frameworks: Design, tuning and functions. *Dalton Trans.* **2015**, *44*, 5258–5275. [CrossRef]
8. Xie, H.; Vignesh, K.R.; Zhang, X.; Dunbar, K.R. From spin-crossover to single molecule magnetism: Tuning magnetic properties of Co(II) bis-ferrocenylterpy cations via supramolecular interactions with organocyanide radical anion. *J. Mater. Chem. C* **2020**, *8*, 8135–8144. [CrossRef]
9. Kurmoo, M. Magnetic metal–organic frameworks. *Chem. Soc. Rev.* **2009**, *38*, 1353–1379. [CrossRef]
10. Vegter, J.G.; Hibma, T.; Kommandeur, J. New phase transitions in simple M-TCNQ-salts. *Chem. Phys. Lett.* **1969**, *3*, 427. [CrossRef]
11. Lépine, Y.; Caillxex, A.; Larochelle, V. Potassium-tetracyanoquinodimethane (K-TCNQ): A spin-Peierls system. *Phys. Rev. B* **1978**, *18*, 3585. [CrossRef]
12. Konno, M.; Ishii, T.; Saito, Y. The crystal structures of the low- and high-temperature modifications of potassium 7,7,8,8-tetracyanoquinodimethanide. *Acta Cryst.* **1977** *B33*, 7603–7770. [CrossRef]
13. Tanner, D.B.; Jacobsen, C.S.; Bright, A.A.; Heeger, A.J. Infrared studies of the energy gap and electron-phonon interaction in potassium-tetracyanoquinodimethane (K-TCNQ). *Phys. Rev. B* **1977**, *16*, 3283. [CrossRef]
14. Berlie, A.; Terry, I.; Szablewski, M.; Giblin, S.R. Tuneability and criticality in a three-dimensional stacked molecular system. *Phys. Rev. B* **2016**, *93*, 054422. [CrossRef]
15. Hillier, A.D.; Blundell, S.J.; McKenzie, I.; Umegaki, I.; Shu, L.; Wright, J.A.; Prokscha, T.; Bert, F.; Shimomura, K.; Berlie, A.; et al. Muon Spin Spectroscopy. *Nat. Rev. Methods Prim.* **2022**, *2*, 4. [CrossRef]
16. Blundell, S.J.; Renzi, R.D.; Lancaster, T.; Pratt, F.L. (Eds.) *Muon Spectroscopy: An Introduction*; Oxford University Press: Oxford, UK, 2022.
17. Berlie, A.; Terry, I.; Cottrell, S.; Pratt, F.L.; Szablewski, M. Magnetic ordering of defects in a molecular spin-Peierls system. *J. Phys. Condens. Matter* **2017**, *29*, 025809. [CrossRef]
18. Wheland, R.C.; Martin, E.L. Synthesis of Substituted 7,7,8,8-Tetracyanoquinodimethanes. *J. Org. Chem.* **1975**, *40*, 3101. [CrossRef]
19. Melby, L.R.; Harder, R.J.; Hertler, W.R.; Mahler, W.; Benson, R.E.; Mochel, W.E. Substituted Quinodimethans. II. Anion-radical Derivatives and Complexes of 7,7,8,8-Tetracyanoquinodimethan. *J. Am. Chem. Soc.* **1962**, *84*, 3374–3387. [CrossRef]
20. Bonner, J.C.; Fisher, M.E. Linear magentic chains with anisotropic coupling. *Phys. Rev.* **1964**, *135*, A640. [CrossRef]
21. Berlie, A.; Terry, I.; Szablewski, M.; Giblin, S.R. Separating the ferromagnetic and glassy behavior within the metal-organic magnet Ni(TCNQ)$_2$. *Phys. Rev. B* **2015**, *92*, 184431. [CrossRef]
22. Blundell, S.J.; Lancaster, T.; Brooks, M.L.; Pratt, F.L.; Taliaferro, M.L.; Miller, J.S. A μSR study of the metamagnetic phase transition in the electron-transfer salt [Fe(Cp)$_2$] [TCNQ]. *Phys. B Condens. Matter* **2006**, *374–375*, 114–117. [CrossRef]

23. Berlie, A.; Terry, I.; Szablewski, M. A 3D antiferromagnetic ground state in a quasi-1D π-stacked charge-transfer system. *J. Mater. Chem. C* **2018**, *6*, 12468–12472. [CrossRef]
24. Lovett, B.W.; Blundell, S.J.; Pratt, F.L.; Jestädt, T.; Hayes, W.; Tagaki, S.; Kurmoo, M. Spin fluctuations in the spin-Peierls compound MEM(TCNQ)$_2$ studied using muon spin relaxation. *Phys. Rev. B* **2000**, *61*, 12241. [CrossRef]
25. Frandsen, B.A.; Petersen, K.A.; Ducharme, N.A.; Shaw, A.G.; Gibson, E.J.; Winn, B.; Yan, J.; Zhang, J.; Manley, M.E.; Hermann, R.P. Spin dynamics and a nearly continuous magnetic phase transition in an entropy-stabilized oxide antiferromagnet. *Phys. Rev. Mater.* **2020**, *4*, 074405. [CrossRef]

Disclaimer/Publisher's Note: The statements, opinions and data contained in all publications are solely those of the individual author(s) and contributor(s) and not of MDPI and/or the editor(s). MDPI and/or the editor(s) disclaim responsibility for any injury to people or property resulting from any ideas, methods, instructions or products referred to in the content.

Communication

Two One-Dimensional Copper-Oxalate Frameworks with the Jahn–Teller Effect: [(CH$_3$)$_3$NH]$_2$[Cu(μ-C$_2$O$_4$)(C$_2$O$_4$)]·2.5H$_2$O (I) and [(C$_2$H$_5$)$_3$NH]$_2$[Cu(μ-C$_2$O$_4$)(C$_2$O$_4$)]·H$_2$O (II)

Bin Zhang [1,*], Yan Zhang [2], Zheming Wang [3,*], Yang Sun [4], Tongling Liang [4], Mei Liu [4] and Daoben Zhu [1,*]

[1] Organic Solid Laboratory, BNLMS, CMS & Institute of Chemistry, Chinese Academy of Sciences, Beijing 100190, China
[2] Department of Physics, Institute of Condensed Material Physics, Peking University, Beijing 100871, China; zhang_yan@pku.edu.cn
[3] State Key Laboratory of Rare Earth Materials Chemistry and Applications, BNLMS, College of Chemistry and Molecular Engineering, Peking University, Beijing 100871, China
[4] BNLMS, CMS & Institute of Chemistry, Chinese Academy of Sciences, Beijing 100190, China
* Correspondence: zhangbin@iccas.ac.cn (B.Z.); zmw@pku.edu.cn (Z.W.); zhudb@iccas.ac.cn (D.Z.)

Citation: Zhang, B.; Zhang, Y.; Wang, Z.; Sun, Y.; Liang, T.; Liu, M.; Zhu, D. Two One-Dimensional Copper-Oxalate Frameworks with the Jahn–Teller Effect: [(CH$_3$)$_3$NH]$_2$ [Cu(μ-C$_2$O$_4$)(C$_2$O$_4$)]·2.5H$_2$O (I) and [(C$_2$H$_5$)$_3$NH]$_2$[Cu(μ-C$_2$O$_4$) (C$_2$O$_4$)]·H$_2$O (II). *Magnetochemistry* 2023, 9, 120. https://doi.org/10.3390/magnetochemistry9050120

Academic Editor: Carlos J. Gómez García

Received: 20 March 2023
Revised: 27 April 2023
Accepted: 27 April 2023
Published: 29 April 2023

Copyright: © 2023 by the authors. Licensee MDPI, Basel, Switzerland. This article is an open access article distributed under the terms and conditions of the Creative Commons Attribution (CC BY) license (https://creativecommons.org/licenses/by/4.0/).

Abstract: Two one-dimensional oxalate-bridged Cu(II) ammonium salts, [(CH$_3$)$_3$NH]$_2$[Cu(μ-C$_2$O$_4$) (C$_2$O$_4$)]·2.5H$_2$O (**I**) and [(C$_2$H$_5$)$_3$NH]$_2$[Cu(μ-C$_2$O$_4$)(C$_2$O$_4$)]·H$_2$O (**II**) were obtained and characterized. They were composed of ammonium: (CH$_3$)$_3$NH$^+$ in (**I**), (C$_2$H$_5$)$_3$NH$^+$ in (**II**), [Cu(μ-C$_2$O$_4$)(C$_2$O$_4$)$^{2-}$]$_n$ and H$_2$O. The Jahn–Teller-distorted Cu(II) is octahedrally coordinated by six O atoms from three oxalates and forms a one-dimensional zigzag chain. The hydrogen bonds between ammonium, the anion and H$_2$O form a three-dimensional network. There is no hydrogen bond between the anion chains. They were insulated at 20 °C with a relative humidity of 40%. Ferromagnetic and weak-ferromagnetic behaviors were observed in **I** and **II**, separately. No long-range ordering was observed above 2 K.

Keywords: Jahn–Teller effect; Cu(II); oxalate; crystal structure; conductivity; magnetism

1. Introduction

The Jahn–Teller effect plays an important role in inorganic superconductors and colossal magneto-resistance materials [1–9]. Interesting conductivity and magnetic behaviors are expected when the Jahn–Teller effect exists in molecular crystals, such as metal–organic frameworks. Oxalate (C$_2$O$_4$$^{2-}$), as one of the most commonly used short connectors, plays a key role in molecular-based magnets [10–12]. Long-range ordering has been reported in metal–oxalate framework compounds from one-dimensional zigzag chains (K$_2$Fe(μ-C$_2$O$_4$)(C$_2$O$_4$), K$_2$Co(μ-C$_2$O$_4$)(C$_2$O$_4$), Co(μ-C$_2$O$_4$)(μ-HOC$_3$H$_6$OH), TTF[Fe(μ-C$_2$O$_4$)Cl$_2$] and κ-BETS$_2$[Fe(μ-C$_2$O$_4$)Cl$_2$]) to two-dimensional honeycomb lattices ([(C$_4$H$_9$)$_4$N][CrMn(μ-C$_2$O$_4$)$_3$], A[M^{II}Fe^{III}(μ-C$_2$O$_4$)$_3$] (A = ammonium; M = Mn, Fe), [C$_5$H$_{10}$N$_3$O]$_2$[Fe$_2$(μ-C$_2$O$_4$)$_3$], Fe$_2$(μ-C$_2$O$_4$)$_3$·4H$_2$O) and square lattices ([Fe(μ-C$_2$O$_4$)(CH$_3$OH)]$_n$) to three-dimensional metal–oxalate framework compounds ([Co(bpy)$_3$][Co$_2$(μ-C$_2$O$_4$)$_3$]ClO$_4$, Mn(μ-C$_2$O$_4$)(H$_2$O)$_{0.25}$, [Z^{II}(bpy)$_3$][M^{II}Cr^{III}(μ-C$_2$O$_4$)$_3$][ClO$_4$] (M = Mn, Fe, Co, Ni) and (Me$_4$N)$_6$[Mn$_3$Cr$_4$(μ-C$_2$O$_4$)$_{12}$] ·6H$_2$O), while a single chain magnet [C$_{12}$H$_{24}$O$_6$K]$_{0.5}$[(C$_{12}$H$_{24}$O$_6$)(FC$_6$H$_4$NH$_3$)]$_{0.5}$[Co(H$_2$O)$_2$ Cr(μ-C$_2$O$_4$)(C$_2$O$_4$)$_2$] has been reported [13–29].

Quantum spin liquid is an intriguing magnetic state, where spin ordering or freezing prevents spin frustration in a resonating valence bond (RVB) state. In 1979, P. W. Anderson proposed the RVB state in $S = 1/2$, a two-dimensional triangular lattice [30]. In 1987, he proposed that La$_2$CuO$_4$ is a parent compound of cuprate superconductors. The antiferromagnetic insulator La$_2$CuO$_4$ turns into a diamagnetic superconductor after hole doping, and a quantum spin liquid with Jahn–Teller distortion on Cu(II) is an indispensable magnetic state [31]. The spin-frustrated copper-oxalate framework with Jahn–Teller distortion supports a platform for molecular-based quantum spin liquids [32,33]. Strong antiferromagnetic interactions without a long-range ordering above 2 K with spin frustration were observed in two-dimensional honeycomb lattices: θ^{21}-(BEDT-TTF)$_3$[Cu$_2$(μ-C$_2$O$_4$)$_3$]·2CH$_3$OH, θ^{21}-(BETS-TTF)$_3$[Cu$_2$(μ-C$_2$O$_4$)$_3$]·2CH$_3$OH, [(C$_3$H$_7$)$_3$NH]$_2$[Cu$_2$(μ-C$_2$O$_4$)$_3$]·2.2H$_2$O, hydrogen-bonded square lattice β''-(BEDT-TTF)$_3$[Cu$_2$(μ-C$_2$O$_4$)(C$_2$O$_4$)$_2$(CH$_3$OH)(H$_2$O)], and three-dimensional hyperhoneycomb lattice [(C$_2$H$_5$)$_3$NH]$_2$[Cu$_2$(μ-C$_2$O$_4$)$_3$] [34–38]. [(C$_2$H$_5$)$_3$NH]$_2$[Cu$_2$(μ-C$_2$O$_4$)$_3$], which is a quantum spin liquid with no long-range ordering was observed until 60 mK [39]. In these compounds, the antiferromagnetic behavior depends on the antiferromagnetic interaction between the ferromagnetic couple. The magnetic structure of [(C$_2$H$_5$)$_3$NH]$_2$[Cu$_2$(μ-C$_2$O$_4$)$_3$] is lower than the three-dimensional, with the coexistence of ferromagnetic and antiferromagnetic interactions between Jahn–Teller distorted Cu(II) [37,39]. Researching the magnetic properties of Jahn–Teller distorted one-dimensional copper-oxalate frameworks [Cu(μ-C$_2$O$_4$)(C$_2$O$_4$)$_2{}^{2-}$]$_n$ without the hydrogen bond between anions will help us to quantitatively analyze the magnetic interaction in Jahn–Teller-distorted two-dimensional and three-dimensional copper-oxalate frameworks and design new candidate quantum spin liquids. Two one-dimensional copper-oxalate framework compounds, [(CH$_3$)$_3$NH]$_2$[Cu(μ-C$_2$O$_4$)(C$_2$O$_4$)]·2.5H$_2$O (**I**) and [(C$_2$H$_5$)$_3$NH]$_2$[Cu(μ-C$_2$O$_4$)(C$_2$O$_4$)]·H$_2$O (**II**), have been obtained and characterized. The related work is presented here.

2. Experiment

[(CH$_3$)$_3$NH]$_2$[Cu(μ-C$_2$O$_4$)(C$_2$O$_4$)]·2.5H$_2$O (**I**) and [(C$_2$H$_5$)$_3$NH]$_2$[Cu(μ-C$_2$O$_4$)(C$_2$O$_4$)]·H$_2$O (**II**) were obtained from a methanol solution of Cu(NO$_3$)$_2$·3H$_2$O and H$_2$C$_2$O$_4$·2H$_2$O with (CH$_3$)$_3$N for **I** and (C$_2$H$_5$)$_3$N for **II** in a 1:3:5 ratio at room temperature. Bulk blue plateful crystals of **I** and **II** were obtained after four weeks. The crystal was washed with CH$_3$COOC$_2$H$_5$ and dried. Elemental analysis calculated (%) for C$_{10}$H$_{25}$CuN$_2$O$_{10.50}$ (**I**): C 29.67, H 6.22 and N 6.92 and found C 29.87, H, 6.08 and N 6.97. For C$_{16}$H$_{34}$CuN$_2$O$_9$ (**II**): C 41.60, H 7.42, N 6.06 and found C 42.03, H 7.46 and N 6.11.

Elemental analyses of carbon, hydrogen and nitrogen were performed using the Flash EA 1112 elemental analyzer. The IR spectra were recorded on a Bio-rad FTS6000/UMA500 spectrometer (Figure S1). Thermogravimeter analysis was carried out on a Shimadzu DTG-60 analyzer at a 10 °C/min heating rate from room temperature to 550 °C under N$_2$ gas with an Al bag. **I** remains stable until 40 °C, and **II** remains stable until 80 °C.

X-ray powder diffraction was carried out using a Rigaku RINT2000 diffractometer at room temperature with Cu Kα radiation (λ = 1.54056 Å) in a flat-plate geometry (Figures S2 and S3).

Single-crystal X-ray diffraction was carried out on an Enraf-Nonius KappaCCD diffractometer at room temperature. The crystal structure was solved using the direct method and refined using the full-matrix least square on F^2 using the SHELX program, with anisotropic thermal parameters for all non-hydrogen atoms [40]. The hydrogen atoms on C and N were located through calculation, and on H$_2$O they were located through a difference Fourier map. All of the H were refined isotropically. The crystallographic data are listed in Table S1.

The resistance measurement was performed on a single crystal at Tonghui TH2828. Gold wires were attached to the best developed surfaces of a single crystal with a size of 0.40 $\ast$ 0.30 $\ast$ 0.11 mm (**I**) and 0.71 $\ast$ 0.60 $\ast$ 0.17 mm (**II**) using gold paste. The two-probe conductivity was measured at 20 °C and a relative humidity (RH) of 40%.

Magnetization measurements were performed on a polycrystalline sample tightly packed into a capsule on a Quantum Design MPMS 7XL SQUID system above 2 K. Susceptibility data were corrected for the diamagnetism of the sample by Pascal constants and background by experimental measurement of the sample holder [41]. Temperature-dependent magnetization was performed under an applied field of 1000 G. Isothermal magnetization was measured at 2 K from 0 to 65 kG.

3. Result and Discussion

I crystallizes in a triclinic system with space group $P\bar{1}$. There are two $(CH_3)_3NH^+$, one Cu^{2+}, one oxalate anion, two half-oxalate anions and one-and-a-half H_2O coexisting in an independent unit (Figure 1). Cu^{2+} is coordinated to two O atoms from one bidentate oxalate (O1 and O2) and four O atoms from two disbidentate oxalates in the Q_3 Jahn–Teller distortion mode of a CuO_6 octahedron [8].

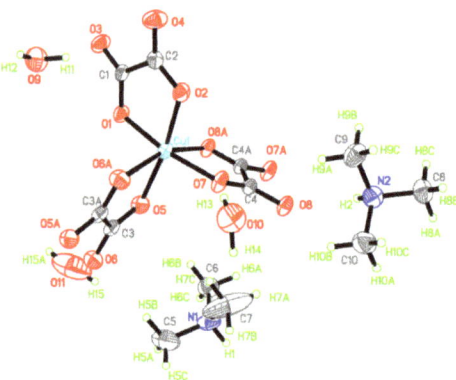

Figure 1. Atomic structure of **I** in an independent unit with scheme label and 50% ellipsoids. Asymmetry code for C3A, O5A, O6A: 2 − x, 2 − y, −z; C4A, O7A, O8A: 1 − x, 1 − y, −z; H15A: −x, 2 − y, −z.

The Cu–O distances are 1.954(2)~1.992(2) Å on the equatorial plane and 2.292(2) Å, 2.329(2) Å from the apex as a result of the Jahn–Teller distortion. The elongated Cu–O bonds (Cu1–O6, Cu1–O8) on the Jahn–Teller-distorted octahedron around Cu^{2+} are highlighted with solid black lines (Figure 2). The cis O–Cu–O angles are 77.73° and 77.12° for the bridged oxalate, 83.56° for the terminal oxalate, and 90.70°, 98.89°, 90.31°, 96.85°, 93.12° and 100.68° among the terminal and bridged oxalate. The trans O–Cu–O angles are in the range of 161.01(6)~171.46(6)°. The axial Cu to oxalate-oxygen angles are 109.17° (Cu1–O6–C3) and 108.27° (Cu1–O8–C4). A one-dimensional zigzag $[Cu(\mu\text{-}C_2O_4)(C_2O_4)^{2-}]_n$ is formed along the b axis.

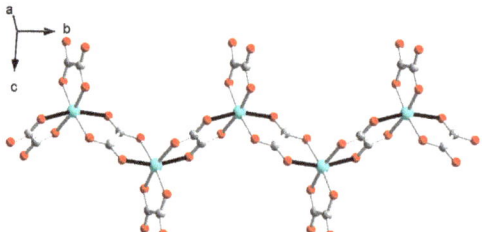

Figure 2. The zigzag anionic chain in **I**.

The one-dimensional $[Cu(\mu\text{-}C_2O_4)(C_2O_4)^{2-}]_n$ zigzag chain running along the b axis is separated by $(CH_3)_3NH^+$ (N1) along the c axis, and there are hydrogen bonds N–H$\cdots$O, C–H$\cdots$O between the cation and the out O atom in the oxalate. The anionic sheets composed of a $[Cu(\mu\text{-}C_2O_4)(C_2O_4)^{2-}]_n$ chain and $(CH_3)_3NH^+$ in a 1:1 ratio are separated by a cation layer composed of a zigzag $(CH_3)_3NH^+$ (N2) chain and a zigzag H_2O chain along the c axis. There are hydrogen bonds between neighboring H_2O molecules. Five H_2O molecules formed a hydrogen-bond $[H_2O]_5$ linear cluster along the c axis (Figure 3).

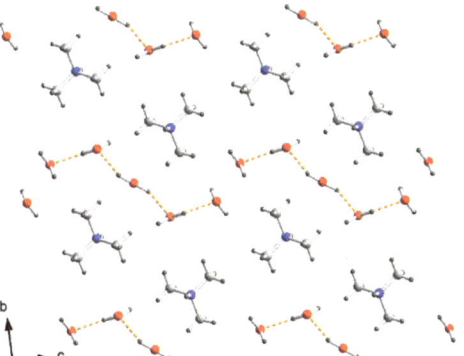

Figure 3. Arrangement of $(CH_3)_3NH^+$ and H_2O in cation layer of **I**. Dashed yellow lines are hydrogen bonds between H_2O in $[H_2O]_5$.

At last, the hydrogen bonds N–H···O and C–H···O between the ammonium and the inner O of the terminal and bridged oxalate, the O–H···O between H_2O and the oxalate, and the O–H···O between the H_2O molecules form a three-dimensional hydrogen-bonded network in crystal (Figure 4). There is no hydrogen bond between the anions.

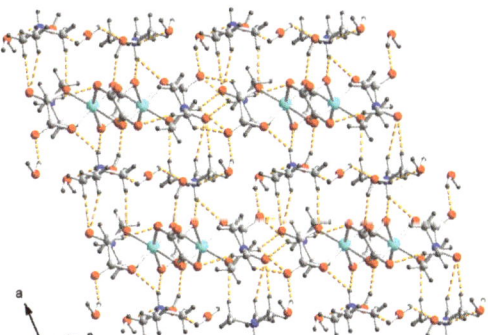

Figure 4. Packing diagram of **I** viewed along the *b* axis. Dash yellow lines are hydrogen bonds. Color code: Cu, cyan; O, red; C, white; N, blue; H, light grey.

II crystallizes in a monoclinic system with the space group $P2_1/c$. There are two $(C_2H_5)_3NH^+$, one Cu^{2+}, one and two half oxalates, and one H_2O in an independent unit (Figure 5). The Cu^{2+} is octahedrally coordinated by six O atoms from two bisbidentate oxalates and one bidentate oxalate, as in **I**, with Cu–O distances of 1.960(2)~1.999(2) Å on the equatorial plane, and Cu1–O8: 2.314(2) Å and Cu1–O7: 2.368(2) Å from the apex. The Cu–O distances of **II** in the equatorial plane are shorter than the direction of the apex as a result of the Q_3 Jahn–Teller distortion mode, as in **I**. The cis–O–Cu–O angles are 75.93° and 77.79° for the bridged oxalate, 88.31° for the terminal oxalate and 92.34°, 93.72°, 95.98°, 89.66°, 95.48° and 102.48° between the terminal and bridged oxalate. The trans-O–Cu–O angles are in the range of 160.21(7)~172.64(7)°. The axial Cu to oxalate-oxygen angles are 108.09° (Cu1–O1–C3) and 108.53° (Cu1–O8–C4). The Cu–O distances and O–Cu–O angles in **I** and **II** are in the same range of the Cu–oxalate coordination polymer [29,30,34,35,37,38]. A one-dimensional oxalate-bridged zigzag $[Cu(\mu-C_2O_4)(C_2O_4)^{2-}]_n$ chain is formed along the *b* axis.

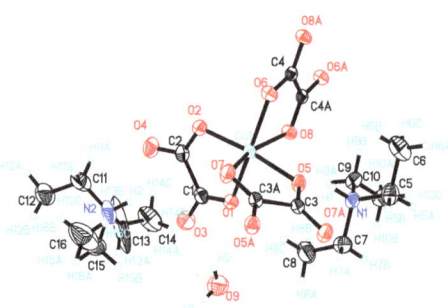

Figure 5. ORTEP drawing of **II** in an independent unit with scheme label and 50% ellipsoids. Asymmetry code for C3A, O5, O7: $-x, -y, 1-z$; C4A, C6A, C8A: $-x, 1-y, 1-z$.

The zigzag chain in **I** and **II** is centrosymmetric with the inversion center located at the middle point of the oxalate bridge; thus, the metal sites have a ΔΔΛΛ configuration along the b axis in **I** and **II** (Figure 6). It is similar to [(CH$_3$)$_4$N]$_2$Cu(C$_2$O$_4$)$_2$(H$_2$O) [42,43]. The hydrogen bonds between ammonium and the anion and H$_2$O and the anion influence the bond length of the CuO$_6$ octahedron due to the Jahn–Teller distortion. Due to the magnetic orbitals of $dx^2 - y^2$ on Cu(II) with the unpaired electrons parallel to each other and the axial Cu to oxalate-oxygen angles, which are sensitive to magnetic interaction and smaller than 109.5°, a ferromagnetic interaction was expected [43,44].

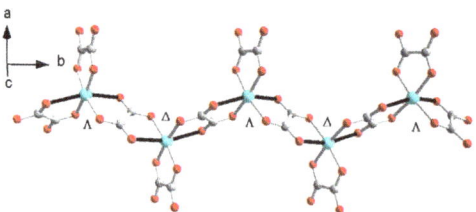

Figure 6. The zigzag anionic chain in **II**.

In **II**, the [Cu(μ-C$_2$O$_4$)(C$_2$O$_4$)$^{2-}$]$_n$ chain is surrounded by zigzag chains of (C$_2$H$_5$)$_3$NH$^+$ and H$_2$O. A pair (C$_2$H$_5$)$_3$NH$^+$ column separates two zigzag chains along the c axis. There are hydrogen bonds between the N of the ammonium and the O on the bridged oxalate: N1–H1···O5 2.10 Å/146.9°, N1–H1···O8 2.44 Å/131.7°. There are hydrogen bonds between the N of the ammonium and the O on the terminal oxalate: N2–H2···O3 2.20 Å/139°, N2–H2···O4 2.21 Å/143.1°. There are hydrogen bonds between H$_2$O and the terminal oxalate: O9–H4···O3 2.11 Å/159°, O9–H3···O4 2.20 Å/164°. There are hydrogen bonds between ammonium and H$_2$O: C8–H8C···O1 2.45 Å/174°; C15–H15A···O8 2.40 Å/155°. The hydrogen bond forms a two-dimensional (2D) network on the (201) plane (Figure 7). There is no hydrogen bond between the one-dimensional [Cu(μ-C$_2$O$_4$)(C$_2$O$_4$)$^{2-}$)]$_n$ chains.

On the basis of the hydrogen bonded cation layer, the resistance, as the proton conductivity under different relative humidities (RH), was measured. Depending on the thermal dynamic analysis, **I** and **II** dehydrate at 40 °C (**I**) and 76 °C (**II**), losing H$_2$O, with a relative weight of 11.2% in **I** and 4% in **II**; therefore, the experiment should be carried out below 40 °C (Figure 8). When the RH increased, the conductivity of **I** and **II** increased. Under a relatively high RH, the surfaces of the crystal were covered with debris at first, which was solvable in gel. Although the sample was restored to a solid state when the RH decreased and reached the same value as the beginning, the sample turned out to be in a polycrystalline state but not a single crystal. When a single crystal of **I** or **II** was exposed to air under a low relative humidity, such as when the relative humidity was lower than 35%, guest molecules, such as H$_2$O, in **I** and **II** would escape from the crystal, leading to crystalline collapse. The crystal surface remained transparent and clear after measurements at 20 °C and an RH of 40%. The resistance comes from the intrinsic behavior of the crystal. The resistance is 1×10^9 Ω·cm in **I** and 1×10^7 Ω·cm in **II**. They are insulators.

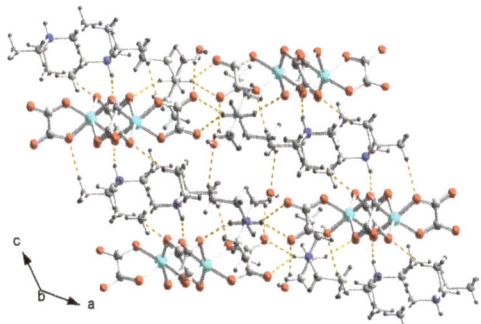

Figure 7. Arrangement of $(C_2H_5)_3NH^+$, zigzag $[Cu(\mu-C_2O_4)(C_2O_4)^{2-}]_n$ chain and H_2O on two-dimensional hydrogen-bond network viewed along the *b* axis in **II**. Dashed yellow lines are hydrogen bonds. Color code: Cu, cyan; O, red; C, white; N, blue; H, light grey.

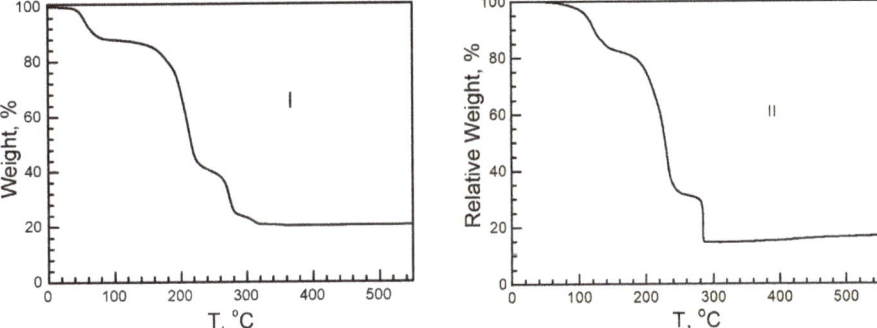

Figure 8. Schematic TGA curves of **I** and **II**. Above 300 °C, the final residue is CuO.

On the basis of the oxalate-bridging and Jahn–Teller distortion of the Cu(II) ion, Cu(II) in an independent unit, and the magnetic properties were studied per Cu^{2+}/mol.

At 300 K, the χT value of **I** was 0.473 cm^3 K mol^{-1} and $g = 2.25$. It is higher than 0.375 cm^3 K mol^{-1} for an isolated, spin only Cu(II) ion with $S = 1/2$, $g = 2.00$ and in the range of Cu^{2+} compounds [34–37,45,46]. The χT value remained stable at 0.478 cm^3 K mol^{-1} at 40 K and increased slowly, reaching 0.71 cm^3 K mol^{-1} at 2 K. No bifurcation is observed from zero-field-cool magnetization and field-cooled magnetization (ZFCM/FCM) measurements from 2 K to 100 K under 100 G (Figure S4). The magnetic data were fitted with the Curie–Weiss law from 2 to 300 K: $C = 0.4711(2)$ cm^{-1}·K/mol, $\theta = 0.61(6)$ K and $R = 1.37 \times 10^{-5}$. It suggests a ferromagnetic interaction in **I** (Figure 9) [47,48].

A one-dimensional Baker–Rushbrooke–Gilbert model was used to fit the temperature-dependent magnetization above 2 K, yielding $J = 0.60(2)$ cm^{-1}, $g = 2.31(1)$ and $R = 9.2 \times 10^{-4}$ (Figure 10) [49]. It shows an intrachain ferromagnetic interaction and corresponds with the Curie–Weiss fitting.

At 2 K, the isothermal magnetization (M) saturated at 1.11 Nβ (N is Avogadro's number and β is the Bohn magneton, 1 Nβ = 5585 cm^{-1} G mol^{-1}) at 65 kG (Figure 11). The average anisotropic *g*-factor calculated from isothermal magnetization at 2 K is 2.22. It is in the range of 2.25 from χT at 300 K and 2.31 from Baker–Rushbrooke–Gilbert model fitting.

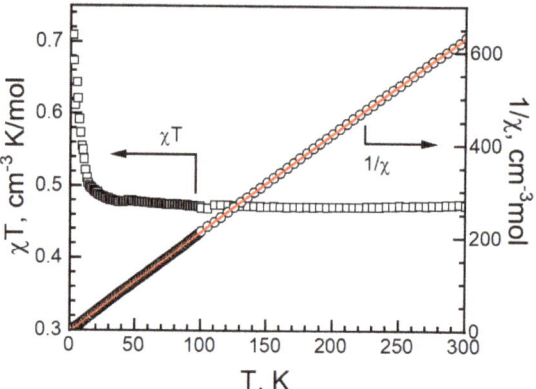

Figure 9. χT vs. T (left, black empty square), $1/\chi$ vs. T (right, black empty circle) and Curie–Weiss fitting data (red solid) of **I**.

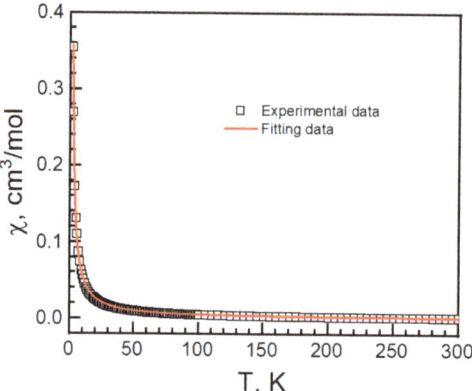

Figure 10. Temperature-dependent susceptibility of **I**. Black empty square is experimental data. Red solid curve is the best fit from the Baker–Rushbrooke–Gilbert model.

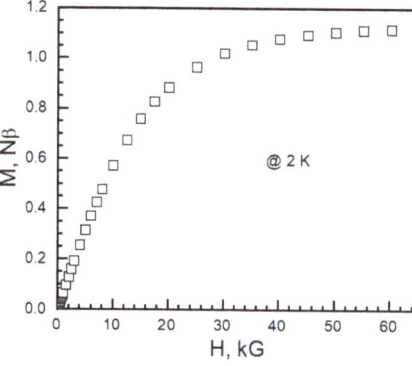

Figure 11. Isothermal magnetization of **I** at 2 K.

In **II**, the χT value was 0.443 cm^3 K mol^{-1} at 300 K with a g-factor of 2.13. It is higher than the 0.375 cm^3 K mol^{-1} of an isolated, spin only Cu(II) ion with $S = 1/2$, $g = 2.00$. It is in the range of Cu^{2+} compounds, as in **I** [34–37,45,46]. As the temperature decreased, the

χT value decreased slowly to 0.393 cm^3 K mol^{-1} around 30 K, and then increased, reaching 0.446 cm^3 K mol^{-1} at 2 K (Figure 12). No bifurcation was observed in the ZFCM/FCM measurement from 2 K to 100 K under 100 G (Figure S5). The magnetic data were fitted with Curie–Weiss law from 80 to 300 K with $C = 0.462(1)$) cm^{-1}·K/mol, $g = -14.2(4)$ K and $R = 3.8 \times 10^{-5}$ (Figure 12).

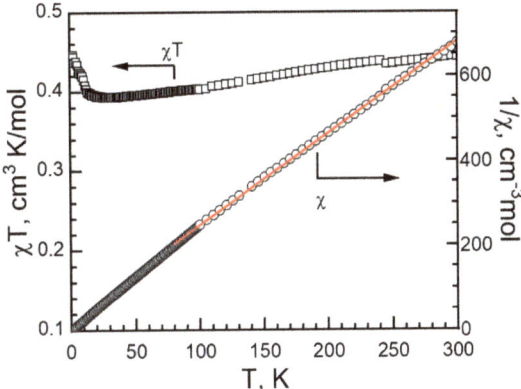

Figure 12. cT vs. T (left, black empty square), $1/c$ vs. T (right, black empty circle) and Curie–Weiss fitting data (red solid) of **II**.

A one-dimensional Baker–Rushbrooke–Gilbert model combined with exchange coupling was used to fit the temperature-dependent magnetization from 2 to 300 K with $J = 0.87(2)$ cm^{-1}, $g = 2.035(3)$, $zJ = -0.65(2)$ cm^{-1} and $R = 6.76 \times 10^{-5}$ (Figure 13) [50]. It shows that intrachain ferromagnetic interaction is stronger than intrachain antiferromagnetic interaction. The g-factor is in the range of 2.13 from χT at 300 K.

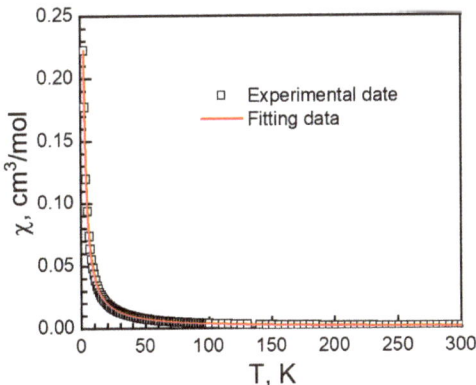

Figure 13. Temperature-dependent susceptibility of **II**. Empty solid square, is experimental data. Red solid curve is best fit from the Baker–Rushbrooke–Gilbert model with exchange coupling.

At 2 K, the magnetization increases with increasing field and is saturated at 0.89 Nβ at 65 kG (Figure 14). The average anisotropic g-factor calculated from isothermal magnetization at 2 K is 1.78. Its magnetic behavior is not the same as expected. This means the Jahn–Teller effect is important to the magnetic property of the copper-oxalate framework. This is different from the compounds [CrMn(C$_2$O$_4$)$_3^-$]$_n$, where in the ferromagnetic order, temperature and isothermal magnetization at 2 K are the same as those taken from ammonium salts to charge-transfer salts [19,51,52]. Depending on the difference in magnetic behaviors between **I** and **II**, the Jahn–Teller effect will help us to obtain molecular-based candidate quantum spin liquid and to look for a new superconductor and colossal magnetoresistance material from copper-oxalate frameworks as cuprate superconductors and colossal magnetoresistance material.

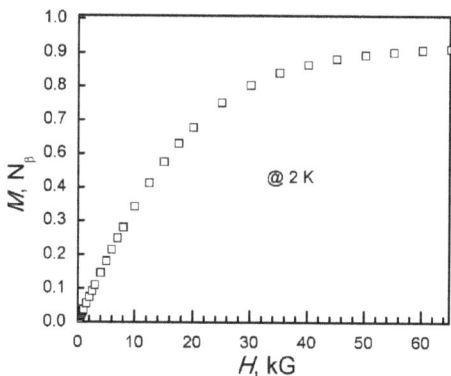

Figure 14. Isothermal magnetization of **II** at 2 K.

4. Conclusions

Two one-dimensional copper-oxalate framework compounds were obtained and characterized. The hydrogen bonds among ammonium, H_2O and the copper-oxalate framework form a three-dimensional hydrogen-bond network, and there is no hydrogen bond between the one-dimensional $[Cu(\mu\text{-}C_2O_4)(C_2O_4)^{2-}]_n$ chains. The Q_3-mode Jahn–Teller distortion of elongated CuO_6 octahedrons is observed. They are insulators. The Jahn–Teller effect results the ferromagnetic and weak-ferromagnetic interaction between Cu(II) in **I** and **II**. No long-range ordering is observed above 2 K.

Supplementary Materials: The following supporting information can be downloaded at: https://www.mdpi.com/article/10.3390/magnetochemistry9050120/s1, Table S1. Crystallographic data of **I** and **II**; Figure S1. IR spectra on crystalline **I** (top) and **II** (bottom); Figure S2. Experimental X-ray powder diffraction pattern of crystalline sample and simulated one based on single crystal structure of **I**. **I** shows preferred orientation; Figure S3. Experimental X-ray powder diffraction pattern of crystalline sample and simulated one based on single crystal structure of **II**; Figure S4. ZFCM/FCM of polycrystal of **I** under 100 G; Figure S5. ZFCM/FCM of polycrystal of **II** under 100 G.

Author Contributions: The manuscript was written with the contributions of all authors. Funding acquisition: B.Z., D.Z. and Y.Z. Synthesized the sample: Y.Z. Performed the magnetic measurements: Z.W. Performed the single-crystal X-ray diffraction experiments: T.L., Y.S. and M.L. Performed the X-ray experiments and data analysis: B.Z. Conducted experiments: B.Z. and Z.W. Analyzed the data: B.Z., Y.Z. and Z.W. Wrote the main manuscript text. All authors have read and agreed to the published version of the manuscript.

Funding: This research was funded by the National Natural Science Foundation of China: 22273109, 22073106, 21573242 and 21173230 and the Strategic Priority Research Program (B) of the Chinese Academy of Sciences (Grant No. XDB12030100).

Institutional Review Board Statement: Not applicable.

Informed Consent Statement: Not applicable.

Data Availability Statement: The IR spectra, powder X-ray diffraction pattern and ZFCM/FCM of **I** and **II** are available in the Supplementary Materials.

Conflicts of Interest: The authors declare no conflict of interest.

References

1. Jahn, H.A.; Teller, E. Stability of polyatomic molecules in degenerate electronic states-I—Orbital degeneracy. *Proc. R. Soc. Lond. Ser. A Math. Phys. Sci.* **1937**, *161*, 220–235.
2. Opik, U.; Pryce, M.H.L. Studies of the Jahn-Teller effect. I. A survey of the static problem. *Proc. R. Soc. Lond. Ser. A Math. Phys. Sci.* **1957**, *238*, 425–427.
3. Longuet-Higgins, H.C.; Opik, U.; Pryce, H.I.; Sack, R.A. Studies of the Jahn-Teller effect. II. The dynamical problem. *Proc. R. Soc. Lond. Ser. A Math. Phys. Sci.* **1958**, *244*, 1–16.
4. Bednorz, J.G.; Muller, K.A. Possible high Tc superconductivity in the Ba−La−Cu−O system. *Z. Für Phys. B Condens. Matter* **1986**, *64*, 189–193. [CrossRef]
5. Radaelli, P.G.; Cox, D.E.; Marezio, M.; Cheong, S.W.; Schiffer, P.E.; Ramirez, A.P. Simultaneous Structural, Magnetic, and Electronic Transitions in La 1−x Cax MnO_3 with x= 0.25 and 0.50. *Phys. Rev. Lett.* **1995**, *75*, 4488–4491. [CrossRef]
6. Dagotto, E.; Hotta, T.; Moreo, A. Colossal magnetoresistant materials: The key role of phase separation. *Phys. Rep.* **2001**, *344*, 1–153. [CrossRef]
7. Dagotto, E. Complexity in strongly correlated electronic systems. *Science* **2005**, *309*, 257–262. [CrossRef] [PubMed]
8. Goodenough, J.B. Jahn-Teller phenomena in solids. *Annu. Rev. Mater. Sci.* **1998**, *28*, 1–27. [CrossRef]
9. Bersuker, I.B. *The Jahn-Teller Effect*; Cambridge University Press, Edinburgh Building: Cambridge, UK, 2006.
10. Staub, U.; Scagnoli, V.; Muleders, A.M.; Janousch, M.; Honda, Z.; Tonnerre, J.M. Charge/orbital ordering vs. Jahn-Teller distortion in $La0.5Sr1.5MnO_4$. *Europhys. Lett.* **2006**, *76*, 926–932. [CrossRef]
11. Clement, R.; Decurtins, S.; Gruselle, M.; Train, C. Polyfunctional two-(2D) and three-(3D) dimensional oxalate bridged bimetallic magnets. *Mon. Chem.* **2003**, *134*, 117–135. [CrossRef]
12. Daze, C.L.F.; Noa, M.A.; Nenwa, J.; Ohrstrom, L. Natural and synthetic metal oxalates–a topology approach. *CrystEngComm* **2019**, *21*, 6156–6164. [CrossRef]
13. Hursthouse, M.B.; Light, M.E.; Price, D.J. One-Dimensional Magnetism in Anhydrous Iron and Cobalt Ternary Oxalates with Rare Trigonal-Prismatic Metal Coordination Environment. *Angew. Chem. Int. Ed.* **2004**, *43*, 472–475. [CrossRef]
14. Zhang, B.; Zhang, Y.; Chang, G.; Wang, Z.; Zhu, D. Crystal-to-Crystal Transformation from K_2 [Co $(C_2O_4)_2(H_2O)_2$]·$4H_2O$ to K_2 [Co (μ-C_2O_4)(C_2O_4)]. *Magnetochemistry* **2021**, *7*, 77. [CrossRef]
15. Duan, Z.; Zhang, Y.; Zhang, B.; Zhu, D. Co(C_2O_4)(HO($CH_2)_3$OH): An antiferromagnetic neutral zigzag chain compound showing long-range ordering of spin canting. *Inorg. Chem.* **2008**, *47*, 9152–9154. [CrossRef]
16. Zhang, B.; Wang, Z.; Fujiwara, H.; Kabayashi, H.; Kurmoo, M.; Inoue, K.; Mori, T.; Gao, S.; Zhang, Y.; Zhu, D. Tetrathiafulvalene [FeIII (C_2O_4)Cl_2]: An Organic–Inorganic Hybrid Exhibiting Canted Antiferromagnetism. *Adv. Mater.* **2005**, *17*, 1988–1991. [CrossRef]
17. Zhang, B.; Wang, Z.; Zhang, Y.; Takahashi, K.; Okano, Y.; Cui, H.; Kobayashi, H.; Inoue, K.; Kurmoo, M.; Pratt, F.; et al. Hybrid Organic—Inorganic Conductor with a Magnetic Chain Anion: κ-$BETS_2$ [Fe^{III}(C_2O_4)Cl_2][BETS = Bis (ethylenedithio) tetraselenafulvalene]. *Inorg. Chem.* **2006**, *45*, 3275–3280. [CrossRef]
18. Tamaki, H.; Zhong, Z.J.; Matsumoto, N.; Kida, S.; Koikawa, M.; Achiwa, N.; Hashimoto, Y.; Okawa, H. Design of metal-complex magnets. Syntheses and magnetic properties of mixed-metal assemblies {NBu4[MCr (ox)$_3$]}$_x$ (NBu$_4^+$ = tetra (n-butyl) ammonium ion; ox^{2-} = oxalate ion; M = Mn^{2+}, Fe^{2+}, Co^{2+}, Ni^{2+}, Cu^{2+}, Zn^{2+}). *J. Am. Chem. Soc.* **1992**, *114*, 6974–6979. [CrossRef]
19. Mathoniere, C.; Nuttall, C.J.; Carling, S.G.; Day, P. Ferrimagnetic mixed-valency and mixed-metal tris (oxalato) iron (III) compounds: Synthesis, structure, and magnetism. *Inorg. Chem.* **1996**, *35*, 1201–1206. [CrossRef] [PubMed]
20. Duan, Z.; Zhang, Y.; Zhang, B.; Zhu, D. Two Homometallic Antiferromagnets Based on Oxalato-Bridged Honeycomb Assemblies:(A)^2[M^{II}(C_2O_4)$_3$](A= Ammonium Salt Derived from Diethylenetriamine; M^{II} = Fe^{2+}, Co^{2+}). *Inorg. Chem.* **2009**, *48*, 2140–2146. [CrossRef] [PubMed]
21. Rousse, G.; Rodriguez-Carvajal, J. Oxalate-mediated long-range antiferromagnetism order in $Fe_2(C_2O_4)_3$ $4H_2O$. *Dalton Trans.* **2016**, *45*, 14311–14316. [CrossRef] [PubMed]
22. Zhang, B.; Zhang, Y.; Zhang, J.; Li, J.; Zhu, D. A neutral molecular-based layered magnet [Fe (C_2O_4)(CH_3OH)]$_n$ exhibiting magnetic ordering at T_N ≈ 23 K. *Dalton Trans.* **2008**, 5037–5040. [CrossRef] [PubMed]
23. Decurtins, S.; Schmalle, H.W.; Schneuwly, P.; Esling, J.; Gutlich, P. A concept for the synthesis of 3-dimensional homo-and bimetallic oxalate-bridged networks [M_2(ox)$_3$]$_n$. Structural, moessbauer, and magnetic studies in the field of molecular-based magnets. *J. Am. Chem. Soc.* **1994**, *116*, 9521–9528. [CrossRef]
24. Hernandez-Molina, M.; Lloret, F.; Ruiz-Perez, C.; Julve, M. Weak Ferromagnetism in chiral 3-dimensional oxalato-bridged cobalt (II) compounds. Crystal structure of [Co(bpy)$_3$][Co$_2$(ox)$_3$]ClO_4. *Inorg. Chem.* **1998**, *37*, 4131–4135. [CrossRef]
25. Coronado, E.; Galan-Mascaros, J.R.; Gomez-Garzia, C.J.; Martinez-Agudo, J.M. Molecule-based magnets formed by bimetallic three-dimensional oxalate networks and chiral tris (bipyridyl) complex cations. The series [Z^{II} (bpy)$_3$][ClO_4][$M^{II}Cr^{III}$ (ox)$_3$](Z^{II}= Ru, Fe, Co, and Ni; M^{II}= Mn, Fe, Co, Ni, Cu, and Zn; ox= oxalate dianion). *Inorg. Chem.* **2001**, *40*, 113–120. [CrossRef]
26. Zhang, B.; Zhang, Y.; Zhang, J.; Hao, X.; Zhu, D. Mn (C_2O_4)(H_2O)$_{0.25}$: An antiferromagnetic oxalate-based cage compound. *Dalton Trans.* **2011**, *40*, 5430–5432. [CrossRef] [PubMed]
27. Mon, M.; Grancha, T.; Verdaguer, M.; Train, C.; Armentano, D.; Pardo, E. Solvent-Dependent Self-Assembly of an Oxalato-Based Three-Dimensional Magnet Exhibiting a Novel Architecture. *Inorg. Chem.* **2016**, *55*, 6845–6847. [CrossRef] [PubMed]
28. Thorarinsdottir, A.E.; Harris, T.D. Metal–organic framework magnets. *Chem. Rev.* **2020**, *120*, 8716–8789. [CrossRef]

29. Coronado, E.; Galan-Mascaros, J.R.; Marti-Gastaldo, C. Single chain magnets based on the oxalate ligand. *J. Am. Chem. Soc.* **2008**, *130*, 14987–14989. [CrossRef]
30. Anderson, P.W. Resonating valence bonds: A new kind of insulator? *Mater. Res. Bull.* **1973**, *8*, 153–160. [CrossRef]
31. Anderson, P.W. The resonating valence bond state in La_2CuO_4 and superconductivity. *Science* **1987**, *235*, 1196–1198. [CrossRef]
32. Ramirez, A.P. Strongly geometrically frustrated magnets. *Annu. Rev. Mater. Sci.* **1994**, *24*, 453–480. [CrossRef]
33. Balent, I. Spin liquids in frustrated magnets. *Nature* **2010**, *464*, 199–208. [CrossRef] [PubMed]
34. Zhang, B.; Zhang, Y.; Zhu, D. $(BEDT-TTF)_3Cu_2(C_2O_4)_3(CH_3OH)_2$: An organic–inorganic hybrid antiferromagnetic semiconductor. *Chem. Commun.* **2012**, *48*, 197–198. [CrossRef]
35. Zhang, B.; Zhang, Y.; Wang, Z.; Gao, S.; Guo, Y.; Liu, F.; Zhu, D. $BETS_3[Cu_2(C_2O_4)_3](CH_3OH)_2$: An organic–inorganic hybrid antiferromagnetic metal (BETS = bisethylene(tetraselenfulvalene)). *CrystEngComm* **2013**, *15*, 3529–3535. [CrossRef]
36. Zhang, B.; Zhang, Y.; Wang, Z.; Wang, D.; Yang, D.; Gao, Z.; Chang, G.; Guo, Y.; Mori, T.; Zhao, Z.; et al. Organic–inorganic hybrid metallic conductors based on bis (ethylenedithio) tetrathiafulvalene cations and antiferromagnetic oxalate-bridged copper (ii) dinuclear anions. *J. Mater. Chem. C* **2022**, *10*, 2845–2852. [CrossRef]
37. Zhang, B.; Zhang, Y.; Wang, Z.; Wang, D.; Baker, P.; Pratt, F.; Zhu, D. Candidate quantum spin liquid due to dimensional reduction of a two-dimensional honeycomb lattice. *Sci. Rep.* **2014**, *4*, 6451. [CrossRef]
38. Zhang, B.; Zhang, Y.; Zhu, D. $[(C_2H_5)_3NH]_2Cu_2(C_2O_4)_3$: A three-dimensional metal–oxalato framework showing structurally related dielectric and magnetic transitions at around 165 K. *Dalton Trans.* **2012**, *14*, 8509–8511. [CrossRef] [PubMed]
39. Zhang, B.; Baker, P.; Zhang, Y.; Wang, D.; Wang, Z.; Su, S.; Zhu, D.; Pratt, F.L. Quantum spin liquid from a three-dimensional copper-oxalate framework. *J. Am. Chem. Soc.* **2018**, *140*, 122–125. [CrossRef] [PubMed]
40. Sheldrick, G.M. *Shelx-97*; University of Göttingen: Göttingen, Germany, 1997.
41. Bain, A.B.; Berry, J.F. Diamagnetic corrections and Pascal's constants. *J. Chem. Educ.* **2008**, *85*, 532–536. [CrossRef]
42. Zhang, B. CCDC843075. Available online: http://www.ccdc.cam.ac.uk (accessed on 5 September 2011).
43. Vilela, R.S.; Oliveira, T.I.; Martins, F.T.; Ellena, J.A.; Lloret, F.; Julve, M.; Cangussu, D. Synthesis, crystal structure and magnetic properties of the helical oxalate-bridged copper(II) chain $\{[(CH_3)_4N]_2[Cu(C_2O_4)_2]\cdot H_2O\}_n$. *Comptes Rendus Chim.* **2012**, *15*, 856–865. [CrossRef]
44. Cano, J.; Alemany, P.; Alvarez, S.; Verdaguer, M.; Ruiz, E. Exchange Coupling in Oxalato-Bridged Copper (ii) Binuclear Compounds: A Density Functional Study. *Chem. A Eur. J.* **1998**, *4*, 476–484. [CrossRef]
45. Carlin, R.; van Duyneveldt, A. *Magnetic Properties of Transition Metal Compounds*; Springer: New York, NY, USA, 1977; p. 69.
46. Zhang, B.; Zhang, Y.; Zhang, J.; Yan, X.; Zhu, D. Step by step crystal-to-crystal transformation from 1D $K_2Cu(C_2O_4)_2(H_2O)_4$ (1) to 1D $K_2Cu(C_2O_4)_2(H_2O)_2$ (2) and then 1D $K_2Cu(C_2O_4)_2$ (3) by dehydration. *CrystEngComm* **2016**, *18*, 5062–5065. [CrossRef]
47. Kahn, O. *Molecular Magnetism*; VCH Publisher Inc.: New York, NY, USA, 1993; pp. 10–29.
48. Mugiraneza, S.; Hallas, A.M. Tutorial: A beginner's guide to interpreting magnetic susceptibility data with the Curie-Weiss law. *Commun. Phys.* **2022**, *5*, 95. [CrossRef]
49. Baker, G.A.; Rushbrooke, G.S.; Gilbert, H.E. High-temperature series expansions for the spin-$\frac{1}{2}$ Heisenberg model by the method of irreducible representations of the symmetric group. *Phys. Rev.* **1964**, *135*, A1272–A1277. [CrossRef]
50. Carling, R.L. *Magnetochemistry*; Springer: Berlin/Heidelberg, Germany, 1986; pp. 132–133.
51. Coronado, E.; Galan-Mascaros, J.R.; Gomez-Gracia, C.J.; Laukhin, V. Coexistence of ferromagnetism and metallic conductivity in a molecule-based layered compound. *Nature* **2000**, *408*, 447–449. [CrossRef] [PubMed]
52. Alberola, A.; Coronado, E.; Galan-Mascaros, J.R.; Gimenez-Saiz, C.; Gomez-Garcia, C.J. A molecular metal ferromagnet from the organic donor bis (ethylenedithio) tetraselenafulvalene and bimetallic oxalate complexes. *J. Am. Chem. Soc.* **2003**, *125*, 10774–10775. [CrossRef] [PubMed]

Disclaimer/Publisher's Note: The statements, opinions and data contained in all publications are solely those of the individual author(s) and contributor(s) and not of MDPI and/or the editor(s). MDPI and/or the editor(s) disclaim responsibility for any injury to people or property resulting from any ideas, methods, instructions or products referred to in the content.

Article

A Simulation Independent Analysis of Single- and Multi-Component cw ESR Spectra

Aritro Sinha Roy [1,2], Boris Dzikovski [2], Dependu Dolui [3], Olga Makhlynets [4], Arnab Dutta [3] and Madhur Srivastava [1,2,*]

1. Department of Chemistry and Chemical Biology, Cornell University, Ithaca, NY 14853, USA; as836@cornell.edu
2. National Biomedical Resource for Advanced ESR Spectroscopy, Cornell University, Ithaca, NY 14853, USA; bd55@cornell.edu
3. Department of Chemistry, Indian Institute of Technology Bombay, Mumbai 400076, India; arnabdutta@chem.iitb.ac.in (A.D.)
4. Department of Chemistry, Syracuse University, Syracuse, NY 13244, USA; ovmakhly@syr.edu
* Correspondence: ms2736@cornell.edu

Abstract: The accurate analysis of continuous-wave electron spin resonance (cw ESR) spectra of biological or organic free-radicals and paramagnetic metal complexes is key to understanding their structure–function relationships and electrochemical properties. The current methods of analysis based on simulations often fail to extract the spectral information accurately. In addition, such analyses are highly sensitive to spectral resolution and artifacts, users' defined input parameters and spectral complexity. We introduce a simulation-independent spectral analysis approach that enables broader application of ESR. We use a wavelet packet transform-based method for extracting g values and hyperfine (A) constants directly from cw ESR spectra. We show that our method overcomes the challenges associated with simulation-based methods for analyzing poorly/partially resolved and unresolved spectra, which is common in most cases. The accuracy and consistency of the method are demonstrated on a series of experimental spectra of organic radicals and copper–nitrogen complexes. We showed that for a two-component system, the method identifies their individual spectral features even at a relative concentration of 5% for the minor component.

Keywords: ESR spectral analysis; hyperfine decoupling; resolution enhancement; wavelet packet transform; simulation-free spectra analysis

Citation: Sinha Roy, A.; Dzikovski, B.; Dolui, D.; Makhlynets, O.; Dutta, A.; Srivastava, M. A Simulation Independent Analysis of Single- and Multi-Component cw ESR Spectra. *Magnetochemistry* **2023**, *9*, 112. https://doi.org/10.3390/magnetochemistry9050112

Academic Editor: Laura C. J. Pereira

Received: 16 March 2023
Revised: 11 April 2023
Accepted: 17 April 2023
Published: 23 April 2023

Copyright: © 2023 by the authors. Licensee MDPI, Basel, Switzerland. This article is an open access article distributed under the terms and conditions of the Creative Commons Attribution (CC BY) license (https://creativecommons.org/licenses/by/4.0/).

1. Introduction

ESR spectroscopy is a useful and powerful tool for studying biological free radicals and transition metal cofactors in proteins [1–7]. Understanding the electronic structure and local environment in such systems provides important insights into catalytic mechanisms [8–13] and redox processes in biological systems [6,14–16]. However, extracting information from ESR spectra can be challenging, especially in the case of weak and poorly resolved coupling interactions between studied spins. Traditionally, ESR spectral analysis is carried out by the spectral simulations, followed by user interpretation, which not only introduces bias but impairs the consistency of such analysis. The low-intensity signal components are difficult to analyze through standard spectral simulations in a variety of cases [17–19]. Accurate spectral analysis in such cases requires quantum chemical computations or special techniques [20–22], but without sufficient expertise and knowledge in the field, the interpretations rely heavily on the researchers' experience and intuition rather than the robustness of the method. An even more challenging problem, which occurs frequently, is the presence of more than one structurally similar molecule in a system. For closely resembling species, identifying multiple components by ESR alone could be a daunting task, and the standard simulation tools are incapable of extracting such information from the spectra without user manipulation. Additionally, every standard ESR simulation method requires user defined starting configuration for optimization [23,24], which could lead to overfitting and/or unintended manipulation. Hence, the direct extraction of the relevant spin Hamiltonian parameters, namely the g-factors and and hyperfine coupling constant (A) values from an experimental ESR spectrum, would be ideal. The extracted parameter values can be used directly to interpret the electronic structure or, if needed, can be further optimized by standard spectral fitting software.

The capability of the wavelet transform to choose a periodic hyperfine pattern from poorly resolved ESR spectra was highlighted previously [25]. In a series of recent publications [26–28], we have presented an improved version of the wavelet transform based spectral analysis, which can decouple different frequency components in an ESR or nuclear magnetic resonance (NMR) spectrum, enhancing the resolution and providing an opportunity to extract spectral information or parameters in a selective and objective manner. Initially, we have modified the Noise Elimination and Reduction via Denoising (NERD) method [25,29], which is based on the discrete wavelet transform (DWT) method, for the separation of the hyperfine lines in cw ESR spectra and the extraction of spectral parameters [26]. Later, in analyzing ^{1}H NMR spectra of molecular mixtures, which feature highly overlapped resonance lines due to the presence of scalar coupling between intramolecular protons, we recognized that spectral decomposition by the wavelet packet transform (WPT) is superior to DWT in separating the central frequencies from the multiplet structures encompassing them [27,28]. Using the same concept, a cw ESR spectrum is decomposed into its different frequency components by WPT, and at an optimum level of decomposition, a detail component in the wavelet domain is used to extract the hyperfine and/or superhyperfine structure, if any. The process is explained with an illustrative example later in this work (cf. Method). In the case of copper complexes, the analysis of the nitrogen-hyperfine structure is used for the determination of both $g_\perp$ and hyperfine splitting by copper, $A_\perp(Cu)$.

The outcome of the method has been validated by comparing the extracted hyperfine coupling constant values from both partially resolved and unresolved ESR spectra of Tempol and Tempo, recorded in the absence and in the presence of oxygen, respectively. We demonstrate the consistency of the WPT-based spectral analysis by using different wavelets in our analysis, namely Daubechies (Db6 and Db9) and Coiflet (coif3) wavelets. The method is applied across a wide range of copper complexes, and both the copper and nitrogen-hyperfine coupling constants are recovered from partially resolved and unresolved ESR spectra. The analysis did not use any prior knowledge about the structure of the complexes or user defined inputs in deriving the spectral parameters. The derived coordination geometry aligned with the structure predicted from analog studies, independent experiments and/or quantum computation, which further validates the results obtained by the WPT-based spectral analysis. The g and A-values obtained were compared with the optimized parameters obtained from spectral fitting by EasySpin software for a set of selected cases. The comparison shows that our method remains unaffected by artifacts, such as saturation and the passage effect, which affects simulations or spectral fitting significantly. While the performances of both methods were at par for well resolved spectra, unresolved spectral features remained inaccessible to the spectral fitting strategy. In those later cases, to the best of our knowledge, the proposed WPT-based analysis is the only method that can separate the so-called hidden features from poorly resolved spectra and extract the relevant spectral parameters by the direct analysis of an ESR spectrum. In this work, along with describing the extraction of spectral parameters by the WPT analysis of the ESR spectra of nitroxides and copper complexes, we demonstrated the efficiency of the method in identifying and analyzing multi-component spectra.

2. Method

2.1. Overview of the Wavelet Packet Transform Theory

A continuous wavelet transform can be defined as [30]

$$F(\tau,s) = \frac{1}{\sqrt{|s|}} \int_{-\infty}^{+\infty} f(\delta)\psi^*\left(\frac{\delta-\tau}{s}\right) dt \quad (1)$$

where s is the inverse frequency (or frequency range) parameter, τ is the signal localization parameter, δ represents the chemical shift, $f(\delta)$ is the spectrum, $F(\tau,s)$ is the wavelet-transformed signal at a given signal localization and frequency, and $\psi^*\left(\frac{\delta-\tau}{s}\right)$ is the signal probing function called "wavelet". Different wavelets are used to vary the selectivity or sensitivity of adjacent frequencies with respect to signal localization. They are not dependent on *a priori* information of the signal or its characteristics.

Discrete wavelet transform (DWT) is expressed by two sets of wavelet components (detail and approximation) in the following way [30]:

$$D_j[n] = \sum_{m=0}^{p-1} f[\delta_m] 2^{\frac{j}{2}} \psi[2^j \delta_m - n] \quad (2)$$

$$A_j[n] = \sum_{m=0}^{p-1} f[\delta_m] 2^{\frac{j}{2}} \phi[2^j \delta_m - n] \quad (3)$$

where $f[\delta_m]$ is the discrete input spectrum, p is the length of input signal $f[\delta_m]$, $D_j[n]$ and $A_j[n]$ are the detail and approximation components, respectively, at the jth decomposition level, and $\psi[2^j \delta_m - n]$ and $\phi[2^j \delta_m - n]$ are wavelet and scaling functions, respectively. The maximum number of decomposition levels that can be obtained is N, where $N = \log_2 p$, and $1 \leq j \leq N$. The scaling and wavelet functions, at a decomposition level, are orthogonal to each other, as they represent non-overlapping frequency information. Similarly, wavelet functions at different decomposition levels are orthogonal to each other.

The detail component $D_j[n]$ is the discrete form of Equation (1), where j and n are associated with s and τ, respectively. The approximation component $A_j[n]$ represents the remaining frequency bands not covered by the detail components until the jth level. The signal $f[\delta_m]$ can be reconstructed using the inverse discrete wavelet transform as follows:

$$f[\delta_m] = \sum_{k=0}^{p-1} A_{j_0}[k]\phi_{j_0,k}[\delta_m] + \sum_{j=1}^{j_0}\sum_{k=0}^{p-1} D_j[k]\psi_{j,k}[\delta_m] \quad (4)$$

where j_0 is the maximum decomposition level from which an input signal needs to be reconstructed. Compared to that, both the approximation and detail components at each level are further decomposed into a set of approximation and detail components. A schematic diagram of DWT and WPT decomposition against increasing levels are shown for comparison in Figure 1 [27].

2.2. A Case Study: WPT Analysis of cw ESR Spectrum of Tempo

In this section, we explain the details of WPT spectral analysis by using the cw ESR spectrum of Tempo as an example, shown in Figure 2. The ESR spectrum of Tempo is split into three major lines, corresponding to the hyperfine coupling of the ^{14}N nucleus. Each of those lines are further split by the interaction of the ^{1}H nuclei in the molecule, giving rise to what we designate as the superhyperfine splitting. Our goal is to separate the superhyperfine structure in the Tempo spectrum from the other spectral features. The first level of wavelet decomposition by the DB9 wavelet in Figure 2 shows a pair of approximate (A1) and detail (D1) components. It can be seen that D1 comprises noise, and the entire spectrum is represented by A1. Hence, D1 and all the components that derive from D1 during successive decomposition are not used in the spectral analysis. The decomposition of A1 is continued until level-4, where the approximation component of AAA3 shows no superhyperfine splitting, and correspondingly, the approximation component derived from the decomposition of DAA3 contains the superhyperfine splitting pattern. This is why, in this case, decomposition until level-4 is considered to be optimal in separating the hyperfine and superfine splittings of Tempo's ESR spectrum. For cases where the superhyperfine is not resolved or visible, the optimum level of decomposition is chosen to be 4 based on previous work [26]. The WPT analysis is performed using Matlab software, version 9.12.0.1884302 (R2022a), and an illustrative code is given in Appendix A.1.

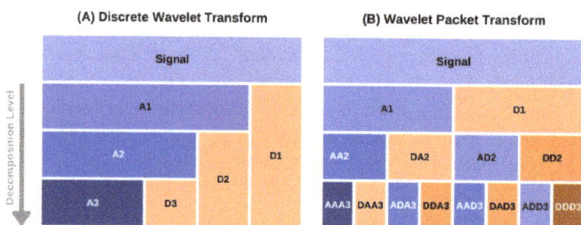

Figure 1. A schematic diagram of data decomposition in discrete (**A**) and packet wavelet transform (**B**) methods. The approximation and detail components at level k are denoted as A_k and D_k in (**A**). In case of wavelet packet transform, the approximation and detail components at a decomposition level are denoted by the component name of the previous level followed by A_k or D_k, respectively [27]. Copyright, 2022, The Journal of Physical Chemistry.

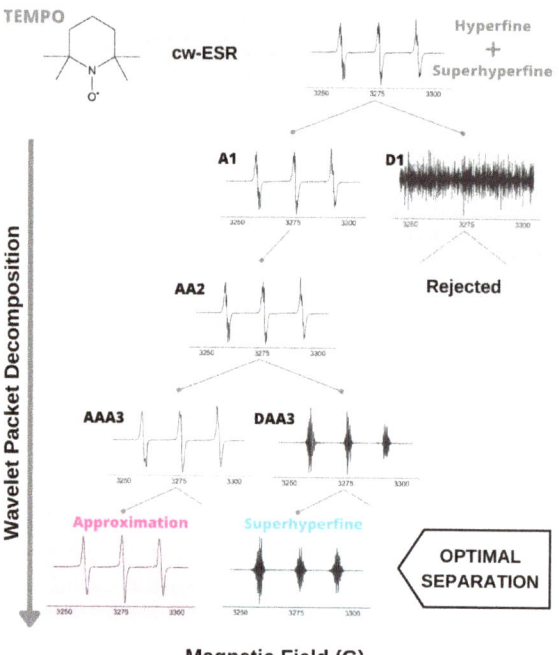

Figure 2. Separation of hyperfine and superhyperfine components in Tempo's ESR spectrum by WPT decomposition using Db9 wavelet. Given that at level-1, A1 contains all the spectral information, D1 and all the components derived from D1 are rejected. Complete separation of superhyperfine structure from the approximate component occurred after the decomposition level-3. Pure hyperfine (approximation) and superfine components are obtained from the decomposition of AAA3 and DAA3.

3. Materials

A series of experimental X-band ESR spectra were used in our analysis, and the corresponding molecular structures are given in Table 1. ESR spectra of Tempol and Tempo are very well studied, and we presented their spectral analysis by WPT for validation purposes. The mutation of superoxide dismutase-1 (SOD1) is arguably correlated with the incidence of significant fractions of familial and spontaneous cases of amyotrophic lateral schlerosis (ALS) disease [31,32]. For one of its mutants, SOD1:H48Q, the cw ESR spectrum was collected at 9.26 GHz, at a temperature of 30 K and concentration in the range of 300 to 350 µM using 50 µL 0.4 mm glass capillaries. No cryoprotectant was added given the viscous nature of the sample. Due to the presence of a glutamine residue in place of a histidine in the mutant, the nitrogen-hyperfine structure along the $g_\perp$ became partially resolved as a result of the reduction of the number of nitrogen in the copper coordination sphere. Cu-AHAHARA spectra were collected at 9.39 GHz for two different temperatures, 10 K and 100 K. ESR measurements of CuQu and CuQuA were performed in DMSO solutions at 100 K with an ESR frequency of 9.32 GHz.

Table 1. Molecular structures and ESR frequencies corresponding to the experimental cw ESR spectra used in this work.

Molecule	Structure	ESR Frequency (GHz)
Tempo		9.33
Tempol		9.33
SOD1:H48Q	SOD1 mutant, histidine (48) replaced with glutamine	9.26
Cu-AHAHARA	A complex of Cu(II) and AHAHARA peptide: C-terminus as amide and N-terminus as acetyl group	9.39
CuQu		9.316
CuQuA		9.316

3.1. Experimental Section
Synthesis

Copper quinoline (CuQuA): 160.0 mg (1.0 mM) 2-amino-8-quinolinol was taken in an 80.0 mL Schlenk flask with a magnetic stirrer bar, and 15.0 mL methanol was added to make it a homogeneous yellow color solution. Then, 85.0 mg (0.49 mM) CuCl$_2$·2H$_2$O was taken separately with 10 mL MeOH in another Schlenk flask under steam of N$_2$. Then, the two methanolic solutions were charged under nitrogen flow. An immediate dark brown solution appeared after adding the metal salt to the ligand. Reducing the solvent with continuous N$_2$/Ar flashing, whitish deep brownish precipitate was observed. The reaction mixture was further stirred for two hours for completion with N$_2$ purging. Then, the precipitate was collected, washed with hexane and diethyl ether and dried under a high vacuum (yield = 137.0 mg, 76% w.r.t. 2-amino-8-quinolinol). XRD-suitable crystals (CCDC 2127156) were grown from slow diffusion from methanol/diethyl ether solution.

Copper quinone (CuQu): Synthesized as reported above, where 8-hydroxyquinoline was utilized as the precursor ligand (yield = 120.0 mg).

Copper AHAHARA: The AHAHARA peptide was synthesized by manual Fmoc solid-phase synthesis at elevated temperature using Amide Rink resin, Fmoc-protected amino acids and previously reported protocols [33].

SOD1:H48Q: The recombinant SOD1:H48Q protein, replacing a histidine with glutamine, was expressed and purified as described in a previous work [34].

3.2. ESR Experiments

Oxygen-free samples of Tempol and Tempo were prepared by flame-sealing after a triple repeat of the freeze–thaw cycle under vacuum. Oxygen-saturated samples were prepared by passing oxygen gas through the radical solution in water for 10 min and kept under oxygen atmosphere during measurements. Samples with an intermediate oxygen concentration were prepared and handled in air. The nitroxide concentration in all the samples was 100 µM. The ESR spectra were recorded at 293 K at a microwave (MW) frequency of 9.33 GHz, power of 0.1 mW, modulation frequency of 100 KHz and modulation amplitude of 0.1 G. The ESR spectrum of SOD1:H48Q was recorded at 30 K at an MW frequency of 9.26 GHz, power of 0.06325 mW, modulation frequency of 100 kHz and modulation amplitude of 4 G. The copper concentration was estimated to be 50 µM. The ESR spectra of Cu-AHAHARA complex were acquired in Wilmad tubes using a Bruker Elexsys E500 EPR spectrometer equipped with a cryostat. Peptide stocks at pH 2 were prepared fresh by adding lyophilized solid (>90%) to 10 mM HCl until peptide concentration reached 1 mM. First, Cu(II) solution in water (1 mM, 75 µL) and peptide stock (1 mM in 10 mM HCl, 150 µL) were mixed, and then buffer (91 mM Hepes, pH 8, 275 µL) was added to make a solution with 150 µM Cu(II) and 300 µM peptide (500 µL final volume). The mixture was incubated at room temperature for 2 h, and then glycerol was added to a final concentration of 10%, transferred into an ESR tube and flash frozen in liquid nitrogen. The final pH of the sample was 7.6 as measured by a Spintrode electrode (Hamilton). ESR spectra were acquired at 100 K and 10 K using the following conditions: frequency 9.39 GHz, power 5 mW or 2 mW, modulation frequency 100 kHz, modulation amplitude 8 G (10 K) and 4 G (100 K) and time constant 163.8 ms. The ESR spectra of CuQu and CuQuA were collected using the Bruker cw ESR EMX spectrometer at the National Biomedical Resource for Advanced ESR Spectroscopy (ACERT). About 100 mg of the air-stable copper complexes were dissolved in 1 mL DMSO. The solutions were poured in 4 mm ESR tubes, frozen in liquid nitrogen and kept in liquid nitrogen Dewar until they were inserted in the spectrometer. The spectra were recorded at 100 K, a microwave frequency of 9.316 GHz and attenuation of 30 dB for CuQu and 50 dB for CuQuA.

3.3. ESR Spectral Mix

For multi-component spectral analysis, four mixed spectra were calculated by mixing the ESR spectra of the compounds, CuQu and CuQuA, in proportions of (A) 2:1, (B) 4:1, (C) 10:1 and (D) 20:1 (Figure 3).

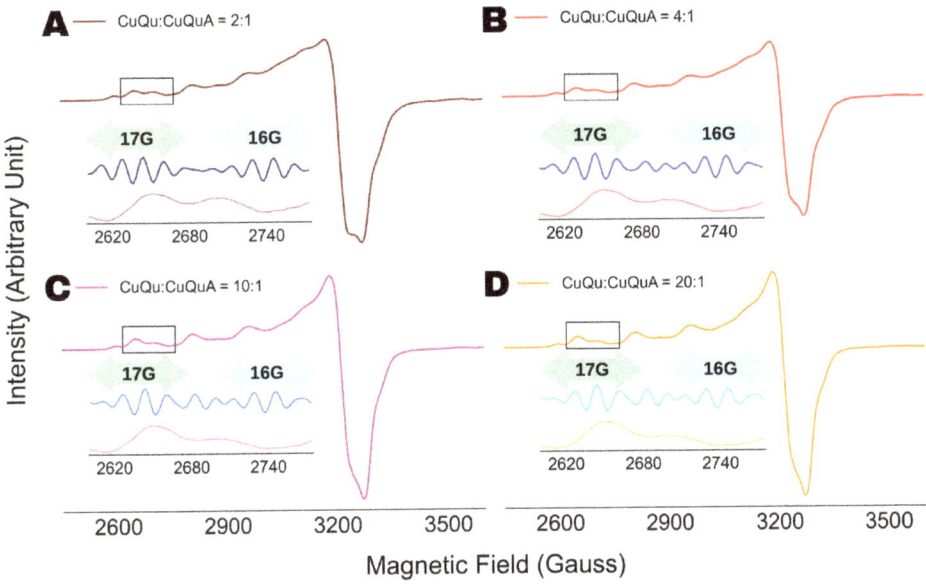

Figure 3. WPT analysis of ESR spectra of mixtures of CuQu and CuQuA, mixed in proportions of (**A**) 2:1, (**B**) 4:1, (**C**) 10:1 and (**D**) 20:1, respectively. The insets show the WPT component for the spectral region between 2600 G and 2775 G. Two components were identified in each of the spectral analyses from the difference in nitrogen-hyperfine splitting of 17 G (green double headed arrow) and 16 G (blue double headed arrow).

4. Results and Discussion

4.1. Validation of the Analysis

The interpretation of partially resolved ESR spectra by fitting has been the standard practice in the study of organic radicals and paramagnetic metal complexes. The approximations and the principles of such optimizations remain unknown to the users, whose interpretation of the results is largely dominated by the goodness of the fit and a priori knowledge about the structure or nature of the molecule under investigation. In contrast, a direct extraction of parameters, a process that lacks visual aids, can raise queries about the accuracy of the outputs. That was our motivation to run the analysis on a series of X-band ESR spectra of Tempol and Tempo under varying oxygen concentrations, as shown in Figure 4. The superhyperfine lines are fully or partially resolved in absence of oxygen or when it is present in very low concentration. However, the features become invisible due to line broadening by increased oxygen concentration. Consequently, while the superhyperfine constant in the former cases could be obtained by simple visual inspection or spectral fitting, none of those are applicable in the latter. In all those cases, the WPT-based method recovered the superhyperfine structure from the corresponding ESR spectra and extracted the value of the coupling constant, illustrating both its advantage over the existing spectral analysis approaches and its consistency.

It should be noted that the hyperfine components contain some features in between the regions of interest, clearly visible in Figure 4(B1). Such artifacts can be discarded by either (i) comparing with the original spectra (which was the case for Figure 4(B1)) or (ii) analyzing the inter-peak spacing, which yields non-uniform splitting in case of artifacts.

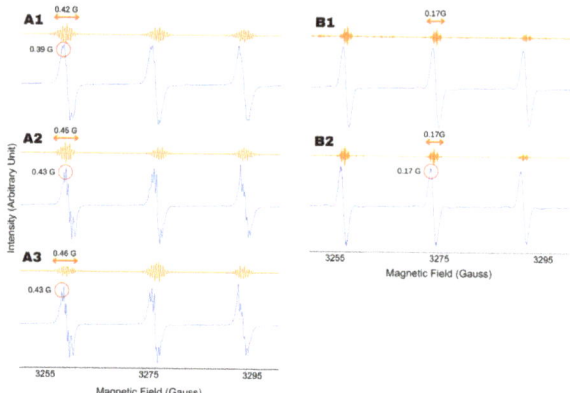

Figure 4. Analysis of X-band cw ESR spectra (blue) of Tempol (**A**) and Tempo (**B**) under varying concentrations of oxygen. The superhyperfine splitting due to the protons is partially resolved in (**A(1–3),B2**). The superfine splitting is invisible in the fully oxygenated Tempo spectrum (**B1**) due to line broadening. The superfine spectra recovered by the wavelet packet transform-based analysis of the spectra are shown (orange). The similarity between the superfine splitting constant obtained from the WPT analysis and direct analysis of the partially resolved should be noted. The validity of the analysis in case of an unresolved spectrum is demonstrated for (**B1,B2**).

4.2. Robustness Against Wavelet Selection

An important and non-trivial variable in the WPT analysis of ESR spectra is the selection of a wavelet, or in other words, understanding the robustness of the analysis against the types of wavelets used in the analysis. We repeated the analysis shown in Figure 4 with three different types of wavelets, Coiflet-3 (Coif3), Daubechies-6 (Db6) and Daubechies-9 (Db9), and the results are summarized in Figure 5. The extracted superhyperfine constant values for Tempol and Tempo, 0.44 ± 0.01 G and 0.20 ± 0.01 G, showed insignificant variation for all three wavelets while producing the same amount of splitting across all the analyses, consistent with the molecular structure. This observation illustrated the robustness of the method while further validating the accuracy of the extracted parameters. For the rest of the analysis in this work, we used the Db9 wavelet.

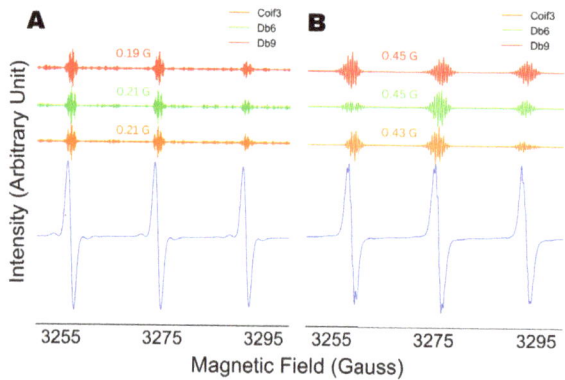

Figure 5. Analysis of unresolved and partially resolved X-band cw ESR spectra (blue) of Tempo (**A**) and Tempol (**B**). The superfine spectra recovered by the wavelet packet transform-based analysis of the spectra are shown using three different wavelets, Coiflet-3 (orange), Daubechies-6 (green) and Daubechies-9 (red). Superfine coupling constants obtained are consistent across the analysis involving three different wavelets.

4.3. Spectral Analysis of Partially Resolved ESR Spectra

We started our analysis by fitting the X-band cw ESR spectra of SOD1:H48Q recorded at 30 K and Cu-AHAHARA, recorded at 10 K and 100 K by using the EasySpin software. The optimized simulations along with the experimental spectra are shown in Figure 6. For SOD1:H48Q, the EasySpin analysis yielded nitrogen-hyperfine constant values of 15.6 G (axial) and 9.0 G (parallel), while the simulations for Cu-AHAHARA spectra did not require any nitrogen-hyperfine parameter in fitting the spectra. It should be noted that the Cu-AHAHARA spectrum at 10 K shows no nitrogen-hyperfine splitting; however, partial nitrogen-hyperfine splitting is evident for the spectrum at 100 K. This apparent anomaly might have resulted from the partial saturation and passage effect [35,36] in the case of the former, as evidenced by the behavior of the first integral of the spectrum, and consequently, the EasySpin fit for the 10 K spectrum performed poorly compared to that of the 100 K spectrum. In addition, the simulation presented for SOD1:H48Q in Figure 6A emphasized the potential presence of a second component, with a slightly shifted $g_\parallel$ and/or different hyperfine coupling constants.

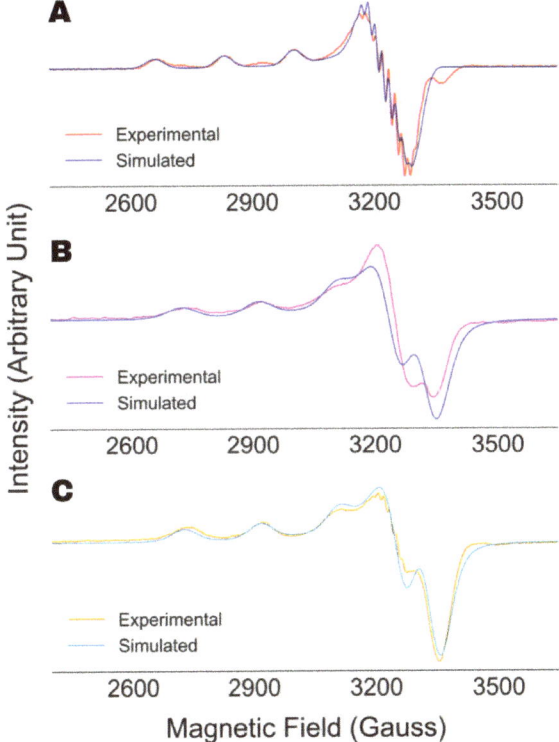

Figure 6. EasySpin simulation results along with experimental ESR spectra of SOD1:H48Q (**A**), Cu-AHAHARA spectrum recorded at 10 K (**B**) and 100 K (**C**).

In Figure 7, we present the WPT analysis of the ESR spectra of (A) SOD1:H48Q, (B) Cu-AHAHARA at 10 K and (C) Cu-AHAHARA at 100 K. For all the analyses, the spectra were decomposed to level-4 using the DB9 wavelet. For SOD1:H48Q, our analysis yielded a nitrogen-hyperfine coupling constant of 14.1 G for the axial peaks in the range of 3200 to 3320 G. However, an analysis of the $g_\parallel$ splitting (Figure 7(A.2)) revealed two different nitrogen-hyperfine splitting constants: 12.5 G along the dominant Cu parallel hyperfine splitting and 16.5 G along the minor component. Following that, we analyzed the splitting along the small axial peak between 3220 and 3400 G (not shown), which emphasized the presence of four nitrogens and a nitrogen-hyperfine constant of A_N = 16.5 G. From this analysis, we could infer a spherical electron density distribution in the second component, suggesting a tetrahedral copper complex with a large hyperfine splitting of ~205 G. Similar analysis for Cu-AHAHARA yielded nitrogen-hyperfine coupling constants of 13.9 G (10 K) and 14.0 G (100 K) along the $g_\perp$ signal component. The consistency of the results demonstrates the robustness of the WPT-based spectral analysis against artifacts, such as an over-saturation effect in this case, which is a major advantage over the standard spectral fitting methods.

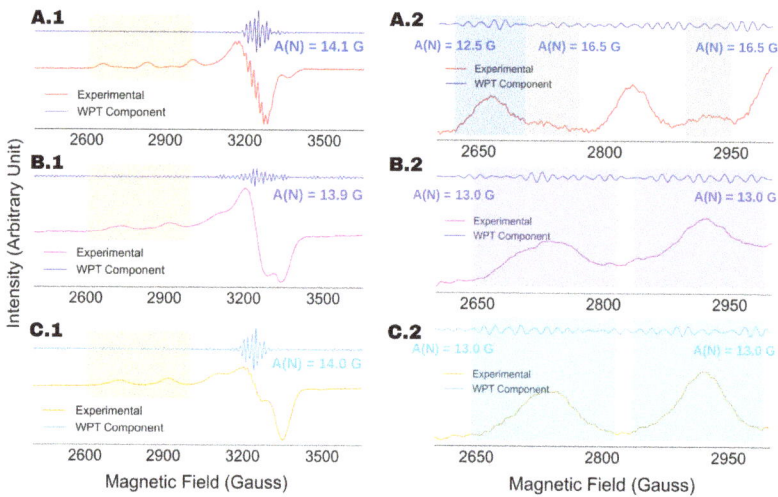

Figure 7. Recovery of nitrogen-hyperfine features by WPT analysis from experimental cw ESR spectra of SOD1:H48Q (**A**), Cu-AHAHARA at 10 K (**B**) and Cu-AHAHARA at 100 K (**C**). The left panel illustrates analysis of the $g_\perp$ component, while the $g_\parallel$ component is highlighted in yellow and the corresponding analysis is shown on the right. For SOD1:H48Q (**A**), resolving the nitrogen-hyperfine splitting along $g_\parallel$, a dominating component with A_N = 12.5 G (shaded blue) and a minor component with A_N = 16.5 G (shaded gray) were obtained.

It should be noted that the copper coordination geometry of the Cu-AHAHARA complex could not be confirmed in a previous work from the ESR spectral analysis [33]. The WPT analysis presented in Figure 7 suggests a strong overlap of nitrogen-hyperfine components in the $g_\perp$ region of the ESR spectra at 10 K and 100 K, while the former seems to contain spectral artifacts, making analysis by standard procedure error-prone. Hence, we analyzed the WPT components originating from the $g_\parallel$ region for the both the cases, which is summarized in Figure 8. In the case of the spectrum collected at 10 K, the WPT component originating from the ESR peak centered at 2730 G showed nine evenly spaced lines at 12.5 G apart from each other, indicating an N_4-coordination for the copper center in the Cu-AHAHARA complex. In addition, a smaller coupling constant in the $g_\parallel$ region in comparison to the value obtained in the $g_\perp$ region (13.9 G) suggests four equivalent nitrogens in the equatorial plane. This interpretation aligns well with a series of independent experimental and theoretical studies conducted to elucidate the previous geometry of the complex [33]. For the spectrum recorded at 100 K, a similar analysis in the $g_\parallel$ region did not reproduce the exact same results because of unresolved spectral overlapping in the WPT component. However, upon close inspection, we resolved half of the nitrogen-hyperfine splitting window for the $g_\parallel$ peaks centered at 2734 G and 2901 G. In this regard, it should be noted that only the even spacing between the peaks in a WPT component around a $g_\parallel$ peak was used as the criteria for recovering nitrogen-hyperfine splitting. The WPT analysis is highly accurate in recovering spectral information with respect to their location but not necessarily the intensity. Factors such as the partial overlapping of resonance lines, residual noise and spectral artifacts affect the intensities of the peaks in a WPT component, and hence the recovered superfine splitting is unlikely to reproduce the relative intensity pattern expected for perfectly resolved spectra. The analysis of the Cu-AHAHARA ESR spectrum at 100 K yielded the same nitrogen-hyperfine coupling constant of 12.5 G and suggested an N_4 coordination for the copper center.

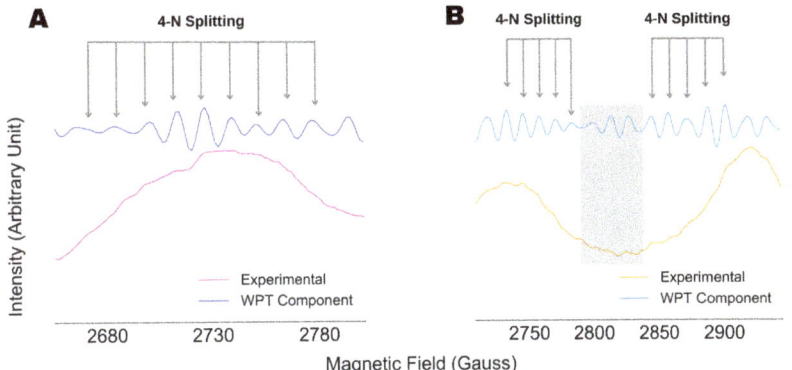

Figure 8. Analysis of the $g_\parallel$ components in the ESR spectra of Cu-AHAHARA complex at (**A**) 10 K and (**B**) 100 K for probing the copper coordination. For (**A**), the splitting pattern observed in the WPT component between 2650 G and 2800 G revealed nine equally spaced lines centered at 2727 G with an inter-peak spacing of 12.5 G, indicating four nitrogens in the copper coordination sphere. In the case of (**B**), the entire ranges of nitrogen-hyperfine splitting for none of the $g_\parallel$ components were visible due to overlapping in the WPT component. However, for both the $g_\parallel$ components centered at 2734 G and 2901 G, half of the nitrogen-hyperfine splitting windows were resolved, indicating four nitrogen coordination geometry for the copper center with a coupling constant of 12.5 G. The WPT component in the shaded region (gray) was not considered in the analysis because of varying inter-peak spacing in the region.

The WPT analysis for the X-band ESR spectra of two copper–nitrogen complexes, CuQu and CuQuA, are shown in Figure 9. The detail component in the optimal level of decomposition, which was three for both the spectra, revealed an unresolved hyperfine structure due to nitrogens coordinated to the copper centers in the complexes. For example, Cu(II) splits the resonance line at $g_\perp$ into four lines—we call them h-1, h-2, h-3 and h-4 for explanation purposes. Each of these lines split further due to the n interacting nitrogens into $m = (2 \times n + 1)$ lines. Given the small magnetic field window between h-1 and h-4, as well as the nitrogen-hyperfine coupling, most of the resonance lines cannot be resolved due to complete or partial overlapping. However, for most of the cases, it might be possible to identify the first $m/2$ lines originating from h-1 and the last $(m+1)/2$ lines originating from h-4, where the extent of overlapping of resonance lines is the least. Using this logic, from the detail component analysis for CuQu in Figure 9A, we identified three lines with a total separation of 53 G, which corresponded to an $m = 5$ or two coupled nitrogens with h-1 and h-4 at 3203 G and 3314 G. Further, we calculated the nitrogen-hyperfine coupling constant to be $A(N) = 53/2$ G or 26.5 G, the Cu(II) hyperfine coupling $A(Cu) = (3314 - 3203)/3$ G or 37 G and the $g_\perp = (3314 + 3203)/2$ G $\equiv 2.0427$. The calculated spin-Hamiltonian parameters are in good agreement with previously reported analogous N_2O_2-coordinated copper(II) complexes [37,38]. With a similar analysis for CuQuA, Figure 9B, we inferred that three nitrogens were coordinating with the Cu(II) center with $A(N) = 21.3$ G, while the $g_\perp$ and $A(Cu)$ were calculated to be 2.0411 and 23.3 G. The original N_2O_2 coordination geometry for the copper center observed in CuQu complex can be expanded in CuQuA due to the presence of peripheral amine in the QuA ligand scaffold. The potential involvement of one amine group is supported by the EPR data, indicating an N_3O_2 coordination.

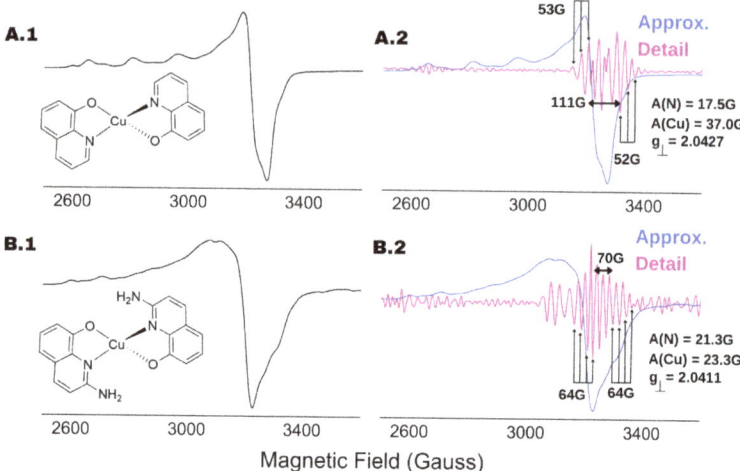

Figure 9. Recovery of nitrogen-hyperfine structure and coordination of Cu(II) center in (**A**) CuQu and (**B**) CuQuA. Both the ESR spectra (left panel) were recorded at 100 K and an ESR frequency of 9.316 GHz with attenuation of (**A**) 30 db and (**B**) 50 db, respectively. The approximation (blue) and detail (magenta) spectral components originated at the optimal decomposition level of 3 are shown in the right panel (scaled arbitrarily for visualization purpose). For both the cases, the nitrogen superfine structure were identified, showing (**A**) 2-N and (**B**) 3-N coordination to the copper centers in the two cases, along with the position of $g_\perp$ and hyperfine coupling due to Cu(II).

4.4. Analysis of Multi-Component ESR Spectra

It has already been shown in Figure 7 that the spectral analysis of the $g_\parallel$ components in the ESR spectrum of SOD1:H48Q revealed the presence of two components. In this section, we present further proof of how the WPT-based spectral analysis can be utilized efficiently to identify multi-component spectra, even when the components are present in highly disproportionate amounts. A set of four mixed spectra was produced for the analysis by mixing the ESR spectra of CuQu and CuQuA in proportions of (A) 2:1, (B) 4:1, (C) 10:1 and (D) 20:1, shown in Figure 3. The WPT spectral analysis was conducted using Db9 wavelet at decomposition level of 4, and the WPT detail components for the spectral region between 2600 and 2775 G are shown in the insets of Figure 3. In all the cases, five lines with a splitting constant of 17 G were recovered between 2610 and 2670 G, which corresponds to CuQu, which is expected because of the abundance of the complex in the mixtures. However, a second component with a splitting constant of 16 G was identified in between 2700 and 2748 G. For the latter, only four or three lines were visible due to spectral overlap, but it can be separated from artifacts by the consistent positions of the peaks appearing the detail WPT component. It should be noted that this strategy clearly identifies a second minor component directly from the spectral analysis, in this case CuQuA, but it may not be possible to determine the structure of the component solely from the analysis due to spectral overlaps, leading to unrecoverable information loss. However, we would like to emphasize that we used poorly resolved X-band ESR spectra of the individual complexes, and given such constraints, we believe that it is a major achievement to detect a second component solely from ESR spectral analysis, even when the second component is present at as low a relative concentration as 5%. In general, such findings from the WPT spectral analysis will help researchers to decide on further structural analysis by employing high-frequency ESR and/or complementary techniques.

5. Conclusions

In this work, we have provided the complete recipe for a simulation-free analysis of cw ESR spectra using a wavelet packet transform-based algorithm and experimental X-band ESR spectra of organic free radicals (Tempol and Tempo) and metal complexes (copper (II) inorganic and biological complexes). We showcased three major accomplishments of the WPT-based spectral analysis over standard simulations: (1) direct extraction of nitrogen-hyperfine structure from poorly resolved/unresolved spectra, (2) insensitivity toward spectral artifacts and (3) identification of multi-component samples. A consistent extraction of the superhyperfine coupling constant due to protons from both partially resolved and unresolved spectra for Tempol is presented, validating the accuracy of the new spectral analysis technique. The limitations of standard ESR spectral analysis are illustrated by running EasySpin simulations for partially resolved and poorly resolved spectra, with one of the cases containing some experimental artifacts. While the accuracy and efficiency of the simulations varied drastically in those cases, the WPT-based analysis extracted spectral parameters, revealing hyperfine and superhyperfine splittings not recoverable by EasySpin simulations, and the analysis was unperturbed by spectral artifacts. For the copper–nitrogen complexes, the resolved hyperfine structures confirmed the number of nitrogen atoms coordinating with copper centers as well as their coordination geometry. Finally, by close examination of the detail WPT components originating from the $g_{\parallel}$-regions of the spectra, the presence of two different components in the case of SOD1:H48Q, a mutant of SOD1, has been confirmed along with their nitrogen-hyperfine coupling constants. The strength of the method in resolving multi-component spectra has been displayed further by analyzing four mixed spectra of two copper–nitrogen complexes, CuQu and CuQuA. The spectral analysis revealed the presence of the two components even when the proportion of the two components in the spectrum was 20:1. In addition, three different wavelets were used for selected cases, emphasizing the robustness of the method against the choice of wavelets. The comprehensive analysis presented in this work is expected to motivate broad adoption of the technique in analyzing ESR spectra, and the spectral parameters obtained from the analysis can be further optimized using simulations and/or other computational methods.

Author Contributions: Conceptualization, A.S.R., B.D. and M.S.; methodology, A.S.R., B.D. and M.S.; formal analysis, A.S.R. and B.D.; material synthesis, A.D., B.D., D.D. and O.M.; spectral data collection, A.S.R., B.D. and O.M.; investigation, A.S.R.; writing—original draft preparation, A.S.R.; writing—review and editing, O.M., A.D. and M.S.; visualization, A.S.R.; supervision, M.S.; project administration, M.S.; funding acquisition, M.S. and A.D. All authors have read and agreed to the published version of the manuscript.

Funding: Data were collected at the ACERT facility for ESR on Cornell's campus (the National Institute of General Medical Sciences/National Institutes of Health under grant R24GM146107). This research was funded in part by the DST-SERB ore research grant (CRG/2020/001239).

Institutional Review Board Statement: Not applicable.

Informed Consent Statement: Not applicable.

Data Availability Statement: The data used in this paper can be accessed on 15 April 2023 via https://github.com/Signal-Science-Lab/Simulation_Independent_ESR_Spectral_Analysis.

Conflicts of Interest: The authors declare no conflict of interest.

Abbreviations

The following abbreviations are used in this manuscript:

ESR	Electron spin resonance
WPT	Wavelet packet transform
NERD	Noise Elimination and Reduction via Denoising
SOD1	Superoxide dismutase-1

Appendix A. Analysis of 100 K Cu-AHAHARA ESR Spectrum

Appendix A.1. Matlab Code for WPT Decomposition Shown in Figure A1

```
% set directory
cd "user directory";

% wavelet decomposition level
nlevs = [3,4,5];

% load data
fname = 'AHAHARA_100K.txt';
tag = strtok(fname, '.');
df = readtable(fname);

for i = 1:3
    % apply wavelet packet transform
    n = nlevs(i);
    uwpt = wpdec(data.Var2, n, 'db9');

    % approximation component
    rwpc0 = wprcoef(uwpt, 2^n - 1);
    % hyperfine component
    rwpc1 = wprcoef(uwpt, 2^n + 1);

    % save the results
    writetable(table(df.Var1, rwpc0),
               strjoin([tag,"_aL",int2str(n),".txt"],''));
    writetable(table(df.Var1, rwpc1),
               strjoin([tag,"_hL",int2str(n),".txt"],''));
end
```

Appendix A.2. Spectral Analysis

We plot the approximation and hyperfine components from the decomposition of Cu-AHAHARA ESR spectrum collected at 100 K in Figure A1, following the procedure described in the case study of Section 2.2. It can be seen that in successive wavelet decomposition of the spectrum using the Db9 wavelet, the approximation component at level-5 develops spectral distortion in comparison to the original spectrum. Therefore, we select level-4 as the optimal decomposition level and analyze the hyperfine component at that level to probe nitrogen hyperfine splitting. The next steps in the analysis and the results are given in Section 4.3.

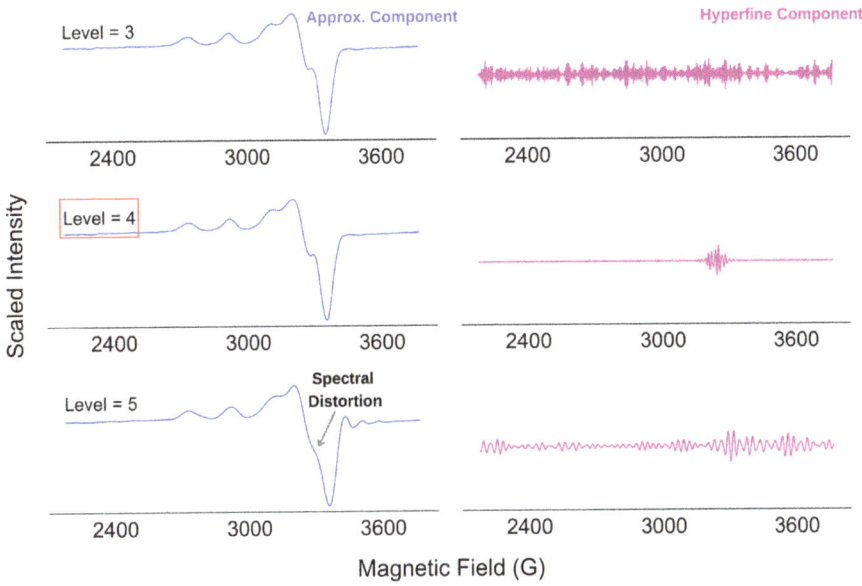

Figure A1. WPT analysis of ESR spectra of Cu-AHAHARA complex recorded at 100 K. The approximation and hyperfine components are shown at decomposition levels of 3, 4 and 5 using the Db9 wavelet. In this case, the approximation component at level-5 showed spectral distortion, implying that level-4 is the optimal decomposition level in this case.

References

1. Mabbs, F.E.; Collison, D. *Electron Paramagnetic Resonance of d Transition Metal Compounds*; Elsevier: Amsterdam, The Netherlands, 2013.
2. Fukuzumi, S.; Ohkubo, K. Quantitative evaluation of Lewis acidity of metal ions derived from the g values of ESR spectra of superoxide: Metal ion complexes in relation to the promoting effects in electron transfer reactions. *Chem.—A Eur J* **2000**, *6*, 4532–4535. [CrossRef]
3. Matsuda, K.; Takayama, K.; Irie, M. Photochromism of metal complexes composed of diarylethene ligands and Zn(II), Mn(II), and Cu(II) hexafluoroacetylacetonates. *Inorg. Chem.* **2004**, *43*, 482–489. [CrossRef] [PubMed]
4. Raman, N.; Dhaveethu Raja, J.; Sakthivel, A. Synthesis, spectral characterization of Schiff base transition metal complexes: DNA cleavage and antimicrobial activity studies. *J. Chem. Sci.* **2007**, *119*, 303–310. [CrossRef]
5. Lund, A.; Shiotani, M.; Shimada, S. *Principles and Applications of ESR Spectroscopy*; Springer Science & Business Media: New York, NY, USA, 2011.
6. Kohno, M. Applications of electron spin resonance spectrometry for reactive oxygen species and reactive nitrogen species research. *J. Clin. Biochem. Nutr.* **2010**, *47*, 1–11. [CrossRef] [PubMed]
7. Khramtsov, V.V.; Volodarsky, L.B. Use of imidazoline nitroxides in studies of chemical reactions ESR measurements of the concentration and reactivity of protons, thiols, and nitric oxide. In *Biological Magnetic Resonance*; Springer: Berlin/Heidelberg, Germany, 2002; pp. 109–180.
8. Lazreg, F.; Nahra, F.; Cazin, C.S. Copper–NHC complexes in catalysis. *Coord. Chem. Rev.* **2015**, *293*, 48–79. [CrossRef]
9. Ali, A.; Prakash, D.; Majumder, P.; Ghosh, S.; Dutta, A. Flexible Ligand in a Molecular Cu Electrocatalyst Unfurls Bidirectional O2/H2O Conversion in Water. *ACS Catal.* **2021**, *11*, 5934–5941. [CrossRef]
10. Chen, Z.; Meyer, T.J. Copper (II) catalysis of water oxidation. *Angew. Chem. Int. Ed.* **2013**, *52*, 700–703. [CrossRef]
11. Weng, Z.; Wu, Y.; Wang, M.; Jiang, J.; Yang, K.; Huo, S.; Wang, X.F.; Ma, Q.; Brudvig, G.W.; Batista, V.S.; et al. Active sites of copper-complex catalytic materials for electrochemical carbon dioxide reduction. *Nat. Commun.* **2018**, *9*, 415. [CrossRef]
12. Bolm, C.; Martin, M.; Gescheidt, G.; Palivan, C.; Neshchadin, D.; Bertagnolli, H.; Feth, M.; Schweiger, A.; Mitrikas, G.; Harmer, J. Spectroscopic investigations of bis (sulfoximine) copper (II) complexes and their relevance in asymmetric catalysis. *J. Am. Chem. Soc.* **2003**, *125*, 6222–6227. [CrossRef]
13. Bonke, S.A.; Risse, T.; Schnegg, A.; Brückner, A. In situ electron paramagnetic resonance spectroscopy for catalysis. *Nat. Rev. Methods Prim.* **2021**, *1*, 33. [CrossRef]
14. Sánchez-Moreno, C. Methods used to evaluate the free radical scavenging activity in foods and biological systems. *Food Sci. Technol. Int.* **2002**, *8*, 121–137. [CrossRef]

15. Okano, H. Effects of static magnetic fields in biology: Role of free radicals. *Front. Biosci.-Landmark* **2008**, *13*, 6106–6125. [CrossRef]
16. Yin, H.; Xu, L.; Porter, N.A. Free radical lipid peroxidation: Mechanisms and analysis. *Chem. Rev.* **2011**, *111*, 5944–5972. [CrossRef]
17. Eaton, G.R.; Eaton, S.S.; Barr, D.P.; Weber, R.T. *Quantitative EPR*; Springer Science & Business Media: New York, NY, USA, 2010.
18. Murphy, D.M.; Farley, R.D. Principles and applications of ENDOR spectroscopy for structure determination in solution and disordered matrices. *Chem. Soc. Rev.* **2006**, *35*, 249–268. [CrossRef] [PubMed]
19. Golombek, A.P.; Hendrich, M.P. Quantitative analysis of dinuclear manganese (II) EPR spectra. *J. Magn. Reson.* **2003**, *165*, 33–48. [CrossRef] [PubMed]
20. Drew, S.C.; Young, C.G.; Hanson, G.R. A density functional study of the electronic structure and spin hamiltonian parameters of mononuclear thiomolybdenyl complexes. *Inorg. Chem.* **2007**, *46*, 2388–2397. [CrossRef]
21. Cox, N.; Jin, L.; Jaszewski, A.; Smith, P.J.; Krausz, E.; Rutherford, A.W.; Pace, R. The semiquinone-iron complex of photosystem II: Structural insights from ESR and theoretical simulation; evidence that the native ligand to the non-heme iron is carbonate. *Biophys. J.* **2009**, *97*, 2024–2033. [CrossRef]
22. Trukhan, S.N.; Yakushkin, S.S.; Martyanov, O.N. Fine-tuning simulation of the ESR spectrum—Sensitive tool to identify the local environment of asphaltenes in situ. *J. Phys. Chem. C* **2022**, *126*, 10729–10741. [CrossRef]
23. Stoll, S.; Schweiger, A. EasySpin, a comprehensive software package for spectral simulation and analysis in EPR. *J. Magn. Reson.* **2006**, *178*, 42–55. [CrossRef]
24. Khairy, K.; Budil, D.; Fajer, P. Nonlinear-least-squares analysis of slow motional regime EPR spectra. *J. Magn. Reson.* **2006**, *183*, 152–159. [CrossRef]
25. Srivastava, M.; Dzikovski, B.; Freed, J.H. Extraction of Weak Spectroscopic Signals with High Fidelity: Examples from ESR. *J. Phys. Chem. A* **2021**, *125*, 4480–4487. [CrossRef]
26. Roy, A.S.; Srivastava, M. Hyperfine Decoupling of ESR Spectra Using Wavelet Transform. *Magnetochemistry* **2022**, *8*, 32. [CrossRef]
27. Sinha Roy, A.; Srivastava, M. Analysis of Small-Molecule Mixtures by Super-Resolved 1H NMR Spectroscopy. *J. Phys. Chem. A* **2022**, *126*, 9108–9113. [CrossRef]
28. Sinha Roy, A.; Srivastava, M. Unsupervised Analysis of Small Molecule Mixtures by Wavelet-Based Super-Resolved NMR. *Molecules* **2023**, *28*, 792. [CrossRef] [PubMed]
29. Srivastava, M.; Anderson, C.L.; Freed, J.H. A New Wavelet Denoising Method for Selecting Decomposition Levels and Noise Thresholds. *IEEE Access* **2016**, *4*, 3862–3877. [CrossRef]
30. Addison, P. *The Illustrated Wavelet Transform Handbook: Introductory Theory and Applications in Science, Engineering, Medicine and Finance*, 2nd ed.; CRC Press: London, UK, 2016.
31. Wang, Q.; Johnson, J.L.; Agar, N.Y.; Agar, J.N. Protein aggregation and protein instability govern familial amyotrophic lateral sclerosis patient survival. *PLoS Biol.* **2008**, *6*, e170. [CrossRef]
32. Pratt, A.J.; Shin, D.S.; Merz, G.E.; Rambo, R.P.; Lancaster, W.A.; Dyer, K.N.; Borbat, P.P.; Poole, F.L.; Adams, M.W.; Freed, J.H.; et al. Aggregation propensities of superoxide dismutase G93 hotspot mutants mirror ALS clinical phenotypes. *Proc. Natl. Acad. Sci. USA* **2014**, *111*, E4568–E4576. [CrossRef] [PubMed]
33. Makhlynets, O.V.; Gosavi, P.M.; Korendovych, I.V. Short Self-Assembling Peptides Are Able to Bind to Copper and Activate Oxygen. *Angew. Chem. Int. Ed.* **2016**, *55*, 9017–9020. [CrossRef] [PubMed]
34. Merz, G.E.; Borbat, P.P.; Pratt, A.J.; Getzoff, E.D.; Freed, J.H.; Crane, B.R. Copper-based pulsed dipolar ESR spectroscopy as a probe of protein conformation linked to disease states. *Biophys. J.* **2014**, *107*, 1669–1674. [CrossRef]
35. Ernst, R.; Anderson, W. Sensitivity enhancement in magnetic resonance. II. Investigation of intermediate passage conditions. *Rev. Sci. Instrum.* **1965**, *36*, 1696–1706. [CrossRef]
36. Portis, A. Rapid passage effects in electron spin resonance. *Phys. Rev.* **1955**, *100*, 1219. [CrossRef]
37. Noël, S.; Perez, F.; Pedersen, J.T.; Alies, B.; Ladeira, S.; Sayen, S.; Guillon, E.; Gras, E.; Hureau, C. A new water-soluble Cu(II) chelator that retrieves Cu from Cu (amyloid-β) species, stops associated ROS production and prevents Cu (II)-induced Aβ aggregation. *J. Inorg. Biochem.* **2012**, *117*, 322–325. [CrossRef] [PubMed]
38. Bunda, S.; May, N.V.; Bonczidai-Kelemen, D.; Udvardy, A.; Ching, H.V.; Nys, K.; Samanipour, M.; Van Doorslaer, S.; Joo, F.; Lihi, N. Copper (II) complexes of sulfonated salan ligands: Thermodynamic and spectroscopic features and applications for catalysis of the Henry reaction. *Inorg. Chem.* **2021**, *60*, 11259–11272. [CrossRef] [PubMed]

Disclaimer/Publisher's Note: The statements, opinions and data contained in all publications are solely those of the individual author(s) and contributor(s) and not of MDPI and/or the editor(s). MDPI and/or the editor(s) disclaim responsibility for any injury to people or property resulting from any ideas, methods, instructions or products referred to in the content.

Communication

A New Organic Conductor of Tetramethyltetraselenafulvalene (TMTSF) with a Magnetic Dy(III) Complex

Qingyun Wan [1,2,*], Masanori Wakizaka [1,*], Haitao Zhang [1], Yongbing Shen [3], Nobuto Funakoshi [1], Chi-Ming Che [2], Shinya Takaishi [1,*] and Masahiro Yamashita [1,4,*]

[1] Department of Chemistry, Graduate School of Science, Tohoku University, Aramaki-Aza-Aoba, Aoba-ku, Sendai 980-8578, Japan
[2] Department of Chemistry, State Key Laboratory of Synthetic Chemistry, HKU-CAS Joint Laboratory on New Materials, The University of Hong Kong, Pokfulam Road, Hong Kong, China
[3] Frontier Institute of Science and Technology (FIST), State Key Laboratory for Mechanical Behavior of Materials, MOE Key Laboratory for Nonequilibrium Synthesis of Condensed Matter, Xi'an Jiaotong University, 99 Yanxiang Road, Xi'an 710054, China
[4] School of Materials Science and Engineering, Nankai University, Tianjin 300350, China
* Correspondence: qywan@connect.hku.hk (Q.W.); masanori.wakizaka.a7@tohoku.ac.jp (M.W.); shinya.takaishi.d8@tohoku.ac.jp (S.T.); yamasita@agnus.chem.tohoku.ac.jp (M.Y.); Tel.: +81-22-765-6547 (M.Y.)

Abstract: A new molecular conductor of $(TMTSF)_5[Dy(NCS)_4(NO_3)_2]CHCl_3$ was prepared using the electrochemical oxidation method. The complex crystallizes in the $Cmc2_1$ (36) space group, where the partially-oxidized TMTSF molecules form a 1D (one-dimensional) column structure. The crystal shows a semiconducting behavior with a room temperature conductivity of 0.2 S·cm^{-1} and an activation energy of 34 meV at ambient pressure.

Keywords: organic conductor; electro-crystallization; Dysprosium(III); TMTSF

1. Introduction

The TMTSF cation constituted the first organic superconductor of $(TMTSF)_2PF_6$ and was reported in 1980 [1], describing a quasi 1D charge-transfer salt system with a superconducting transition temperature (T_c) of 0.9 K under 12 kbar by the suppression of spin density wave (SDW) state. Since then, over 100 organic superconductors have been reported and studied [2–5]. For example, β'-(BEDT-TTF)$_2$ICl$_2$ [BEDT-TTF = bis(ethylenedithio)tetrathiafulvalene] was reported to have a high T_c of 14.2 K under 82 kbar [6], and its superconducting state is obtained under high pressure to suppress the antiferromagnetic Mott insulating state. Other organic conducting systems include potassium-doped para-terphenyl, which shows step-like transitions at about 125 K in the temperature dependent magnetization curve [7].

The search for new organic superconductors and conductors is still ongoing [8–17], and it is interesting to investigate the effects of 4f electrons on the conductivity properties of TMTSF molecules [18,19]. Our group has been working on functional molecular conductors and single-molecule magnets (SMMs) for a long time, reporting various hybrid systems by combining different conductors of TTF (tetrathiafulvalene), BEDT-TTF, M(dmit)$_2$ (dmit = 4,5-dimercapto-1,3-dithiole-2-dithione), and BEDO-TTF (bis(ethylenedioxy)tetrathiafulvalene) with different single-molecule magnets (SMMs) such as [Co(pdms)$_2$]$^{2-}$, [Dy(NCS)$_7$]$^{4-}$, [Mn$_2$]$^{2+}$ clusters, and so on [20–26]. The 4f electrons are well known to have large anisotropic magnetic moments due to strong spin-orbit coupling, which is distinct from 3d electrons. The use of a polyvalent 4f metal complex as a counter-anion also indicates a different degree of conduction band filling in the radical of TMTSF molecules, compared to that of monovalent anions such as PF$_6^-$, Cl$^-$, I$^-$, and so on [2]. Such a change in the filling in the conduction band may lead to new physical properties of the molecular conductors.

Herein, we used a polyvalent 4f metal complex of $[Dy(NCS)_4(NO_3)_2]^{3-}$ as the counter-anion to prepare a new quasi-1D magnetic molecular conductor of $(TMTSF)_5[Dy(NCS)_4(NO_3)_2]CHCl_3$ (**1**, Scheme 1) using an electro-crystallization method. Synthesis, crystal structure, conductivity, optical, magnetic properties, and band structure calculations of **1** have been investigated and discussed in the present work.

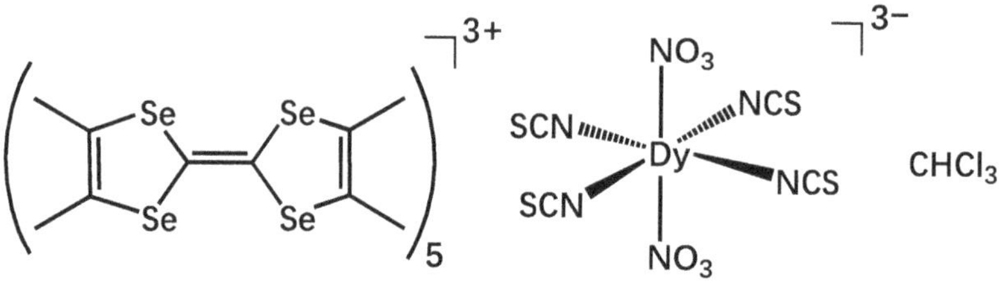

Scheme 1. Chemical structure of compound **1**.

2. Materials and Methods

2.1. Synthesis

TMTSF, $Dy(NO_3)_3 \cdot 6H_2O$, tetrabutylammonium (TBA) thiocyanate salts, and organic solvents were commercially purchased and used without any further purification. $(TBA)_3Dy(NCS)_4(NO_3)_2$ complexes were obtained by following reported procedures [27].

Crystals of **1** were synthesized using an electro-crystallization method of TMTSF (10 mg) and $(TBA)_3Dy(NCS)_4(NO_3)_2$ (80 mg) in $CHCl_3$ (12 mL), with an addition of EtOH (3 mL) on an ITO electrode under galvanostatic conditions (I = 0.5–2 µA) at 25 °C. The crystals of complex **1** grew for 2–4 days depending on the applied current as thin black needles of different sizes.

2.2. Physical Measurements

We measured the temperature-dependent resistivity of compound **1** by using a Quantum Design PPMS 6000 (Quantum Design, San Diego, CA, USA) and Keithley 2611 System Source Meter (Keithley Instruments, Solon, OH, USA). The four-probe method was used, and the measurement was performed under ambient pressure. Gold wires (30 µm diameter) were used to attach the crystal, and carbon paste was used as the electrode.

Single-crystal X-ray crystallographic measurements were performed by using a Rigaku Saturn 70 CCD Diffractometer at 120 K. Graphite-monochromated Mo Kα radiation (λ = 0.71073 Å) was generated by a VariMax microfocus X-ray rotating anode source. We used the CrystalClear crystallographic software package for data processing. The structures were solved and refined by using direct methods included in SIR-92 and SHELXL-2013, respectively [28–30]. The non-H atoms were refined anisotropically, and H atoms were refined by a riding model and were attached to the C atoms using idealized geometries.

We performed the magnetic measurements on compound **1** using MPMS3 (Quantum Design) in the direct current (dc) mode and the alternating current (ac) mode, respectively. We filled the powders sample of compound **1** into a gelatin capsule. Eicosane with a melting point of 310 K was used to fix the sample in a plastic straw.

2.3. Computational Methodology

The band structure of compound **1** was calculated by VASP (Vienna Ab initio Simulation Package) [31,32] using the Perdew-Burke-Ernzerhof (PBE) exchange-correlation functional [33] with a kinetic energy cutoff of 640 eV. PAW pseudopotentials were applied to describe the Dy, Se, C, N, H, O, and Cl atoms [34], where the f electrons of Dy are kept frozen in the core and described by the selected pseudopotentials. A $2 \times 4 \times 2$ Monkhorst-Pack k-mesh was employed for the self-consistent calculation to obtain a converged charge density for the further band structure calculation.

To estimate the charge transfer integrals between two TMTSF units, the energy-splitting-in-dimer (ESID) method was applied [35]. The wavefunction of the dimers was obtained using the Gaussian16 program package [36] under a PBE0/def2-TZVP level [37,38]. The tight convergence threshold (10^{-8} for the root mean square change in the density matrix) was used for the SCF procedure.

3. Results and Discussion
3.1. Crystal Structures

Compound (**1**) crystallized in the Cmc2$_1$ (36) space group with five TMTSF units, one [Dy(NCS)$_4$(NO$_3$)$_2$]$^{3-}$ unit, and one CHCl$_3$ molecule. TMTSF molecules form a quasi-1D π-π stacking column structure along the b axis as shown in Figure 1d. We checked the intermolecular π-π distance between neighboring TMTSF molecules in the 1D column. A small difference of intermolecular distance of 3.44(1) Å and 3.50(1) Å has been observed between each TMTSF molecule in the 1D column, suggesting a dimerization process of TMTSF molecules in the 1D column. Among five TMTSF molecules, four out of them (TMTSF-b1 and TMTSF-b2) were located in the 1D column, while one TMTSF (TMTSF-a) shows orthogonal (T-shaped) packing form with the 1D column structure (Figure 1a). The distance between the TMTSF-a and TMTSF-b2 molecules is 3.71 Å, and the close distance indicates a T-type packing interaction between these two molecules (Figure 1c) [39]. Along the c-axis, the layer is constituted by radical cations of TMTSF-a and [Dy(NCS)$_4$(NO$_3$)$_2$]$^{3-}$ units alternatively. A close distance of 4.95 Å between TMTSF-b1 and complex [Dy(NCS)$_4$(NO$_3$)$_2$]$^{3-}$ was observed (Figure 1c). Compound **1** shows rectangular cavities in its crystal structure (Figure 1c) with a size of 15.15 Å × 12.16 Å; they are built up by alternating TMTSF-a and [Dy(NCS)$_4$(NO$_3$)$_2$]$^{3-}$ units, and the cavities are occupied exclusively by TMTSF-b column.

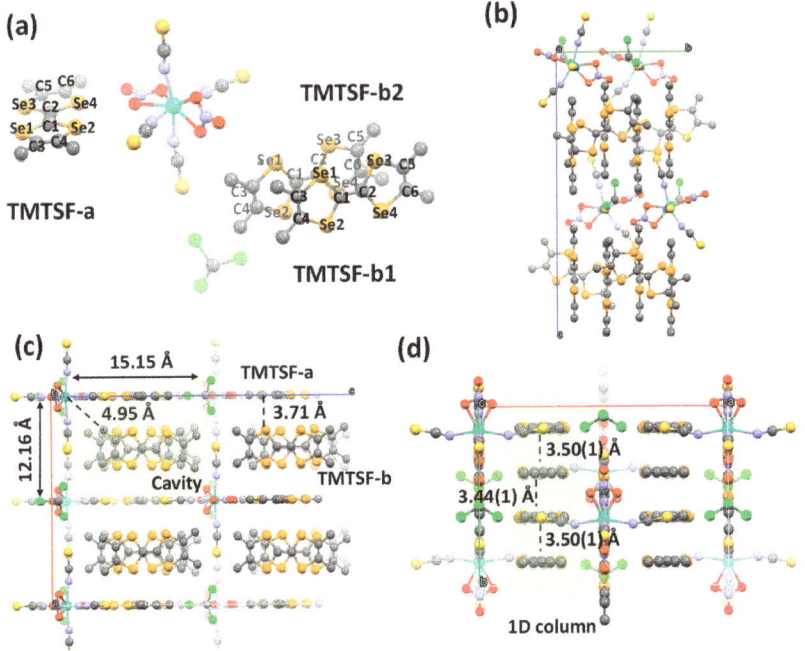

Figure 1. (**a**) X-ray crystal structure of **1**. Crystal packing along the (**b**) a-axis, (**c**) b-axis and (**d**) c-axis.

An examination of intramolecular C-C and C-Se distances in TMTSF-a and TMTSF-b molecules (Figure 1a) was conducted, and the results are summarized in Table 1. A closer C1-C2 is observed in the TMTSF-a molecule (1.31(2) Å) compared to that in TMTSF-b

(1.385(11) and 1.392(10) Å). The different intramolecular bond length indicates a different charge density for TMTSF-a and TMTSF-b molecules in compound **1**. We made a comparison of the C1-C2 distance between TMTSF molecules in compound **1** and other reported TMTSF-type organic conductors, and summarized the results in Table S2. The charge-neutral TMTSF molecule has a C1-C2 distance of 1.347 Å [40]. The TMTSF molecule with an average oxidation ranging from +0.5, to +2/3, to +1 has the C1-C2 distance from 1.430 Å to 1.316 Å [41–47], with no clear relationship between the C1-C2 distance and the charge density of the TMTSF molecule.

Table 1. Intramolecular bond length of TMTSF-a and TMTSF-b molecule in compound **1**.

Bond Length	TMTSF-a	TMTSF-b1	TMTSF-b2
C1-C2	1.31(2) Å	1.385(11) Å	1.392(10) Å
C1-Se1	1.915(13) Å	1.867(7) Å	1.865(7) Å
C1-Se2	1.923(18) Å	1.867(8) Å	1.879(8) Å
C2-Se3	1.917(12) Å	1.885(7) Å	1.871(8) Å
C2-Se4	1.919(19) Å	1.866(8) Å	1.877(7) Å
C3-C4	1.34(3) Å	1.336(13) Å	1.354(12) Å
C5-C6	1.39(2) Å	1.339(13) Å	1.368(13) Å

3.2. Conductivity Properties

Single-crystal temperature-dependent resistivity measurements were performed on compound **1** using the four-probe method along the b-axis of the crystal. The σ-T^{-1} relationship shows a semiconductive behavior in Figure 2a, based on a decreased resistivity upon increasing the temperature. Conductivity of **1** at room temperature (σ_{rt}) was determined to be 0.2 S·cm^{-1}. Analysis of the Ln(σ) versus T^{-1} plot shows a linear curve, as shown in Figure 2b. The curve was fitted using a linear function giving an activation energy (E_a) of 34 meV at ambient pressure, which is the energy difference between the transport level and the Fermi level of compound **1** [48]. The resistivity measurements were performed on another two crystals of compound **1**, giving similar σ_{rt} and E_a values (Figure S1).

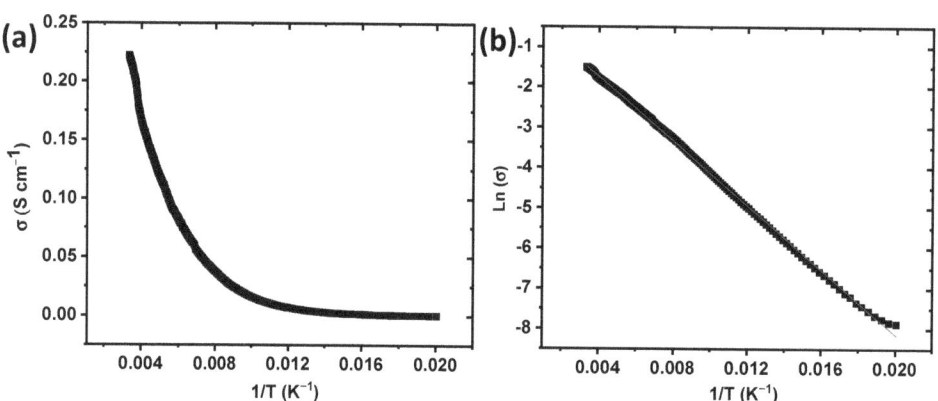

Figure 2. (**a**) Temperature dependence of σ (S·cm^{-1}) for a single crystal of **1**. (**b**) Ln(σ)-T^{-1} and its fitting curve to a linear function (red line).

Since [Dy(III)(NCS)$_4$(NO$_3$)$_2$]$^{3-}$ anion's charge is -3 and the unit cell contains five TMTSF molecules, the valence band made by TMTSF molecules should be partially filling. Such a partially-filled band structure usually leads to metallic behavior instead of semiconductivity [49]. We note that the semiconductive or insulating behavior also shows up in other organic conductors with a partially-filled band, such as (TMTSF)$_2$X (X = PF$_6$, AsF$_6$, SbF$_6$, TaF$_6$, NbF$_6$) where a metal-insulating transitions occurs at 11–17 K [2], and (BEDT-

TTF)$_2$X with a Mott transition near the superconducting state in its phase diagram [3]. The formal oxidation state of TMTSF and BEDT-TTF molecules in these organic conductors is +0.5, indicating a quarter filling of the valence band. The insulating or semiconductive behavior is due to a dimerization of TMTSF or BEDT-TTF molecules, making the charge density +1 per site (Figure 3a). Mott localization occurs subsequently, based on the Coulomb repulsion (U) between dimers, leading to an opening of the gap in their band structures (Figure 3c) [3,10]. In compound **1**, a similar dimerization was observed between two TMTSF-b molecules (Figure 1d). We conceive that two possibilities may lead to the non-metallic behaviors of compound **1**: (a) formation of a dimer-Mott state where an overall charge density of +1 may populate over the TMTSF-b dimer and a +1 charge is assigned for the TMTSF-a molecule (Figure 3a), and (b) formation of a charge-ordering state due to the charge disproportion for the TMTSF-b molecules, as shown in Figure 3b. Under these two conditions, the TMTSF-b dimer would form a half-filled band in the 1D column structure. The hopping integral (t) would be small due to a relatively long intermolecular distance of 3.50 Å between two TMTSF-b dimers. The Mott insulating phase shows up [3], leading to the semiconductive behavior of compound **1**.

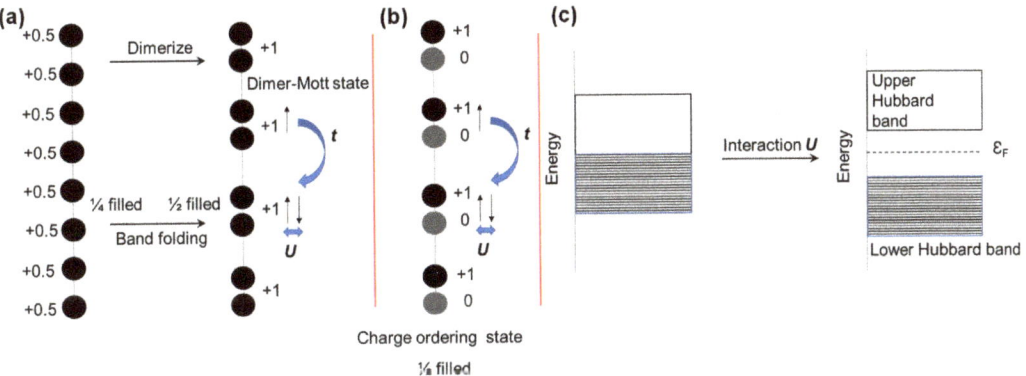

Figure 3. Schematic drawing of possible mechanisms to account for the non-metallic behaviors of compound **1**: the formation of a (**a**) dimer-Mott state, (**b**) a charge-ordering state, or (**c**) formation of upper Hubbard band and lower Hubbard band in the Mott-Hubbard insulators.

A comparison was made among compound **1** and other organic conductors containing 4f metal complexes regarding the conductivity properties. The reported (BEDT-TTF)$_5$[Ln(NCS)]$_6$ (Ln = Ho, Er, Yb and Y) compound has a semiconductive behavior with a large activation energy of ~1.5 eV and the resistivity of 4–6 Ω·cm at 280 K [50]. Two BEDT-TTF molecules co-exist in the crystal structure of (BEDT-TTF)$_5$[Ln(NCS)]$_6$; one is BEDT-TTF$^+$ and the other is BEDT-TTF$^{+0.5}$. The BEDT-TTF$^{+0.5}$ molecule forms a chain structure, and a dimerization of BEDT-TTF molecules leads to Mott localization and the semiconductive behavior of (BEDT-TTF)$_5$[Ln(NCS)]$_6$. The reported (BEDT-TTF)$_5$Dy(NCS)$_7$(KCl)$_{0.5}$ compound has a comparable value of σ_{rt} (1.7 S·cm^{-1}) with compound **1** [20]. (BEDT-TTF)$_5$Dy(NCS)$_6$(NO$_3$)C$_2$H$_5$OH compound is a semiconductor and has a smaller σ_{rt} of 0.01–0.1 S·cm^{-1} and 1–7 × 10^{-5} S·cm^{-1} along two axes of the crystal structure compared to **1** [18]. In the literature, room-temperature conductivities of (BEDT-TTF)$_2$[HoCl$_2$ (H$_2$O)$_6$]Cl$_2$(H$_2$O)$_2$, (BEDT-TTF)$_2$Ln'Cl$_4$(H$_2$O)$_n$(Ln' = Dy, Tb, Ho) crystals were measured and determined to be 0.004, 0.007, 0.0008, and 0.035 S/cm, respectively, with semiconductor behavior and an activation energy of conductivity of 220 meV, 300 meV, 320, and 290 meV, respectively [51]. Several factors can influence the conductivity properties of molecular conductors, including the degree of charge transfer, dimensionality, and conformation variations in the radical cations [3,18]. Moreover, these factors are considered to lead to the discrepancy of conductivity properties among these 4f-π organic conductors. A high degree

3.3. Optical Properties

To further investigate the electronic structure of **1**, a polarized IR (infrared) reflectance spectrum was recorded for the crystal of **1**. The excitation light was polarized along (// direction) and perpendicular (⊥ direction) to the long axis of the crystal, respectively. As shown in Figure 4, the intensity of the reflection spectrum recorded at the // direction is stronger than that along the ⊥ direction, indicating an anisotropic 1D electronic structure of compound **1**. A broad peak around 100 meV was observed in the reflectance spectrum, which is attributed to the existence of a small energy band gap of crystal **1** in its band structure.

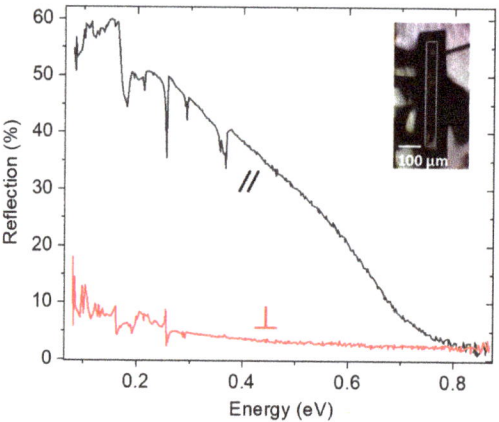

Figure 4. Polarized IR reflectance spectrum of crystal **1** at room temperature. Inset is the optical image of the crystal. Black line: excitation along the direction of the crystal. Red line: excitation perpendicular to the direction of the crystal.

3.4. Magnetic Properties

The magnetic field dependence of static normalized magnetization (M/M_s) was measured on polycrystalline samples of **1** at 1.8 K (Figure 5). The magnetization process is considered to be mostly contributed by a Dy(III) complex having a large magnetic moment ($J = 15/2$, $S = 5/2$, $L = 5$), with minor contributions from the TMTSF radical cations. No hysteresis was observed for compound **1** at 1.8 K (Figure 5a). The temperature dependence of the magnetization curves was simulated on the PHI program using the following spin Hamiltonian (Figure 5b) [52]:

$$\hat{\mathcal{H}}_{SO} = \lambda \hat{L} \cdot \hat{S} \qquad (1)$$

$$\hat{\mathcal{H}}_{ZEE} = \mu_B \hat{S} \cdot g_J \cdot \hat{B} \qquad (2)$$

$$\hat{\mathcal{H}}_{ZFS} = D\left\{\hat{S}_z^2 - \frac{1}{3}S(S+1)\right\} \qquad (3)$$

where λ, L, S, and B with hats, μ_B, g_J, D, and S refer to spin–orbit coupling constant, operators of orbit and spin, magnetic field, the Bohr magneton, g-factor for lanthanide, axial zero-field splitting (ZFS) constant, and total spin on the metal ion, respectively. The simulation curves were applied typical values of $g_J = 4/3$ and $\lambda = -360$ cm^{-1} [52,53]. Without the ZFS parameter, the simulation curves (dotted lines) do not match with the experiment plots. In contrast, applying $D = -0.9$ cm^{-1} matches well with the experimental plots, suggesting that Dy(III) centers in **1** have a small negative D term. A negative D term is necessary for SMMs with uniaxial anisotropy. However, D of -0.9 cm^{-1} is too small to induce hysteresis. As expected, the dynamic susceptibility exhibited no significant signals at 1.8 K in the measurement range in Figure S2. The susceptibility of the out-of-phase component (χ'') rises at a higher frequency region in Figure S2, suggesting that the peak would be out of measurement range (>1000 Hz) and a very fast magnetic relaxation of SMMs of compound **1**. Among the reported 4f-π system, SMM behavior was observed in the (BEDT-TTF)$_5$Dy(NCS)$_7$(KCl)$_{0.5}$ system [20], while it is absent in (BEDT-TTF)$_2$[HoCl$_2$(H$_2$O)$_6$]Cl$_2$(H$_2$O)$_2$ and (BEDT-TTF)$_2$Ln'Cl$_4$(H$_2$O)$_n$(Ln' = Dy, Tb, Ho) compounds [51]. Molecular symmetry is found to be closely related to SMM properties [54], and is considered to be a probable reason for the existence of strong and weak SMM properties in (BEDT-TTF)$_5$Dy(NCS)$_7$(KCl)$_{0.5}$ and compound **1**, respectively, as well as the absence of SMM properties in (BEDT-TTF)$_2$[HoCl$_2$(H$_2$O)$_6$]Cl$_2$(H$_2$O)$_2$ and (BEDT-TTF)$_2$Ln'Cl$_4$(H$_2$O)$_n$ compounds.

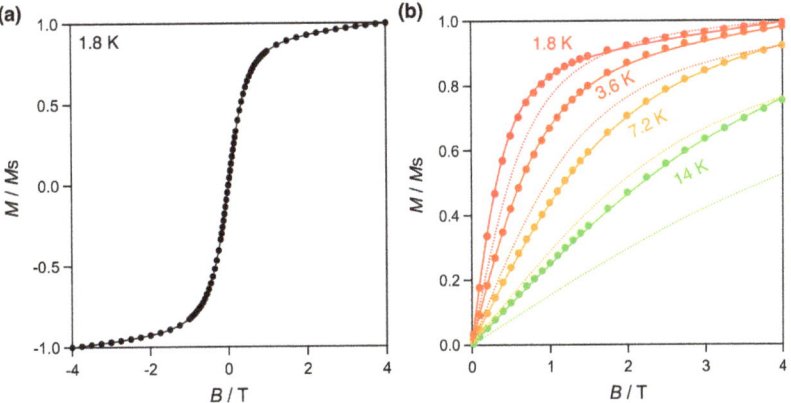

Figure 5. Magnetic field-dependent magnetization of compound **1**. (**a**) Loop at 1.8 K and (**b**) curves at 1.8 K (red line), 3.6 K (orange line), 7.2 K (yellow line), and 14 K (lime-green line). The solid-colored lines show the simulation curves using $S = 5/2$, $L = 5$, $g_J = 4/3$, $\lambda = -360$ cm^{-1}, $D = -0.9$ cm^{-1}, whereas the dotted lines are fitting curves without the D term.

We further examined the magnetoresistance (MR) of compound **1** and the results are shown in Figure 6. Negative MR was observed for compound **1**, where the MR approaches −3.5% under a magnetic field of 9 T at 30 K.

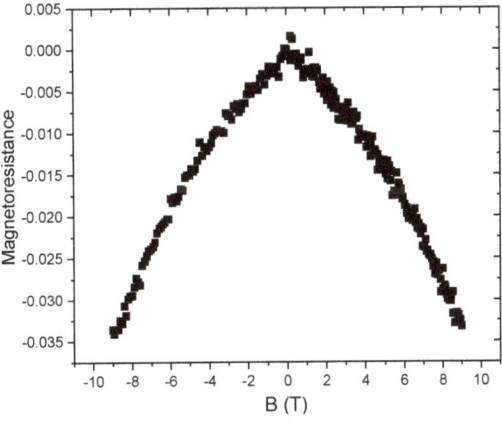

Figure 6. Magnetoresistance of compound **1** under a sweet rate of 20 Oe s^{-1} at the temperature of 30 K.

3.5. Band Structure Calculations

To understand the high conductivity of compound **1** (0.2 S·cm^{-1}), we calculated its band structure, which was shown in Figure 7. The Fermi level was calculated to be at ~1.47 eV which crosses the bands at XS, YG, UR, and TZ directions. The band dispersion is much more significant along the *b* direction compared to that of other two directions, which indicates that the TMTSF units have a stronger orbital overlap along the *b* axis than the other two directions. The partially-filled bands are consistent with the fact that TMTSF molecules in compound **1** are partially oxidized. Notably, the PBE functional lacks accuracy to describe a localized electronic structure; hence, the on-site Coulomb repulsion calculations have not been involved and considered here to describe the Mott-insulating characteristics of compound **1**. We further calculated the charge transfer integral (*t*) between the TMTSF dimer of compound **1** and summarized the results in Figure S3. Larger *t* was calculated and observed for TMTSF-b molecules along the 1D column *b* direction, compared to that of TMTSF-a molecules, indicating the anisotropic 1D nature of the crystal **1**.

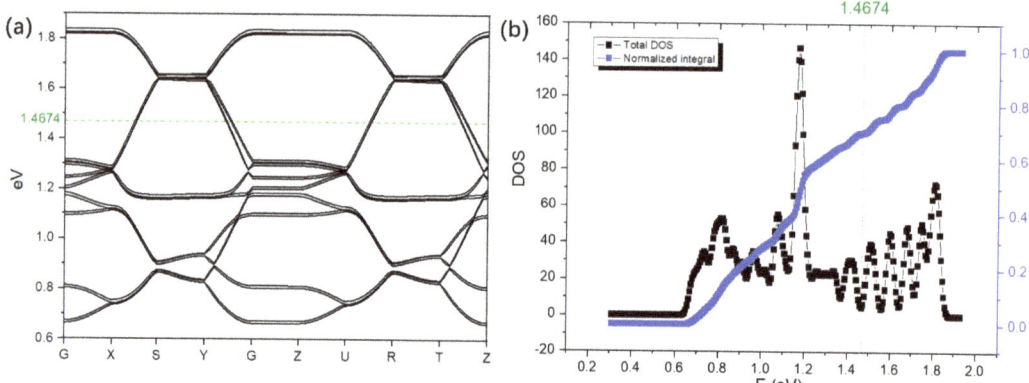

Figure 7. (**a**) The band structure of compound **1**. The corresponding Fermi level is represented by the green dash line. G (0, 0, 0); X (0.5, 0, 0); Y (0, 0.5, 0); Z (0, 0, 0.5); S (0.5, 0.5, 0); U (0.5, 0, 0.5); T (0, 0.5, 0.5); R (0.5, 0.5, 0.5). (**b**) Density of states (DOS) of compound **1**.

4. Conclusions

A new organic conductor (**1**) composed of a 1D cationic TMTSF column and 4f metal complexes of [Dy(III)(NCS)$_4$(NO$_3$)$_2$]$^{3-}$ has been prepared. Its conductivity at room temperature was determined to be 0.2 S·cm^{-1} with an activation energy of 34 meV. This preliminary study provides information for designing new hybrid materials based on molecular conductors and polyvalent magnetic 4f metal complexes.

Supplementary Materials: The following supporting information can be downloaded at: https://www.mdpi.com/article/10.3390/magnetochemistry9030077/s1, Figure S1: (a and c) Temperature dependence of σ (S·cm^{-1}) for two single crystals of **1**. (b and d) Ln(σ)-T^{-1} and its fitting curve to a linear function (red line) of panel a and c, respectively; Figure S2: Frequency dependence of (a) the in-phase and (b) the out-of-phase magnetic susceptibility at 1.8 K as a function of the magnetic field of compound **1**; Figure S3: Calculation of charge transfer integral of *t(hole)* and *t(electron)* in the TMTSF dimers of compound **1**; Table S1: Summary of the crystal data of compound **1**; Table S2: A summary of C1-C2 distance in TMTSF-type molecules.

Author Contributions: M.Y. conceived and designed the experiments; M.W., Q.W. and S.T. performed the experiments; H.Z. performed the calculations; M.W., Q.W., M.Y., N.F., Y.S., S.T. and C.-M.C. analyzed the data; Q.W. wrote the paper. All authors have read and agreed to the published version of the manuscript.

Funding: This work was supported by the JSPS KAKENHI Grant Number JP19H05631 and the National Natural Science Foundation of China (NSFC, 22150710513). M.Y. Thanks the 111 projects (B18030) from China.

Acknowledgments: Thanks all the projects from China and Japan.

Conflicts of Interest: The authors declare no conflict of interest.

References

1. Jérome, D.; Mazaud, A.; Ribault, M.; Bechgaard, K. Superconductivity in a synthetic organic conductor (TMTSF)$_2$PF$_6$. *J. Phys. Lett.* **1980**, *41*, 95–98. [CrossRef]
2. Mori, H. Materials viewpoint of organic superconductors. *J. Phys. Soc. Jpn.* **2006**, *75*, 051003. [CrossRef]
3. Ardavan, A.; Brown, S.; Kagoshima, S.; Kanoda, K.; Kuroki, K.; Mori, H.; Ogata, M.; Uji, S.; Wosnitza, J. Recent topics of organic superconductors. *J. Phys. Soc. Jpn.* **2011**, *81*, 011004. [CrossRef]
4. Williams, J.M.; Schultz, A.J.; Geiser, U.; Carlson, K.D.; Kini, A.M.; Wang, H.G.; Kwok, W.-K.; Whangbo, M.-H.; Schirber, J.E. Organic superconductors—New benchmarks. *Science* **1991**, *252*, 1501–1508. [CrossRef]
5. Enoki, T.; Miyazaki, A. Magnetic TTF-based charge-transfer complexes. *Chem. Rev.* **2004**, *104*, 5449–5478. [CrossRef]
6. Taniguchi, H.; Miyashita, M.; Uchiyama, K.; Satoh, K.; Môri, N.; Okamoto, H.; Miyagawa, K.; Kanoda, K.; Hedo, M.; Uwatoko, Y. Superconductivity at 14.2 K in layered organics under extreme pressure. *J. Phys. Soc. Jpn.* **2003**, *72*, 468–471. [CrossRef]
7. Liu, W.; Lin, H.; Kang, R.; Zhu, X.; Zhang, Y.; Zheng, S.; Wen, H.-H. Magnetization of potassium-doped p-terphenyl and p-quaterphenyl by high-pressure synthesis. *Phys. Rev. B* **2017**, *96*, 224501. [CrossRef]
8. Kobayashi, H.; Kobayashi, A.; Cassoux, P. BETS as a source of molecular magnetic superconductors (BETS = bis (ethylenedithio) tetraselenafulvalene). *Chem. Soc. Rev.* **2000**, *29*, 325–333. [CrossRef]
9. Enomoto, M.; Miyazaki, A.; Enoki, T. Magnetic Properties of (C1TEX-TTF) FeBr$_4$ (X=S, Se). *Mol. Cryst. Liq. Cryst. Sci. Technol. Sect. A Mol. Cryst. Liq. Cryst.* **1999**, *335*, 293–302. [CrossRef]
10. Anderson, P.; Lee, P.; Saitoh, M. Remarks on giant conductivity in TTF-TCNQ. *Solid State Commun.* **1973**, *13*, 595–598. [CrossRef]
11. Cui, H.; Otsuka, T.; Kobayashi, A.; Takeda, N.; Ishikawa, M.; Misaki, Y.; Kobayashi, H. Structural, electrical, and magnetic properties of a series of molecular conductors based on BDT-TTP and lanthanoid nitrate complex anions (BDT-TTP = 2, 5-bis (1,3-dithiol-2-ylidene)-1,3,4,6-tetrathiapentalene). *Inorg. Chem.* **2003**, *42*, 6114–6122. [CrossRef]
12. Kushch, N.; Bardin, A.; Buravov, L.; Glushakova, N.; Shilov, G.; Dmitriev, A.; Morgunov, R.; Kulikov, A. Synthesis particularities, structure and properties of the radical cation salts ω-(BEDT-TTF)$_5$M (SCN)$_6$·C$_2$H$_5$OH, M=Mn, Ni. *Synth. Met.* **2014**, *195*, 75–82. [CrossRef]
13. Kushch, N.D.; Buravov, L.I.; Kushch, P.P.; Shilov, G.V.; Yamochi, H.; Ishikawa, M.; Otsuka, A.; Shakin, A.A.; Maximova, O.V.; Volkova, O.S. Multifunctional compound combining conductivity and single-molecule magnetism in the same temperature range. *Inorg. Chem.* **2018**, *57*, 2386–2389. [CrossRef] [PubMed]
14. Zhang, X.; Xie, H.; Ballesteros-Rivas, M.; Woods, T.J.; Dunbar, K.R. Conducting Molecular Nanomagnet of DyIII with Partially Charged TCNQ Radicals. *Chem.–A Eur. J.* **2017**, *23*, 7448–7452. [CrossRef] [PubMed]
15. Sato, T.; Breedlove, B.K.; Yamashita, M.; Katoh, K. Electro-Conductive Single-Molecule Magnet Composed of a Dysprosium (III)-Phthalocyaninato Double-Decker Complex with Magnetoresistance. *Angew. Chem. Int. Ed.* **2021**, *60*, 21179–21183. [CrossRef]
16. Mroweh, N.; Mézière, C.; Pop, F.; Auban-Senzier, P.; Alemany, P.; Canadell, E.; Avarvari, N. In Search of Chiral Molecular Superconductors: κ-[(S, S)-DM-BEDT-TTF]$_2$ClO$_4$ Revisited. *Adv. Mater.* **2020**, *32*, 2002811. [CrossRef]
17. Pop, F.; Auban-Senzier, P.; Canadell, E.; Avarvari, N. Anion size control of the packing in the metallic versus semiconducting chiral radical cation salts (DM-EDT-TTF)$_2$XF$_6$ (X=P, As, Sb). *Chem. Commun.* **2016**, *52*, 12438–12441. [CrossRef]
18. Kushch, N.; Kazheva, O.; Gritsenko, V.; Buravov, L.; Van, K.; Dyachenko, O. Novel packing type of ET radical cation layers in a new organic conductor (ET)$_5$[Dy(NCS)$_6$NO$_3$]·C$_2$H$_5$OH with a metal-complex lanthanide anion. *Synth. Met.* **2001**, *123*, 171–177. [CrossRef]
19. Shvachko, Y.N.; Starichenko, D.; Korolyov, A.; Kushch, N. Temperature evolution of BEDT-TTF$^{+1/2}$ and Dy^{3+} spin systems in novel organic conductor (BEDT-TTF)$_2$Dy(NO$_3$)$_4$: EPR and SQUID studies. *Synth. Met.* **2008**, *158*, 315–319. [CrossRef]
20. Shen, Y.; Cosquer, G.; Zhang, H.; Breedlove, B.K.; Cui, M.; Yamashita, M. 4f-π Molecular Hybrid Exhibiting Rich Conductive Phases and Slow Relaxation of Magnetization. *J. Am. Chem. Soc.* **2021**, *143*, 9543–9550. [CrossRef]
21. Shen, Y.; Ito, H.; Zhang, H.; Yamochi, H.; Cosquer, G.; Herrmann, C.; Ina, T.; Yoshina, S.K.; Breedlove, B.K.; Otsuka, A. Emergence of metallic conduction and cobalt (II)-based single-molecule magnetism in the same temperature range. *J. Am. Chem. Soc.* **2021**, *143*, 4891–4895. [CrossRef]
22. Shen, Y.; Cosquer, G.; Ito, H.; Izuogu, D.C.; Thom, A.J.; Ina, T.; Uruga, T.; Yoshida, T.; Takaishi, S.; Breedlove, B.K. An Organic-Inorganic Hybrid Exhibiting Electrical Conduction and Single-Ion Magnetism. *Angew. Chem.* **2020**, *132*, 2420–2427. [CrossRef]
23. Hiraga, H.; Miyasaka, H.; Nakata, K.; Kajiwara, T.; Takaishi, S.; Oshima, Y.; Nojiri, H.; Yamashita, M. Hybrid molecular material exhibiting single-molecule magnet behavior and molecular conductivity. *Inorg. Chem.* **2007**, *46*, 9661–9671. [CrossRef]
24. Hiraga, H.; Miyasaka, H.; Takaishi, S.; Kajiwara, T.; Yamashita, M. Hybridized complexes of [MnIII2] single-molecule magnets and Ni dithiolate complexes. *Inorg. Chim. Acta* **2008**, *361*, 3863–3872. [CrossRef]

25. Cosquer, G.; Shen, Y.; Almeida, M.; Yamashita, M. Conducting single-molecule magnet materials. *Dalton Trans.* **2018**, *47*, 7616–7627. [CrossRef] [PubMed]
26. Yamashita, M. Next generation multifunctional nano-science of advanced metal complexes with quantum effect and nonlinearity. *Bull. Chem. Soc. Jpn.* **2021**, *94*, 209–264. [CrossRef]
27. Mullica, D.F.; Bonilla, B.M.; David, M.a.C.; Farmer, J.M.; Kautz, J.A. Synthesis, characterization, and structural analyses of three high-coordination tetra-n-butylammonium lanthanide (III) complexes. *Inorg. Chim. Acta* **1999**, *292*, 137–143. [CrossRef]
28. Altomare, A.; Burla, M.C.; Camalli, M.; Cascarano, G.L.; Giacovazzo, C.; Guagliardi, A.; Moliterni, A.G.; Polidori, G.; Spagna, R. SIR97: A new tool for crystal structure determination and refinement. *J. Appl. Crystallogr.* **1999**, *32*, 115–119. [CrossRef]
29. Farrugia, L.J. WinGX and ORTEP for Windows: An update. *J. Appl. Crystallogr.* **2012**, *45*, 849–854. [CrossRef]
30. Sheldrick, G.M. Crystal structure refinement with SHELXL. *Acta Crystallogr. Sect. C Struct. Chem.* **2015**, *71*, 3–8. [CrossRef]
31. Kresse, G.; Furthmüller, J. Efficiency of ab-initio total energy calculations for metals and semiconductors using a plane-wave basis set. *Comput. Mater. Sci.* **1996**, *6*, 15–50. [CrossRef]
32. Kresse, G.; Furthmüller, J. Efficient iterative schemes for ab initio total-energy calculations using a plane-wave basis set. *Phys. Rev. B* **1996**, *54*, 11169. [CrossRef] [PubMed]
33. Perdew, J.P.; Burke, K.; Ernzerhof, M. Generalized gradient approximation made simple. *Phys. Rev. Lett.* **1996**, *77*, 3865. [CrossRef] [PubMed]
34. Kresse, G.; Joubert, D. From ultrasoft pseudopotentials to the projector augmented-wave method. *Phys. Rev. B* **1999**, *59*, 1758. [CrossRef]
35. Valeev, E.F.; Coropceanu, V.; da Silva Filho, D.A.; Salman, S.; Brédas, J.-L. Effect of electronic polarization on charge-transport parameters in molecular organic semiconductors. *J. Am. Chem. Soc.* **2006**, *128*, 9882–9886. [CrossRef]
36. Frisch, M.; Trucks, G.; Schlegel, H.; Scuseria, G.; Robb, M.; Cheeseman, J.; Scalmani, G.; Barone, V.; Mennucci, B.; Petersson, G. G09; Revison D. 01; Gaussian Inc.: Wallingford, CT, USA, 2010.
37. Adamo, C.; Barone, V. Toward reliable density functional methods without adjustable parameters: The PBE0 model. *J. Chem. Phys.* **1999**, *110*, 6158–6170. [CrossRef]
38. Weigend, F.; Ahlrichs, R. Balanced basis sets of split valence, triple zeta valence and quadruple zeta valence quality for H to Rn: Design and assessment of accuracy. *Phys. Chem. Chem. Phys.* **2005**, *7*, 3297–3305. [CrossRef]
39. Grimme, S. Do special noncovalent π–π stacking interactions really exist? *Angew. Chem. Int. Ed.* **2008**, *47*, 3430–3434. [CrossRef]
40. Rani, P.; Rajput, G.; Srivastava, M.; Yadav, R. Structural and vibrational characteristics and vibronic coupling of tetramethyltetraselenafulvalene. *J. Mol. Struct.* **2019**, *1175*, 1–12. [CrossRef]
41. Wu, L.; Coppens, P.; Bu, X. Crystal structure of tetramethyltetraselenafulvalene nitrate,($C_{10}H_{12}Se_4$)NO_3. *Z. Krist.-New Cryst. Struct.* **1997**, *212*, 101–102. [CrossRef]
42. Rosokha, S.V.; Stern, C.L.; Ritzert, J.T. π-Bonded molecular wires: Self-assembly of mixed-valence cation-radical stacks within the nanochannels formed by inert tetrakis [3,5-bis (trifluoromethyl) phenyl] borate anions. *CrystEngComm* **2013**, *15*, 10638–10647. [CrossRef]
43. Sakata, M.; Yoshida, Y.; Maesato, M.; Saito, G.; Matsumoto, K.; Hagiwara, R. Preparation of superconducting (TMTSF)$_2$NbF$_6$ by electrooxidation of TMTSF using ionic liquid as electrolyte. *Mol. Cryst. Liq. Cryst.* **2006**, *452*, 103–112. [CrossRef]
44. Wudl, F. Three-dimensional structure of the superconductor (TMTSF)$_2$AsF$_6$ and the spin-charge separation hypothesis. *J. Am. Chem. Soc.* **1981**, *103*, 7064–7069. [CrossRef]
45. Beno, M.; Blackman, G.; Williams, J.M.; Bechgaard, K. Synthetic metals based on tetramethyltetraselenafulvalene (TMTSF): Synthesis, structure (T = 298 and 125 K), and novel properties of (TMTSF)$_2$H$_2$F$_3$. *Inorg. Chem.* **1982**, *21*, 3860–3862. [CrossRef]
46. Emge, T.J.; Beno, M.A.; Daws, C.A.; Wang, H.H.; Williams, J.M. Novel Structural Features, and their Relationship to the Electrical Properties, of the Organic Conductor (TMTSF)$_2$NO$_3$ at 298 K and 125 K. *Mol. Cryst. Liq. Cryst.* **1984**, *116*, 153–171. [CrossRef]
47. Kazheva, O.N.; Kushch, N.D.; Dyachenko, O.A.; Canadell, E. Rare-earth elements in molecular conductors: Crystal and electronic structures. *J. Solid State Chem.* **2002**, *168*, 457–463. [CrossRef]
48. Mao, L.-F.; Ning, H.; Hu, C.; Lu, Z.; Wang, G. Physical modeling of activation energy in organic semiconductor devices based on energy and momentum conservations. *Sci. Rep.* **2016**, *6*, 24777. [CrossRef] [PubMed]
49. Friedel, J.; Jérome, D. Organic superconductors: The (TMTSF)$_2$X family. *Contemp. Phys.* **1982**, *23*, 583–624. [CrossRef]
50. Tamura, M.; Matsuzaki, F.; Nishio, Y.; Kajita, K.; Kitazawa, T.; Mori, H.; Tanaka, S. Novel BEDT-TTF salts containing rare earth ions, (ET)$_4$Ln (NCS)$_6$·CH$_2$Cl$_2$. *Synth. Met.* **1999**, *102*, 1716–1717. [CrossRef]
51. Flakina, A.M.; Zhilyaeva, E.I.; Shilov, G.V.; Faraonov, M.A.; Torunova, S.A.; Konarev, D.V. Layered Organic Conductors Based on BEDT-TTF and Ho, Dy, Tb Chlorides. *Magnetochemistry* **2022**, *8*, 142. [CrossRef]
52. Chilton, N.F.; Anderson, R.P.; Turner, L.D.; Soncini, A.; Murray, K.S. PHI: A powerful new program for the analysis of anisotropic monomeric and exchange-coupled polynuclear d- and f-block complexes. *J. Comput. Chem.* **2013**, *34*, 1164–1175. [CrossRef] [PubMed]

53. Feltham, H.L.; Brooker, S. Review of purely 4f and mixed-metal nd-4f single-molecule magnets containing only one lanthanide ion. *Coord. Chem. Rev.* **2014**, *276*, 1–33. [CrossRef]
54. Liu, J.-L.; Chen, Y.-C.; Tong, M.-L. Symmetry strategies for high performance lanthanide-based single-molecule magnets. *Chem. Soc. Rev.* **2018**, *47*, 2431–2453. [CrossRef] [PubMed]

Disclaimer/Publisher's Note: The statements, opinions and data contained in all publications are solely those of the individual author(s) and contributor(s) and not of MDPI and/or the editor(s). MDPI and/or the editor(s) disclaim responsibility for any injury to people or property resulting from any ideas, methods, instructions or products referred to in the content.

Article

Superconductivity and Fermi Surface Studies of β''-(BEDT-TTF)$_2$[(H$_2$O)(NH$_4$)$_2$Cr(C$_2$O$_4$)$_3$]·18-Crown-6

Brett Laramee [1,†], Raju Ghimire [1,†], David Graf [2], Lee Martin [3], Toby J. Blundell [3] and Charles C. Agosta [1,*]

1. Department of Physics, Clark University, Worcester, MA 01610, USA
2. National High Magnetic Field Lab, Tallahassee, FL 32310-3706, USA
3. School of Science and Technology, Nottingham Trent University, Nottingham NG11 8NS, UK
* Correspondence: cagosta@clarku.edu
† These authors contributed equally to this work.

Abstract: We report rf-penetration depth measurements of the quasi-2D organic superconductor β''-(BEDT-TTF)$_2$[(H$_2$O)(NH$_4$)$_2$Cr(C$_2$O$_4$)$_3$]·18-crown-6, which has the largest separation between consecutive conduction layers of any 2D organic metal with a single packing motif. Using a contactless tunnel diode oscillator measurement technique, we show the zero-field cooling dependence and field sweeps up to 28 T oriented at various angles with respect to the crystal conduction planes. When oriented parallel to the layers, the upper critical field, $H_{c2} = 7.6$ T, which is the calculated paramagnetic limit for this material. No signs of inhomogeneous superconductivity are seen, despite previous predictions. When oriented perpendicular to the layers, Shubnikov–de Haas oscillations are seen as low as 6 T, and from these we calculate Fermi surface parameters such as the superconducting coherence length and Dingle temperature. One remarkable result from our data is the high anisotropy of H_{c2} in the parallel and perpendicular directions, due to an abnormally low $H_{c2\perp} = 0.4$ T. Such high anisotropy is rare in other organics and the origin of the smaller $H_{c2\perp}$ may be a consequence of a lower effective mass.

Keywords: organic conductors; 2D metals; anisotropic superconductivity

Citation: Laramee, B.; Ghimire, R.; Graf, D.; Martin, L.; Blundell, T.J.; Agosta, C.C. Superconductivity and Fermi Surface Studies of β''-(BEDT-TTF)$_2$[(H$_2$O)(NH$_4$)$_2$Cr(C$_2$O$_4$)$_3$]·18-Crown-6. *Magnetochemistry* **2023**, *9*, 64. https://doi.org/10.3390/magnetochemistry9030064

Academic Editors: Laura C. J. Pereira and Dulce Belo

Received: 26 January 2023
Revised: 14 February 2023
Accepted: 20 February 2023
Published: 24 February 2023

Copyright: © 2023 by the authors. Licensee MDPI, Basel, Switzerland. This article is an open access article distributed under the terms and conditions of the Creative Commons Attribution (CC BY) license (https://creativecommons.org/licenses/by/4.0/).

1. Introduction

Many crystalline organic conductors are highly anisotropic, consisting of alternating conducting cation layers and quasi-insulating anion layers [1–3]. The organic metal β''-(BEDT-TTF)$_2$[(H$_2$O)(NH$_4$)$_2$Cr(C$_2$O$_4$)$_3$]·18-crown-6 (hereafter β''(ET)Cr, for short) is a quasi-2D (Q2D) superconductor that has an extraordinarily wide anion layer spacing, $s = 27.38$ Å [4], compared to other well-known organics such as κ-(ET)$_2$Cu(NCS)$_2$, where $s = 15$ Å [5]. Anisotropic metals, such as these Q2D crystals, tend to support many correlated electron states such as charge density waves, spin density waves, and superconductivity, because the constrained motion of the carriers enhances the electron–electron interactions. It is therefore useful to study these materials to learn about the origin and stability of correlated electron states. Although they tend to only support these correlated states at low temperatures, <10 K, they are generally electronically clean systems with long mean free paths, allowing the use of quantum oscillations to study the Fermi surface and learn about the detailed band structure, in addition to studying the correlated ground states [6]. Moreover, layered anisotropic superconductors can display a wide variety of anisotropic behaviours when subjected to external magnetic fields oriented at different angles with respect to the conducting planes, including highly anisotropic superconducting critical fields [7], inhomogeneous superconductivity [8], and novel vortex effects [9].

The wide anion layer spacing coupled with a reasonable superconducting critical temperature, T_c, make β''(ET)Cr a prime candidate to support the inhomogeneous superconducting FFLO state. Proposed independently in 1964 and 1965 by Fulde and Ferrell [10]

and Larkin and Ovchinnikov [11], the FFLO state is an exotic superconducting state at high fields and low temperatures where superconductivity survives past the Clogston–Chandrasekhar paramagnetic limit, H_P, where the Zeeman splitting energy would ordinarily overcome the binding energy of the Cooper pairs. In this state, the total momentum of the Cooper pairs is non-zero, and the superconducting order parameter is modulated. As a good approximation, $H_P = \sqrt{2}\Delta_s/g\mu_B$, where Δ_s is the superconducting energy gap, g is the Landé g-factor, and μ_B is the Bohr magneton. For $\beta''(ET)Cr$, $\mu_0 H_P = 7.6$ T [4]. In order to reach H_P, however, the orbital destruction of superconductivity needs to be suppressed. For layered organics, magnetic flux lines can penetrate through the quasi-insulating anion layers when the field is aligned parallel to the conduction planes, suppressing the orbital destruction of superconductivity in the cation layers. The wide anion layer of $\beta''(ET)Cr$ should lead to a large Maki parameter $\alpha_M = \sqrt{2}H^0_{orb}/H_P$, where H^0_{orb} is the orbital critical field in the parallel direction, thus favouring the paramagnetic destruction of Cooper pairs [12]. In this case, the FFLO state is favoured for superconductors in the clean limit.

A contactless tunnel diode oscillator measurement technique is used to measure the penetration depth, λ, of the electromagnetic field into the sample. The penetration depth comes from the addition of two parts, λ_L and λ_v, the London penetration depth and the penetration depth due to motion of the vortices, respectively, following the equation

$$\lambda^2 = \lambda_L^2 + \frac{B\Phi_0}{4\pi k_p}, \qquad (1)$$

where B is the magnetic field, Φ_0 is the quantum flux, and k_p is the restoring force on the vortices [13]. In the data, $\Delta f \propto \lambda$, where a lower frequency corresponds to a smaller penetration depth, indicating better superconductivity. Similarly, a lower frequency corresponds to less movement of the vortices or a greater restoring force.

Previous temperature sweeps down to 1 K at constant external magnetic fields showed some signs pointing towards the FFLO state, but no confirmation was given [4]. In this article we show a zero-field cooling curve which confirms the previously reported T_c of roughly 4 K. At a base temperature of 60 mK, we present B-field sweeps both parallel and perpendicular to the conduction layers of the sample. When oriented parallel to the layers, where the magnetic field is in the crystallographic a/b plane, we find no evidence of the FFLO state, contrary to the previous prediction, and the superconductivity does not extend above H_P. When oriented perpendicular to the layers, along the c axis, single frequency Shubnikov–de Haas oscillations are observed up to 28 T. Analysis of the oscillations give the Dingle temperature, T_D [K], and subsequently, the mean free path, l [Å]. The mean free path was calculated to be slightly greater than the superconducting coherence length, but perhaps not enough to stabilize the FFLO state.

The upper critical fields, H_{c2}, of $\beta''(ET)Cr$ in the $B_\parallel$ and $B_\perp$ orientations are surprisingly anisotropic. With $T_c \cong 4$ K, $H_{c2\parallel} = H_P = 7.6$ T is not abnormal, but the observed $H_{c2\perp} = 0.4$ T is unusually low. We compare this anisotropy with other well-known Q2D organics along with other Fermi surface parameters. Further study is needed into the mechanism behind the low perpendicular upper critical field and other potential effects of the high crystal anisotropy on the properties of the material.

2. Materials and Methods

Single crystals of β''-(BEDT-TTF)$_2$[(H$_2$O)(NH$_4$)$_2$Cr(C$_2$O$_4$)$_3$]·18-crown-6 were grown by electrocrystallization as described in [4]. The structure of the layering gives this crystal the widest gap between consecutive conducting (ET) layers of any organic superconductor with a single packing motif (β''), making this material of particular interest for studying Q2D systems.

A contactless tunnel diode oscillator (TDO) penetration depth technique was used to measure the change in frequency of a self-resonant circuit containing the crystal inside an inductor [14]. As the crystal expels the rf field, the complex impedance of the inductor in the resonant circuit changes, which is a function of relative rf-penetration depth, $\Delta\lambda$, a sum of the London penetration depth and penetration due to motion of vortices in a Type-II superconductor, as seen in Equation (1). Therefore, $\Delta f = \Delta f(\lambda)$, where we monitor Δf with a frequency counter or lock-in amplifier. At fields above H_{c2}, the TDO measures the normal-state skin depth, $\Delta \delta$. At cryogenic temperatures, the signal-to-noise ratio of the TDO fundamental frequency results in a resolution of about one part in 10^7, making the TDO technique sensitive to tiny variations in small samples without requiring contacts.

Experiments were performed at the National High Magnetic Field Laboratory (NHMFL) in Tallahassee, FL in a 32 T DC all-superconducting magnet containing a top loading dilution refrigerator. The data presented in this paper comes from two separate single-crystals grown in the same batch, both having a largest in-plane dimension of roughly 750 μm. Each single-crystal was placed inside an 810 μm diameter four-turn coil connected to a TDO circuit board by approximately 2 cm of a twisted wire pair. The coils and TDO circuits were placed on a single-axis rotator probe alongside a calibrated RuO thermometer for temperature measurements. Zero-field cooling conditions were identical for the two crystals as the distance between them was about 1 cm. The fundamental frequencies of the TDOs were 639 and 453 MHz, respectively, which were mixed down to 0.5–10 MHz using a superheterodyne receiver.

After reaching base temperature, the samples were rotated in a finite DC field in order to determine when the orientation of the field was parallel to the conducting layers of the samples. The conducting layers of crystals 1 and 2 were askew by roughly 5.6°. Field-dependent data was collected by rotating to a fixed angle and increasing the field.

3. Results

The zero-field cooling dependence of crystal 2 is shown in Figure 1. The superconducting transition at $T_c = 3.9$ K is broad but comparable to the previous reported value [4], and the transition spans roughly 4 MHz, a sizeable signal strength. Note that the absolute TDO frequency measurement is proportional to relative rf-penetration depth, so the absolute scale for λ is arbitrary. Crystal 1, not shown, has a much less obvious superconducting transition spanning only 0.8 MHz at a lower temperature, suggesting that it was a weaker superconductor than crystal 2. Despite the crystal quality discrepancy, crystal 1 yielded results that supported our findings from crystal 2 throughout the duration of the experiment.

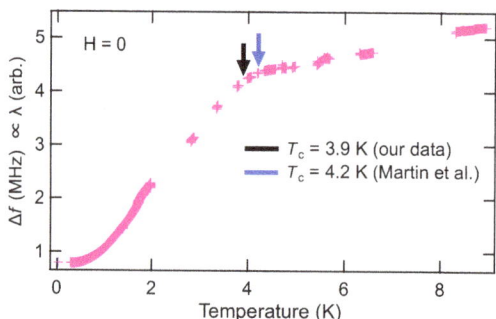

Figure 1. Rf-penetration depth as a function of temperature in a zero magnetic field. The critical temperature, defined by the minimum of $d^2\lambda/dT^2$, is similar to previous work [4].

After reaching a base temperature of 60 mK, a series of field sweeps at different angles tracked $H_{c2}(\theta)$. A subset of these field sweeps is shown in Figure 2a between $\theta = -14°$ and $+2°$. Near $B_\parallel$ ($\theta = 0°$), the superconducting transition is broad, with the highest H_{c2} recorded to be 7.6 T at $\theta = \pm 0.25°$ (green). Away from $B_\parallel$, the superconducting transition becomes sharper and decreases in field. Despite the wide anion layer being a promising indicator that the crystal might be able to harbour the FFLO state, no evidence was found from our field sweep data. The field sweeps nearest to $\theta = 0°$ showed superconductivity persisting only up to $H_P = 7.6$ T, and no anomalous bumps in the 2nd derivatives indicating a transition to the FFLO state were found.

Though the anion layer spacing in $\beta''(ET)Cr$ is wide compared to other organics, the cryogenic temperatures at which the experiment was performed should be well below any dimensional crossover transition. Therefore, $H_{c2}(\theta)$ in Figure 2b follows the Tinkham thin-film formula for a 2D layered Josephson-coupled superconductor,

$$\left| \frac{H_{c2}(\theta)\cos(\theta)}{H_{c2\perp}} \right| + \left[\frac{H_{c2}(\theta)\sin(\theta)}{H_{c2\parallel}} \right]^2 = 1 \qquad (2)$$

as opposed to anisotropic Ginzburg–Landau theory [7,15]. Though Equation (2) is based on the orbital destruction of superconductivity only, it gives reasonable agreement to the data. One remarkable result is the high anisotropy in the ultimate critical fields in the parallel vs. perpendicular orientations, with $H_{c2\parallel}/H_{c2\perp} = 19$, coming from a low $H_{c2\perp}$.

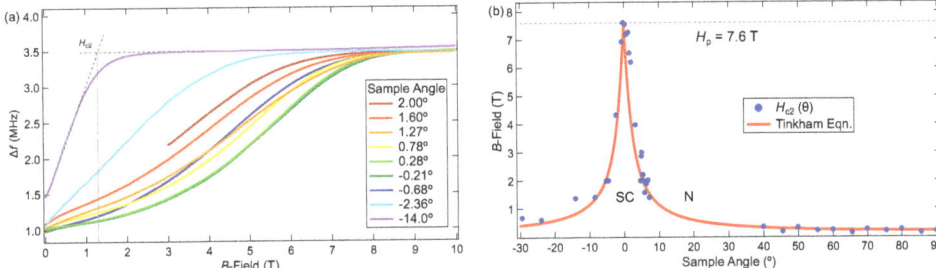

Figure 2. (a) Rf-penetration depth as a function of the magnetic field oriented at different angles with respect to the conducting planes of the sample, with $B_\parallel = 0°$. Field sweeps were performed at $T < 100$ mK. We define H_{c2} as the crossing point of the linear extrapolations above and below the superconducting transition, shown in grey. (b) $H_{c2}(\theta)$ as taken from (a) and from additional data not shown. The solid trace is the 2D Tinkham thin-film equation, Equation (2).

Due to trapped flux in the 32 T superconducting magnet, the given magnetic field, which is reported based on the current in the magnet, may be lower than the real field experienced by the crystals. Therefore, $H_{c2\perp} = 0.2$ T seen in the raw data is a lower bound. We estimate the range of the magnitude of trapped flux could be 0.15–0.25 T at low fields, based on discussion with the NHMFL. Therefore, $H_{c2\perp} = 0.2 + (0.15 - 0.25)$ T, with the most likely value for $H_{c2\perp} = 0.4$ T. Later, we show orbital critical field calculations that are close to this corrected value.

Field sweeps with the magnetic field oriented far from parallel to the conduction layers of the crystal are shown in Figure 3. At $\theta_\perp = 90°$, strong, pure sinusoidal oscillations are seen. At $\theta \cong 60°$ the amplitude of the SdH oscillations pass through zero and the phase flips, indicating a spin-zero. This occurs when the Landau level spacing $\Delta E_L = \hbar eB/m^*$ is an integer multiple of the Zeeman splitting energy $\Delta E_Z = g\mu_B B$, where the effective mass $m^*(\theta) = m^*_\perp/\sin(\theta)$ in our convention. By fitting the SdH oscillation amplitudes as a function of the angle to the Lifshitz–Kosevich (L–K) formula [16], we determine $g^* = 1.89$. According to McKenzie [17], this value of g is lower than expected due to the highly correlated nature of the quasi-particles in these lower-dimensional materials, not unlike how the effective mass m^* is different from the band mass as calculated from the band structure. Therefore, the g-factor as measured by electron spin resonance, which we will refer to as g, is probably still very close to two. The measured g-factor from the location of the spin-zero, g^*, will be enhanced or diminished by many body effects. In this case, $g^* = 1.89$, and the ratio $g^*/g = 0.95$, a value comparable to other organic conductors, as seen in McKenzie's paper [17].

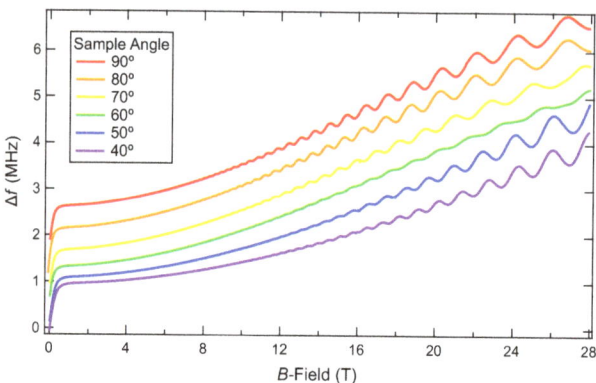

Figure 3. Field sweeps up to 28 T at different angles with respect to the layers of the sample. At $B_\perp = 90°$, quantum oscillations begin as low as 6 T. The data are offset vertically for clarity.

As shown in Figure 4, Shubnikov–de Haas oscillations can be seen as low as 6 T (a) and are of a single frequency of $\alpha = 253$ T (b). There is no evidence of breakdown orbits, and the next highest peak is the 2nd harmonic, which is almost four orders of magnitude weaker than the fundamental. $F_{SdH}(\theta)$ follows the expected behaviour for a 2D metal with a corresponding cylindrical Fermi surface (c). From the SdH oscillation amplitudes we construct the Dingle plot shown in (d). The Dingle temperature is a measure of the scattering (purity) in a crystal; $T_D = X/14.7m^*(\theta)$, where X is the slope of the log of the oscillation amplitudes vs. B^{-1}. Using the previously reported $m^*_\perp = 1.4m_e$ [4], we find $T_D = 2.42$ K for crystal 2 and $T_D > 4$ K for crystal 1, confirming that crystal 2 is a cleaner superconductor.

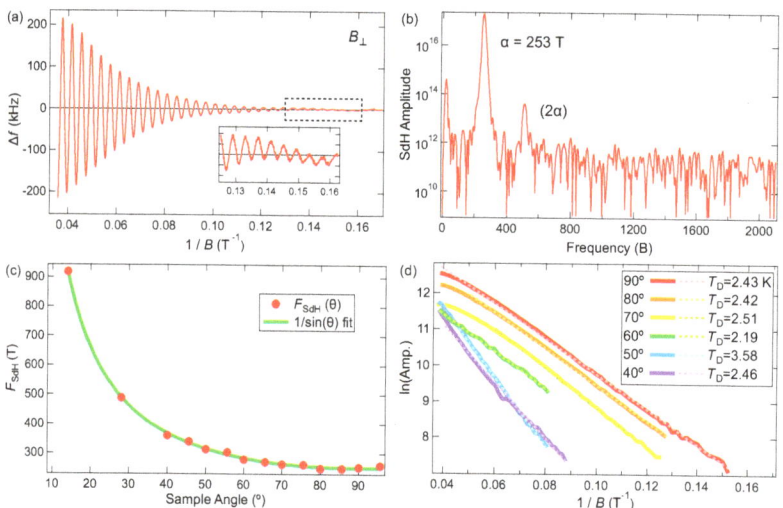

Figure 4. (a) Shubnikov–de Haas oscillations ($\theta = 90°$ from Figure 3) with the background subtracted and plotted against $1/B$. Inset: zoom of main figure; oscillations can be seen down to 6 T. (b) FFT of (a). A singular α frequency can be observed, as well as its weaker 2nd harmonic. (c) $F_{SdH}(\theta)$ as gathered by field sweeps at different angles. The data follows the expected $1/\cos(\theta)$ dependence for a cylindrical Fermi surface, noting that our convention of $B_\perp = 90°$ is reverse to the traditional $B_\perp = 0°$. (d) Dingle plot produced by the data shown in Figure 3, reduced as in (a). Linear fits are overlaid on the data with corresponding Dingle temperatures.

4. Discussion

The temperature sweep data in Figure 1 yields a similar $T_c \cong 4$ K to previous work [4]. We previously found that cooling rate could affect whether or not the crystal exhibits superconductivity, perhaps due to a structural transition at higher temperatures. Because two of our crystals underwent identical experimental cooling, however, the variation in their properties due to cooling rate may not be the only factor that determines the quality of samples. It is now clearer that there may be intrinsic disorder in the samples that also determines the strength of the superconductivity, and crystal variation could be a challenge in future experiments.

Field sweeps up to 28 T were performed below 100 mK, extending the reach of previous work in both field and temperature. Previous results down to 1 K [4] suggested the possibility of forming the FFLO state at low temperatures and high magnetic fields above H_P. Despite this, the $\lambda(B)$ data close to $B_\parallel$, Figure 2, show H_{c2} as not surpassing the Clogston–Chandrasekhar paramagnetic limit and no FFLO state is observed. The 2D Tinkham thin-film equation produces an agreeable fit to our $H_{c2}(\theta)$ plot.

The anisotropy of $H_{c2\parallel}/H_{c2\perp}$ is noteworthy being higher than most other well-known 2D organic superconductors, and a comparison with other similar superconductors in Table 1 shows that the low $H_{c2\perp}$ is anomalous. Using the corrected value of 0.4 T for $H_{c2\perp}$, $H_{c2\parallel}/H_{c2\perp} \cong 19$, which is only less than κ-(ET)$_2$I$_3$ in terms of anisotropy magnitude, and higher than the others.

Table 1. A comparison of crystal properties from experimental results to other well-known quasi-2D organic superconductors. The title compound and its corresponding values are shown in bold.

Crystal Name [Citations]	T_c (K)	$H_{c2\parallel}$ (T)	$H_{c2\perp}$ (T)	H_{orb}^0 (T)	H_P (T)	m^*/m_e	l (Å)	$\xi_\parallel$ (Å)
κ-(ET)$_2$Cu(NCS)$_2$ [18–20]	9.6	28	6	17.6 †	21.6	3.5	900	74
β''-(ET)$_2$SF$_5$CH$_2$CF$_2$SO$_3$ [6,21,22]	4.5	11	1.6	2.07 †	9.2	2.0	520	122
λ-(BETS)$_2$GaCl$_4$ [23–25]	4.3	11	2.9	1.6 †	8.3	3.6	170	107
β''(ET)Cr [4]	**4.1**	**7.6**	**0.4**	**0.49**	**7.6**	**1.4**	**367**	**286**
κ-(ET)$_2$I$_3$ [26–28]	3.5	6.7	0.2	1.37	5.7	3.9	953	410
α-(ET)$_2$NH$_4$Hg(SCN)$_4$ [7,29,30]	0.96	2.1	0.12	0.05 †	2.1	2.5	663	605

† H_{orb}^0 with daggers use Equation (5) with Δ_s from specific heat data. Others use Equation (4), the BCS result.

Away from $B_\parallel$, we show Shubnikov–de Haas oscillations, Figure 3, at multiple field angles with respect to the conduction layers of the crystal. Up to 28 T and as low as 6 T, we find strong, harmonically pure oscillations. At $B_\perp$, the fundamental frequency is 253 T, slightly higher than the previously reported $F_{SdH} = 231.1$ T [4]. The oscillation frequencies follow the expected $1/\sin(\theta)$ behaviour as we rotate the sample (often reported as $1/\cos(\theta)$ when $B_\perp = 0°$). We report the Dingle temperature $T_D = 2.4$ K in our main crystal and $T_D > 4$ K in the other, noisier crystal.

Using the Shubnikov–de Haas oscillations and the effective mass, we can calculate the Fermi surface parameters such as the Fermi velocity, $v_F = \sqrt{2\hbar e F_{SdH}}/m^*$, the scattering time, which is related to the Dingle temperature, $\tau = \hbar/2\pi k_B T_D$, and the mean free path $l = v_F \tau$. For the superconducting coherence length, ξ, we use Equation (3), with $H_{c2\perp} = 0.4$ T at $T < 100$ mK. These parameters are shown in Table 1 alongside comparisons to other Q2D organic superconductors. One of the conditions for forming the FFLO state is for the crystal to be in the clean limit, i.e., $r = l/\xi > 1$, which is barely satisfied for β''(ET)Cr, though this alone is no guarantee of FFLO stabilization, as seen in the paramagnetically limited α-(ET)$_2$NH$_4$Hg(SCN)$_4$ [7].

Considering that the remarkably low $H_{c2\perp}$ is well below H_P, we would expect the orbital effect to dominate the destruction of superconductivity, following the equation

$$H_{c2\perp} \cong H_{orb}^0 = \frac{\Phi}{2\pi \xi_\parallel^2}, \tag{3}$$

where $\xi = \hbar v_F / \pi \Delta_s$, v_F is the Fermi velocity, and the superconducting energy gap is given by the BCS result, $\Delta_s = 1.75 k_B T_c$, resulting in

$$H_{\text{orb}}^0 = \frac{\Phi}{2\pi} \left(\frac{1.75\pi k_B T_c}{\hbar v_F} \right)^2. \qquad (4)$$

With v_F as calculated above, $T_c = 3.9$ K, and using the BCS result, we arrive at $H_{\text{orb}}^0 = 0.49$ T, very close to our measured $H_{c2\perp}$. We note that, once specific heat data are taken, the equation for H_{orb}^0 could be made more accurate using Δ_s calculated empirically via the Alpha model [20,31,32] rather than the BCS result, using the equation

$$H_{\text{orb}}^0 = \frac{\Phi \pi \Delta_s^2 m^*}{4\hbar^2 E_F}, \qquad (5)$$

where $E_F = \hbar e F_{\text{SdH}}$ is the Fermi energy.

Orbital critical field calculations do agree moderately well with $H_{c2\perp}$ for most of the organics in Table 1, with the exceptions of κ-(ET)$_2$Cu(NCS)$_2$ and κ-(ET)$_2$I$_3$. The former can be justified by taking into account Zeeman pair breaking and strong coupling. The calculated H_{orb}^0 for κ-(ET)$_2$Cu(NCS)$_2$ is close to the value of H_P, where one would expect a large contribution of Zeeman pair breaking to limit the upper critical field, as calculated by the the WHH formula [33], and is seen as a downward curvature in $H_{c2\perp}(T)$ [2]. The rest of the materials in Table 1 have relatively linear $H_{c2\perp}(T)$ phase diagrams, given that H_{orb}^0 is much less than H_P for all of them. However, new measurements may be needed to account for the discrepancy between $H_{c2\perp}$ and the calculated H_{orb}^0 for λ-(BETS)$_2$GaCl$_4$ and α-(ET)$_2$NH$_4$Hg(SCN)$_4$, as one would not expect H_{orb}^0 to be less than the measured critical field. Finally, the extremely low $H_{c2\perp}$ for κ-(ET)$_2$I$_3$ cannot be accounted for by measurement error, and another mechanism may be at play in that material.

In fact, many layered organics have $H_{c2\perp}(T)$ phase diagrams that do not follow the simple model suggested by Equation (3) or even more complicated models such as the WHH formula. Although there have been some attempts to build separate theories for $H_{c2}(T)$ in layered superconductors, they have tended to focus on the parallel orientation with the magnetic field along the layers [15,34] or address special cases such as the diverging perpendicular critical fields of some cuprates or dichalcogenides, motivating proposed mechanisms based on magnetic impurities [35] or quantum critical behaviour [36], neither of which are suggested for organic materials. There are possible instances of 3D-to-2D crossover in the perpendicular critical fields of layered organic superconductors [25,37], but there is no theory that supports 3D-to-2D crossover in perpendicular magnetic fields.

5. Conclusions

In this brief report we present high-field, low-temperature measurements on β''(ET)Cr, a layered crystalline organic superconductor with novel crystal geometry, using a tunnel diode oscillator measurement technique. Despite prior predictions, inhomogeneous superconductivity was not found, and the superconductor may be paramagnetically limited. From a series of strong Shubnikov–de Haas oscillations, we calculated Fermi surface parameters confirming the crystal to be in the clean limit, though only just. We found that this material exhibits higher critical field anisotropy than other well-known organics, with H_{c2} in the perpendicular-to-the-layers orientation being surprisingly low. Though, calculation of the expected orbital critical field matches the measured value. This raises further questions as to why other organics do not necessarily follow this behaviour.

Future additional field sweep experiments would help clarify multiple points regarding the properties of β''(ET)Cr presented in this work. In the $B_\perp$ direction, collecting SdH oscillation data at multiple temperatures is of utmost importance to confirm the effective mass value, $m^* = 1.4$ K, reported by Martin et al. [4], which is used in numerous calculations of Fermi surface parameters, the orbital critical field, Equation (5), etc.

In particular, the estimated g-factor, $g^* = 1.89$, could be made more accurate both by confirming m^* and by gathering additional SdH amplitude data at finer angle intervals. Fitting the oscillation amplitudes to the L–K formula produced error bars that were significant, due to limited angular sample size, and only one spin-zero was seen. As an alternative to field sweeps at fixed angles, since $\alpha = 253$ T is the only visible SdH frequency, rotating the sample in a high DC field would immediately reveal at what spin-zero angles the oscillations vanished, which uniquely determine the product m^*g^* [17,38].

The H/T phase diagrams of the upper critical fields in both the $B_\parallel$ and $B_\perp$ orientations would benefit from being completed, as this present work only adds data at $T = 60$ mK, a temperature well below the previously published data [4], leaving a gap in the $0.1 < T < 1$ K range. A saturation of $H_{c2\parallel}(T)$ to a constant value in this region would help confirm whether the crystal is paramagnetically limited [7] and determine the degree to which many-body effects might enhance H_P [17]. The suspected linearity of $H_{c2\perp}(T)$ in this temperature range should also be investigated, as discussed above, to deepen our understanding of perpendicular critical field phase diagrams in layered organics, where the extent of the current theory is incomplete.

Author Contributions: Conceptualization, C.C.A. and L.M.; formal analysis, B.L.; investigation, B.L., R.G. and D.G.; writing, B.L. and C.C.A.; sample synthesis, T.J.B. and L.M.; funding acquisition, C.C.A. All authors have read and agreed to the published version of the manuscript.

Funding: This research was funded by the NSF Agreement No. DMR-1905950. L.M. and T.J.B. thank the Leverhulme Trust for funding for synthesis of crystals (LT170022). A portion of this work was performed at the National High Magnetic Field Laboratory, which is supported by National Science Foundation Cooperative Agreement No. DMR-1644779 and the State of Florida.

Institutional Review Board Statement: Not Applicable.

Informed Consent Statement: Not Applicable.

Data Availability Statement: The data presented in this study are available on request from the corresponding author.

Conflicts of Interest: The authors declare no conflict of interest.

References

1. Ishiguro, T.; Yamaji, K.; Saito, G. *Organic Superconductors*; Springer Series in Solid-State Sciences; Springer: Berlin/Heidelberg, Germany, 1998.
2. Singleton, J.; Mielke, C. Quasi-two-dimensional organic superconductors: A review. *Contemp. Phys.* **2002**, *43*, 63–96. [CrossRef]
3. Lebed, A. *The physics of Organic Superconductors and Conductors*; Springer Series in Materials Sciences; Springer: Berlin/Heidelberg, Germany, 2008; Volume 110.
4. Martin, L.; Lopez, J.R.; Akutsu, H.; Nakazawa, Y.; Imajo, S. Bulk Kosterlitz–Thouless Type Molecular Superconductor β″-(BEDT-TTF)$_2$[(H$_2$O)(NH$_4$)$_2$Cr(C$_2$O$_4$)$_3$]·18-crown-6. *Inorg. Chem.* **2017**, *56*, 14045–14052. [CrossRef] [PubMed]
5. Urayama, H.; Yamochi, H.; Saito, G.; Nozawa, K. A new ambient pressure organic superconductor based on BEDT-TTF with T_c higher than 10 K (T_c= 10.4 K). *Chem. Lett.* **1988**, *17*, 55–58. [CrossRef]
6. Wosnitza, J.; Wanka, S.; Hagel, J.; Häussler, R.; Löhneysen, H.V.; Schlueter, J.A.; Geiser, U.; Nixon, P.G.; Winter, R.W.; Gard, G.L. Shubnikov—De Haas effect in the superconducting state of an organic superconductor. *Phys. Rev. B* **2000**, *62*, R11973–R11976. [CrossRef]
7. Coffey, T.; Martin, C.; Agosta, C.C.; Kinoshota, T.; Tokumoto, M. Bulk two-dimensional Pauli-limited superconductor. *Phys. Rev. B* **2010**, *82*, 212502. [CrossRef]
8. Agosta, C.C. Inhomogeneous Superconductivity in Organic and Related Superconductors. *Crystals* **2018**, *8*, 285. [CrossRef]
9. Mansky, P.; Danner, G.; Chaikin, P. Vortex pinning and lock-in effect in a layered superconductor with large in-plane anisotropy. *Phys. Rev. B (Condens. Matter)* **1995**, *52*, 7554–7563. [CrossRef] [PubMed]
10. Fulde, P.; Ferrell, R.A. Superconductivity in a Strong Spin-Exchange Field. *Phys. Rev.* **1964**, *135*, A550–A563. [CrossRef]
11. Larkin, A.I.; Ovchinnikov, Y.N. Inhomogeneous State of Superconductors. *Sov. Phys. JETP* **1965**, *20*, 762.
12. Maki, K. Effect of Pauli Paramagnetism on Magnetic Properties of High-Field Superconductors. *Phys. Rev.* **1966**, *148*, 362–369. [CrossRef]
13. Mansky, P.A.; Chaikin, P.M.; Haddon, R.C. Evidence for Josephson vortices in κ-(BEDT-TTF)$_2$Cu(NCS)$_2$. *Phys. Rev. B* **1994**, *50*, 15929–15944. [CrossRef]

14. Van Degrift, C.T. Tunnel diode oscillator for 0.001 ppm measurements at low temperatures. *Rev. Sci. Instrum.* **1975**, *46*, 599. [CrossRef]
15. Schneider, T.; Schmidt, A. Dimensional crossover in the upper critical field of layered superconductors. *Phys. Rev. B* **1993**, *47*, 5915–5921. [CrossRef] [PubMed]
16. Shoenberg, D. *Magnetic Oscillations in Metals*; Cambridge University Press: Cambridge, UK, 1984.
17. McKenzie, R.H. Wilson's ratio and the spin splitting of magnetic oscillations in quasi-two-dimensional metals. *arXiv* **1999**, arXiv.9905044.
18. Taylor, O.J.; Carrington, A.; Schlueter, J.A. Specific-Heat Measurements of the Gap Structure of the Organic Superconductors κ-(BEDT-TTF)$_2$-Cu[N(CN)$_2$]Br and κ-(ET)$_2$Cu(NCS)$_2$. *Phys. Rev. Lett.* **2007**, *99*, 057001. [CrossRef]
19. Mihut, I.; Agosta, C.; Martin, C.; Mielke, C.; Coffey, T. Incoherent Bragg reflection and Fermi-surface hot spots in a quasi-two-dimensional metal. *Phys. Rev. B* **2006**, *73*, 125118. [CrossRef]
20. Müller, J.; Lang, M.; Helfrich, R.; Steglich, F.; Sasaki, T. High-resolution ac-calorimetry studies of the quasi-two-dimensional organic superconductor κ-(BEDT-TTF)$_2$Cu(NCS)$_2$. *Phys. Rev. B* **2002**, *65*, 140509.
21. Zuo, F.; Su, X.; Zhang, P.; Brooks, J.S.; Wosnitza, J.; Schlueter, J.A.; Williams, J.M.; Nixon, P.G.; Winter, R.W.; Gard, G.L. Anomalous low-temperature and high-field magnetoresistance in the organic superconductor β''-(BEDT-TTF)$_2$SF$_5$CH$_2$CF$_2$SO$_3$. *Phys. Rev. B* **1999**, *60*, 6296–6299. [CrossRef]
22. Sugiura, S.; Terashima, T.; Uji, S.; Yasuzuka, S.; Schlueter, J.A. Josephson vortex dynamics and Fulde-Ferrell-Larkin-Ovchinnikov superconductivity in the layered organic superconductor β''-(BEDT-TTF)$_2$SF$_5$CH$_2$CF$_2$SO$_3$. *Phys. Rev. B* **2019**, *100*, 014515. [CrossRef]
23. Coniglio, W.A.; Winter, L.E.; Cho, K.; Agosta, C.C.; Fravel, B.; Montgomery, L.K. Superconducting phase diagram and FFLO signature in λ-(BETS)$_2$GaCl$_4$ from rf penetration depth measurements. *Phys. Rev. B* **2011**, *83*, 224507. [CrossRef]
24. Tanatar, M.A.; Ishiguro, T.; Tanaka, H.; Kobayashi, H. Magnetic field-temperature phase diagram of the quasi-two-dimensional organic superconductor λ-(BETS)$_2$GaCl$_4$ studied via thermal conductivity. *Phys. Rev. B* **2002**, *66*, 134503. [CrossRef]
25. Mielke, C.; Singleton, J.; Nam, M.S.; Harrison, N.; Agosta, C.C.; Fravel, B.; Montgomery, L.K. Superconducting properties and Fermi-surface topology of the quasi-two-dimensional organic superconductor λ-(BETS)$_2$GaCl$_4$ (BETS=; bis(ethylenedithio)tetraselenafulvalene). *J. Phys. Condens. Matter* **2001**, *13*, 8325–8345. [CrossRef]
26. Wanka, S.; Beckmann, D.; Wosnitza, J.; Balthes, E.; Schweitzer, D.; Strunz, W.; Keller, H.J. Critical fields and mixed-state properties of the layered organic superconductor κ-(BEDT-TTF)$_2$I$_3$. *Phys. Rev. B* **1996**, *53*, 9301–9309. [CrossRef] [PubMed]
27. Wosnitza, J.; Liu, X.; Schweitzer, D.; Keller, H.J. Specific heat of the organic superconductor κ-(BEDT-TTF)$_2$I$_3$. *Phys. Rev. B* **1994**, *50*, 12747–12751. [CrossRef] [PubMed]
28. Harrison, N.; Mielke, C.H.; Rickel, D.G.; Wosnitza, J.; Qualls, J.S.; Brooks, J.S.; Balthes, E.; Schweitzer, D.; Heinen, I.; Strunz, W. Quasi-two-dimensional spin-split Fermi-liquid behavior of κ-(BEDT-TTF)$_2$I$_3$ in strong magnetic fields. *Phys. Rev. B* **1998**, *58*, 10248–10255. [CrossRef]
29. Wang, H.; Carlson, K.; Geiser, U.; Kwok, W.; Vashon, M.; Thompson, J.; Larsen, N.; McCabe, G.; Hulscher, R.; Williams, J. A New Ambient Pressure Organic Superconductor: (BEDT-TTF)$_2$(NH$_4$)Hg(SCN)$_4$. *Physica C* **1990**, *166*, 57–61. [CrossRef]
30. Brooks, J.S.; Chen, X.; Klepper, S.J.; Valfells, S.; Athas, G.J.; Tanaka, Y.; Kinoshita, T.; Kinoshita, N.; Tokumoto, M.; Anzai, H.; et al. Pressure effects on the electronic structure and low-temperature states in the α-(BEDT-TTF)$_2$MHg(SCN)$_4$ organic-conductor family (M=K, Rb, Tl, NH$_4$). *Phys. Rev. B* **1995**, *52*, 14457–14478. [CrossRef]
31. Padamsee, H.; Neighbor, J.; Shiffman, C. Quasiparticle phenomenology for thermodynamics of strong-coupling superconductors. *J. Low Temp. Phys.* **1973**, *12*, 387–411. [CrossRef]
32. Johnston, D.C. Elaboration of the α-model derived from the BCS theory of superconductivity. *Supercond. Sci. Technol.* **2013**, *26*, 115011. [CrossRef]
33. Werthamer, N.R.; Helfand, E.; Hohenberg, P.C. Temperature and Purity Dependence of the Superconducting Critical Field, H_{c2}. III. Electron Spin and Spin-Orbit Effects. *Phys. Rev.* **1966**, *147*, 295–302. [CrossRef]
34. Klemm, R.A.; Luther, A.; Beasley, M.R. Theory of the upper critical field in layered superconductors. *Phys. Rev. B* **1975**, *12*, 877–891. [CrossRef]
35. Kotliar, G.; Varma, C.M. Low-Temperature Upper-Critical-Field Anomalies in Clean Superconductors. *Phys. Rev. Lett.* **1996**, *77*, 2296–2299. [CrossRef] [PubMed]
36. Ovchinnikov, Y.N.; Kresin, V.Z. Critical magnetic field in layered superconductors. *Phys. Rev. B* **1995**, *52*, 3075–3078. [CrossRef] [PubMed]
37. Agosta, C.; Ivanov, S.; Bayindir, Z.; Coffey, T.; Kushch, N.; Yagubskii, E.; Burgin, T.; Montgomery, L. The anomalous superconducting phase diagram of (BEDO-TTF)$_2$ReO$_4$·H$_2$O. *Synth. Met.* **1999**, *103*, 1795–1796. [CrossRef]
38. Wosnitza, J.; Crabtree, G.W.; Wang, H.H.; Geiser, U.; Williams, J.M.; Carlson, K.D. de Haas–van Alphen studies of the organic superconductors α-(ET)$_2$(NH$_4$)Hg(SCN)$_4$ and κ-(ET)$_2$Cu(NCS)$_2$ [with ET = bis(ethelenedithio)-tetrathiafulvalene]. *Phys. Rev. B* **1992**, *45*, 3018–3025. [CrossRef] [PubMed]

Disclaimer/Publisher's Note: The statements, opinions and data contained in all publications are solely those of the individual author(s) and contributor(s) and not of MDPI and/or the editor(s). MDPI and/or the editor(s) disclaim responsibility for any injury to people or property resulting from any ideas, methods, instructions or products referred to in the content.

Article

Vibronic Relaxation Pathways in Molecular Spin Qubit Na₉[Ho(W₅O₁₈)₂]·35H₂O under Pressure

Janice L. Musfeldt [1,2,*], Zhenxian Liu [3], Diego López-Alcalá [4], Yan Duan [4], Alejandro Gaita-Ariño [4], José J. Baldoví [4] and Eugenio Coronado [4]

[1] Department of Chemistry, University of Tennessee, Knoxville, TN 37996, USA
[2] Department of Physics, University of Tennessee, Knoxville, TN 37996, USA
[3] Department of Physics, University of Illinois Chicago, Chicago, IL 60607-7059, USA
[4] Instituto de Ciencia Molecular, Universitat de Valencia, 46980 Paterna, Spain
* Correspondence: musfeldt@utk.edu

Citation: Musfeldt, J.L.; Liu, Z.; López-Alcalá, D.; Duan, Y.; Gaita-Ariño, A.; Baldoví, J.J.; Coronado, E. Vibronic Relaxation Pathways in Molecular Spin Qubit Na₉[Ho(W₅O₁₈)₂]·35H₂O under Pressure. *Magnetochemistry* **2023**, *9*, 53. https://doi.org/10.3390/magnetochemistry9020053

Academic Editors: Zhao-Yang Li and Quan-Wen Li

Received: 11 January 2023
Revised: 28 January 2023
Accepted: 1 February 2023
Published: 9 February 2023

Copyright: © 2023 by the authors. Licensee MDPI, Basel, Switzerland. This article is an open access article distributed under the terms and conditions of the Creative Commons Attribution (CC BY) license (https://creativecommons.org/licenses/by/4.0/).

Abstract: In order to explore how spectral sparsity and vibronic decoherence pathways can be controlled in a model qubit system with atomic clock transitions, we combined diamond anvil cell techniques with synchrotron-based far infrared spectroscopy and first-principles calculations to reveal the vibrational response of Na₉[Ho(W₅O₁₈)₂]·35H₂O under compression. Because the hole in the phonon density of states acts to reduce the overlap between the phonons and f manifold excitations in this system, we postulated that pressure might move the HoO₄ rocking, bending, and asymmetric stretching modes that couple with the $M_J = \pm 5, \pm 2$, and ± 7 levels out of resonance, reducing their interactions and minimizing decoherence processes, while a potentially beneficial strategy for some molecular qubits, pressure slightly hardens the phonons in Na₉[Ho(W₅O₁₈)₂]·35H₂O and systematically fills in the transparency window in the phonon response. The net result is that the vibrational spectrum becomes less sparse and the overlap with the various M_J levels of the Ho³⁺ ion actually increases. These findings suggest that negative pressure, achieved using chemical means or elongational strain, could further open the transparency window in this rare earth-containing spin qubit system, thus paving the way for the use of device surfaces and interface elongational/compressive strains to better manage decoherence pathways.

Keywords: spin qubit; vibronic coupling; strategies to minimize decoherence; high pressure vibrational spectroscopy

1. Introduction

Single molecule magnets incorporating heavy centers are of foundational importance for exploring orbital localization and chemical bonding, electron correlation vs. spin–orbit coupling, and the different patterns of crystal field energy levels [1–5]. Hundreds have been developed in an effort to control the properties and reveal structure–property relations [6–9]. Several have demonstrated spin qubit behavior [10–14]. Prominent examples include (i) (PPh₄)₂[Cu(mnt)₂] (mnt²⁻ = maleonitriledithiolate) doped into the diamagnetic isostructural host (PPh₄)₂ [Ni(mnt)₂], (ii) [(Cp^{iPr5})Dy(Cp*)]⁺ (where Cp^{iPr5} = penta-iso-propylcyclopentadienyl and Cp* = pentamethylcyclopentadienyl), and (iii) chiral [Zn(OAc)(L)Yb(NO₃)₂] as well as many others [15–18]. In the Ln³⁺-containing family of mononuclear molecular nanomagnets, Na₉[Ho(W₅O₁₈)₂]·35H₂O is attracting considerable attention [15,19–21]. This is because of the 8.4 µs coherence time in diluted systems [20] as well as the fact that the spin qubit dynamics are protected against magnetic noise at favorable operating points known as atomic clock transitions [14,20,22], where this system has also been experimentally found to display magnetoelectric coupling [23]. Of course, even with clock protection, quantum information can be lost through vibrational and thermal processes [4,17,24–30], although very few of these mechanisms have been unraveled in a detailed manner [31–37].

To address this issue, our team recently began exploring decoherence pathways in $Na_9[Ho(W_5O_{18})_2] \cdot 35H_2O$ with the goal of revealing specific vibronic relaxation pathways governing magnetic relaxations [38]. We discovered strong magneto-infrared contrast near 370 and 63 cm^{-1} due to mixing of odd-symmetry vibrations with f-manifold crystal field excitations. Specifically, the $M_J = \pm 7$ crystal field levels couple to the various HoO_4 rocking and bending modes. At the same time, the $M_J = \pm 5$ levels near 63 cm^{-1} (and very likely the $M_J = \pm 2$ levels) are activated by nearby phonons such as asymmetric HoO_8 stretching with cage tilting. Moreover, we reported the first direct evidence for a transparency window in the phonon density of states in a robust clock-like molecular spin qubit. The overall extent of vibronic coupling [17,26–28] in $Na_9[Ho(W_5O_{18})_2] \cdot 35H_2O$ is therefore limited by a modest coupling constant (in the order of 0.25 cm^{-1}) and a transparency window in the phonon density of states that acts to keep the intramolecular vibrations and M_J levels apart [38]. This is different from 3d-containing molecule-based materials such as $Co[N(CN)_2]_2$ where significantly larger spin–phonon coupling constants (2 and 3 cm^{-1}) give rise to avoided crossings and a transfer of oscillator strength from nearby phonons to the localized Co^{2+} electronic excitations and back again under magnetic field [39]. Despite the smaller coupling constant, interaction with phonons is still a significant problem in $Na_9[Ho(W_5O_{18})_2] \cdot 35H_2O$—even at low temperature. Recent simulations in entangled two-qubit gates suggest that increased spin–lattice relaxation time (T_1) is likely with additional cooling [30]. Since in this system T_2 is controlled by T_1, this is relevant for its behavior as a qubit.

Decoherence of quantum states in a qubit can occur when resonances of different types are found in close proximity. The natural spectral sparsity in $Na_9[Ho(W_5O_{18})_2] \cdot 35H_2O$ raises questions about whether even more extreme separations between the magnetic and vibrational excitations can be encouraged and even promoted [40–43]. Besides exploring the consequences of this "hole" on vibronic coupling, we recently proposed several design rules aimed at mitigating decoherence pathways involving vibronic coupling [38]. In addition to isotope effects and studies of chemically analogous Ho^{3+}-containing polyoxometalates, we discussed specific suggestions for alternate rare earth ions and coordination effects as well as stiffer surrounding ligands [4,38]. One promising avenue that was not explored in prior work is the effect of pressure. Compression changes bond lengths and angles [44] and, as a result, tends to harden phonons—perhaps moving them out of the way. Such an approach has the potential to significantly reduce vibronic coupling as a decoherence mechanism by moving these excitations off resonance [29,38,40–42,45].

In order to to test whether pressure can disentangle electronic and vibrational excitations in a molecular magnet with atomic clock transitions, we measured the far infrared response of $Na_9[Ho(W_5O_{18})_2] \cdot 35H_2O$ under compression and compared our findings with complementary first-principles calculations. This work is based upon the premise that both the molecular vibrations and the M_J levels that derive from on-site f-manifold excitations of Ho^{3+} are likely to harden under compression and therefore might be able to be shifted to nullify the spin–spin (T_2) relaxation pathway under small pressures or strain. In other words, pressure is expected to act upon both energy scales, in the same direction (and in both cases, the desired direction), although not necessarily to the same extent. A more compressed molecule will be more rigid, but also closer metal–ligand distances will result in a stronger crystal field. This is what one wants: molecules that are very rigid and with very strong crystal field perturbations. This energy scaling should, in principle, systematically decrease the chance of a close resonance. Rendering the M_J levels ineffective in terms of engaging in vibronic coupling, we find that pressure broadens the low frequency phonons and shifts others more strongly into resonance. At the same time, compression works to systematically fill the hole in the phonon density of states, so rather than making a sparse lattice even more sparse, pressure decreases sparsity by closing the transparency window. These findings, supported by our theoretical calculations, suggest that negative pressures, obtained via crystal engineering or elongational strain, will be more effective to enhance the performance of this molecular spin qubit.

2. Methods

2.1. Experimental Setup

High quality $Na_9[Ho(W_5O_{18})_2]\cdot 35H_2O$ single crystals were grown as described previously [20]. A small, well-shaped piece was selected and loaded into a suitably chosen diamond anvil cell with an annealed ruby ball and hydrocarbon grease (petroleum jelly) as the pressure medium in order to assure quasi-hydrostatic pressure conditions (Figure 1a). In addition to using the ruby ball to determine pressure via fluorescence [46,47], we monitored the shape of the ruby fluorescence spectrum to assure that the sample remained in a quasi-hydrostatic environment (Figure 1b). The synthetic type IIas diamonds in the symmetric diamond anvil cell had 500 μm culets and we employed a 47 μm thick pre-indented stainless steel gasket with a 200 μm hole diameter. Care was taken to optimize optical density in order to reveal the features of interest. In fact, we carried out two different trials using both low and high optical densities. We employed high optical density in the 55–370 cm^{-1} range where the features have smaller intensity, and we used lower optical density in the 300–625 cm^{-1} range where the spectral features are stronger. Taking advantage of the stable, high-brightness beam, synchrotron-based infrared spectroscopy (60–680 cm^{-1}; 4 cm^{-1} resolution; transmittance geometry) was performed using the 22-IR-1 beamline at the National Synchrotron Light Source II at Brookhaven National Laboratory. Absorbance is calculated as $\alpha(\omega) = -\ln(\mathcal{T}(\omega))$, where $\mathcal{T}(\omega)$ is the measured transmittance. Pressure was increased between 0 and 5.2 GPa at room temperature. Our prior work revealed no substantial spectral changes in the phonons or the transparency window with temperature down to 9 K [38], so we did not pursue these effects here. The spectral changes are fully reversible upon release of pressure.

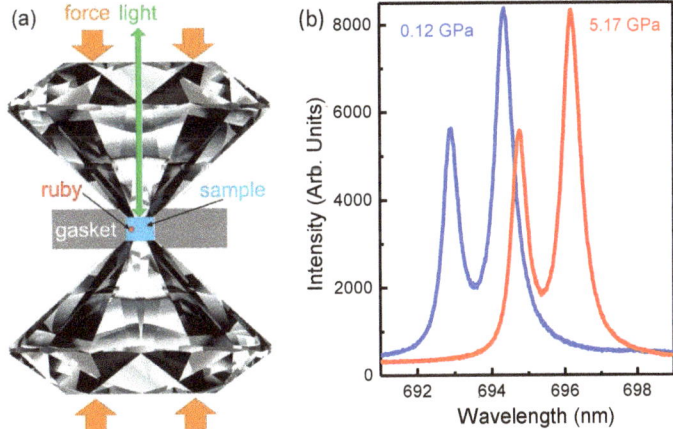

Figure 1. (a) Schematic of the symmetric diamond anvil cell used in these measurements. The ruby ball and the sample crystal share a high pressure environment but are not directly in contact. (b) Fluorescence of the annealed ruby ball when the cell was just closed (0.12 GPa) and at 5.17 GPa. The shape of the fluorescence spectra along with the separation of the two peaks is consistent with quasi-hydrostatic conditions.

2.2. DFT Calculations

The optimization of the molecular geometries and the simulation of vibrational spectra were carried out using the Gaussian09 package in its revision D01 [48]. The PBE0 hybrid-exchange correlation functional was used in all the calculations [49]. Different basis sets were used for each type of atom in the molecule: Stuttgart RSC ANO basis set [50–52] for the Ho^{+3} cation, CRENBL basis set [53] for W, and 6-31G** for O [54,55]. Their corresponding effective core potentials (ECP) for Ho and W atoms were applied. We included Grimme D3BJ dispersion corrections in all the calculations [56]. First, we optimized the crystal structure and then we determined the vibrational spectra. We optimized all the structures until the change in energy of two consecutive steps was lower than 10^{-6} Hartree. The threshold assigned for maximum forces and RMS matrix was $4.5 \cdot 10^{-4}$ and $3 \cdot 10^{-4}$ Hartree/Bohr, respectively. Then, we applied triaxial pressure by decreasing each Cartesian coordinate by 0.5, 1, 1.5, and 2%. This allowed us to simulate the effect of mechanical pressure. Finally, we performed a constrained optimization to maintain the effect of the pressure, thus avoiding negative vibrational frequencies in the spectrum. This constraint consisted of freezing the position of the atoms lying in the Cartesian axes, thus keeping the effect of the triaxial pressure.

2.3. Semi-Empirical Crystal-Field Calculations

To calculate the effect of the applied strain on the energy level scheme of the ground-J manifold, we applied the semi-empirical radial effective charge (REC) model [57], as implemented in the SIMPRE computational package [58]. In order to account for covalency effects, we applied a radial displacement of (D_r = 0.48 Å) and effective point-charges (Z_i = 0.81). These parameters were extracted by fitting the spectroscopic energy levels determined by Vonci et al. [21].

3. Results and Discussion

Figure 2 summarizes the far infrared response of $Na_9[Ho(W_5O_{18})_2] \cdot 35H_2O$ as a function of pressure at room temperature. The main difference between the two trials in panels (a) and (b) is the optical density, which allows us to examine different spectral regions with optimum sensitivity. Molecular materials are well known to be soft and flexible but, unlike a number of other systems [59–61], there is no evidence for a structural phase transition between 0 and 5 GPa in $Na_9[Ho(W_5O_{18})_2] \cdot 35H_2O$. Instead, compression acts to (i) harden the vibrational modes and (ii) reduce spectral sparsity. These findings are discussed below.

As we know, vibrations play an important role in magnetic relaxation processes of molecular spin qubits as they couple to spin states, leading to the loss of quantum information [17,26–28]. In $Na_9[Ho(W_5O_{18})_2] \cdot 35H_2O$, we are primarily interested in vibronic decoherence pathways involving odd-symmetry vibrations near 370 and 63 cm^{-1} that mix with f-manifold crystal field excitations [38]. As expected, pressure hardens the majority of vibrational modes at a rate of approximately 0.9 cm^{-1}/GPa (Figure 2a,c), so it is possible to push a mode that is on resonance and therefore detrimental (such as those near 370 or 63 cm^{-1}) away from resonance with a particular M_J level of the Ho^{3+} ion. In this system, however, the vibrational modes do not harden at significantly different rates so, rather than taking a particular mode off-resonance and leaving the rest unperturbed, the full set of modes hardens systematically. Above approximately 1 GPa, these modes begin to interact with other spin levels in $Na_9[Ho(W_5O_{18})_2] \cdot 35H_2O$—not just the M_J = ±7, ±5, and ±2 levels—so it is really only the smallest pressure (or presumably strain) that is potentially useful.

Our prior analysis suggests that coherence in $Na_9[Ho(W_5O_{18})_2] \cdot 35H_2O$ benefits from the limited frequency overlap between Ho^{3+} crystal field levels and the phonon manifold [38]. The limited overlap is due to a transparency window or "hole" in the phonon density of states that renders many of the M_J levels ineffective in terms of engaging in vibronic coupling. Revealing how pressure affects the transparency window in the phonon density of states is therefore extremely important. For this technique to work well, the hole should stay the same size or expand slightly. Unfortunately, the transparency window in the phonon response closes systematically under compression (Figure 2a). We quantify this effect by integrating the area under the absorption in this frequency window and plotting it as a function of pressure (inset, Figure 2a). Initially, this quantity grows linearly with pressure. It levels off around 5 GPa as the filling saturates and the transparency window closes. Therefore, while the tiniest bit of compression might impact the coherence time in a positive manner, in general, pressure is not an effective external stimulus because it eliminates the transparency window in the vibrational spectrum. Broadening the transparency window with "negative pressure" would be a better approach—at least in this system.

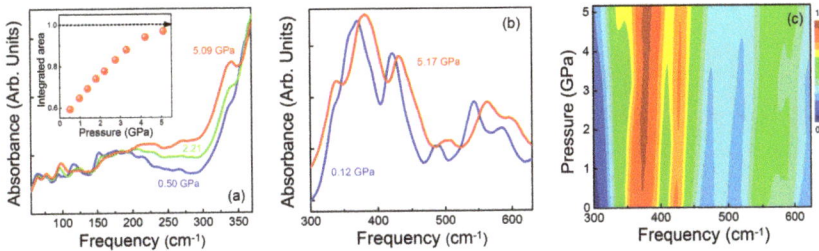

Figure 2. (**a**) Synchrotron-based far infrared spectrum of $Na_9[Ho(W_5O_{18})_2]\cdot 35H_2O$ as a function of pressure in the 55–375 cm^{-1} region. Both intramolecular and intermolecular features are observed. Inset: Integrated area under the curve in the 220–310 cm^{-1} frequency window. The latter corresponds to the transparency window and was therefore used as the integration range. (**b**) Close-up view of the far infrared response of $Na_9[Ho(W_5O_{18})_2]\cdot 35H_2O$ between 300 and 625 cm^{-1}. (**c**) The same data as in panel (**b**) shown as a contour plot. Spectra at 12 different pressures were used to create the contour plot.

In order to rationalize these findings, we performed first-principles calculations on a fully optimized $[Ho(W_5O_{18})_2]^{9-}$ molecule at the density functional theory level. We applied different compressive triaxial strain values ranging from 0.5 to 2% (Figure 3a), which are compatible with the structural changes of the crystallographic coordinates in the experiment, assuming a typical Young's modulus of $\sim$100 GPa for this class of crystals. Our results corroborate the hardening of the molecular vibrations under compression, thus supporting the trend observed in the synchrotron-based far infrared measurements (Figure 2b). This is expected due to the shorter bond lengths in the molecule, which are displaced from their equilibrium positions, thus enhancing the force constant of the bonds within the harmonic oscillator approximation. At the same time, we computed the evolution of the M_J energy level scheme under the same compressive strain values (Figure 2b). As one can observe, the reduction of the bond distances between the Ho^{3+} and the coordinated oxygen centers leads to a linear increase in the crystal field splitting, considering covalency effects by an effective displacement of the electrostatic point charges, under strain. This linear response of the crystal field is the expected behavior at small distortions. The crystal field terms respond to changes in metal–ligand distance as $O_k^q \propto 1/R^{(k+1)}$ but for distortions in the order of 2% these are very well approximated by linear behavior. In Figures S1–S5 of the Supporting Information, we display a combination of both calculated crystal field energy levels and the infrared spectrum for each applied strain.

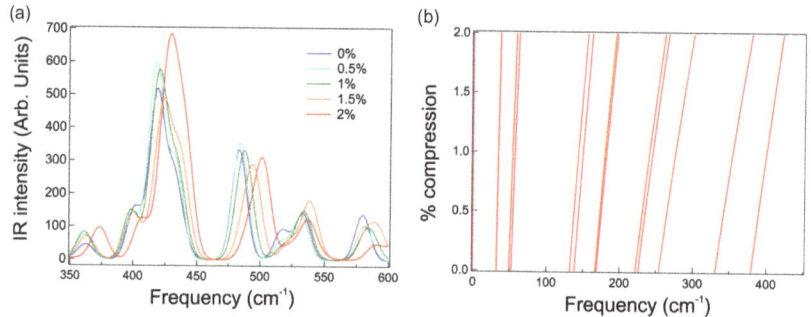

Figure 3. (**a**) Calculated infrared spectra of the building block anion $[Ho(W_5O_{18})_2]^{9-}$ at different compressive strain values (from 0 to 2%) in the 350–600 cm^{-1} region. (**b**) Evolution of the spin energy levels of $Ho(W_5O_{18})_2]^{9-}$ under compressive strain calculated by the radial effective charge model.

$Na_9[Ho(W_5O_{18})_2] \cdot 35H_2O$ already benefits from a relatively sparse lattice (due to the hole in the phonon density of states), and it is likely that it can be made sparser with "chemical pressure" or elongational strain. In our target system, larger more covalent counterions might encourage overall lattice expansion or even just local relaxation sufficient to expand the hole in the phonon density of states. Crystal engineering with chemical pressure is very much akin to chemical control of a host matrix environment [29,38,40–43,45]. At the same time, elongational strain can be tested using a piezostack and incorporated into a device using well-known principles of strain engineering and substrate choice. Again, the idea would be to shield the M_J levels of Ho^{3+} by moving the phonons that engage in the strongest coupling out of the way—without introducing new opportunities for interaction. It is far from certain that the clock transitions will remain near 9 GHz (the X-band) under pressure or strain. Small distortions have already been predicted to change the energy of the clock transitions, indicating the fragility of this parameter [22]. Theory can also test strategies for expanding the lattice and blocking decoherence pathways, simulating the effect of the modified phonon density of states on the vibronic coupling constants and the T_1 relaxation time. This approach to tuning the relative position of states, if successful, has the potential to accelerate the development of molecular spin qubits with improved lifetimes, electric field control, and higher operating temperatures [4,23].

4. Summary and Outlook

$Na_9[Ho(W_5O_{18})_2] \cdot 35H_2O$ is a model spin qubit system with atomic clock transitions and a relatively sparse vibrational spectrum that supports limited overlap between the f manifold levels of Ho^{3+} and the infrared-active vibrational modes, circumstances that favor modest vibronic coupling constants and longer coherence times. In this work, we tested whether an external stimulus in the form of pressure could manipulate this favorable situation even further, perhaps expanding the hole in the phonon density of states to reduce the overlap of these states (and thus the importance of vibronic decoherence pathways) still further. To test this strategy, we measured the far infrared response of $Na_9[Ho(W_5O_{18})_2] \cdot 35H_2O$ under pressure. It turns out that the vibrational modes of this molecular nano-magnet harden under compression. This is in agreement with our simulation of the far infrared spectrum of the evolution of molecular vibrations under compressive strain. On the other hand, the transparency window in the phonon density of states also begins to close, being filled by approximately 5 GPa. These findings suggest that rare earth-containing molecular spin qubits like $Na_9[Ho(W_5O_{18})_2] \cdot 35H_2O$ would instead benefit from negative pressure. While there are a number of efforts to tune similar systems using chemical pressure, tensile strain is under-explored as a technique for disentangling these processes. In this context, device surfaces and interfaces may offer important opportunities in future work.

Supplementary Materials: The following supporting information can be downloaded at: https://www.mdpi.com/article/10.3390/magnetochemistry9020053/s1.

Author Contributions: J.L.M. conceived the study and carried out the high pressure synchrotron far infrared measurements in collaboration with Z.L.; and J.L.M. treated the data, drew the figures, and drafted the manuscript. D.L.-A. performed the theoretical simulations and corresponding figures supervised by J.J.B.; A.G.-A. contributed to the theoretical analysis. Y.D. synthesized the sample under the supervision of E.C. All authors have read and agreed to the published version of the manuscript.

Funding: It is our pleasure to contribute this article to the issue of *Magnetochemistry* in honor of Professor Manuel Almeida on the occasion of their retirement. Thanks for so many useful discussions. Research at the University of Tennessee is supported by the National Science Foundation (DMR-1707846). Work at the National Synchrotron Light Source II at Brookhaven National Laboratory is funded by the Department of Energy (DE-AC98-06CH10886). Use of the 22-IR-1 beamline is supported by COMPRES, the Consortium for Materials Properties Research in Earth Sciences, under NSF Cooperative Agreement EAR 1606856 and CDAC (DE-NA0003975). Research at Universitat de Valencia is supported by the EU (ERC-2018-AdG-788222 MOL-2D) and the QUANTERA project SUMO; the Spanish MCIU (grant CTQ2017-89993 and PGC2018-099568-B-I00 cofinanced by FEDER, grant MAT2017-89528; the Unit of excellence 'María de Maeztu' CEX2019-000919-M); and the Generalitat Valenciana (Prometeo Program of Excellence, SEJI/2018/035 and grant CDEIGENT/2019/022).

Institutional Review Board Statement: None.

Informed Consent Statement: None.

Data Availability Statement: Data are available from the corresponding author upon reasonable request.

Acknowledgments: We thank A. Ullah for useful conversations.

Conflicts of Interest: The authors declare no conflict of interest.

References

1. Woodruff, D.N.; Winpenny, R.E.; Layfield, R.A. Lanthanide single-molecule magnets. *Chem. Rev.* **2013**, *113*, 5110–5148. [CrossRef] [PubMed]
2. McAdams, S.G.; Ariciu, A.-M.; Kostopoulos, A.K.; Walsh, J.P.; Tuna, F. Molecular single-ion magnets based on lanthanides and actinides: Design considerations and new advances in the context of quantum technologies. *Coord. Chem. Rev.* **2017**, *346*, 216–239. [CrossRef]
3. Atzori, M.; Sessoli, R. The second quantum revolution: role and challenges of molecular chemistry. *J. Am. Chem. Soc.* **2019**, *141*, 11339–11352. [CrossRef] [PubMed]
4. Escalera-Moreno, L.; Baldoví, J.J.; Gaita-Ariño, A.; Coronado, E. Design of high-temperature f-block molecular nanomagnets through the control of vibration-induced spin relaxation. *Chem. Sci.* **2020**, *11*, 1593–1598. [CrossRef]
5. Coronado, E. Molecular magnetism: from chemical design to spin control in molecules, materials and devices. *Nat. Rev. Mater.* **2020**, *5*, 87–104. [CrossRef]
6. Baldoví, J.J.; Clemente-Juan, J.M.; Coronado, E.; Duan, Y.; Baldoví, J.J.; Clemente-Juan, J.M.; Coronado, E.; Duan, Y.; Gaita-Ariño, A.; Gimenez-Saiz, C. Construction of a general library for the rational design of nanomagnets and spin qubits based on mononuclear f-block complexes: The polyoxometalate case. *Inorg. Chem.* **2014**, *53*, 9976–9980. [CrossRef]
7. Chakarawet, K.; Atanasov, M.; Ellis, J.E.; Lukens, W.W., Jr.; Young, V.G., Jr.; Chatterjee, R.; Neese, F.; Long, J.R. Effect of spin–orbit coupling on phonon-mediated magnetic relaxation in a series of zero-valent vanadium, niobium, and tantalum isocyanide complexes. *Inorg. Chem.* **2021**, *60*, 18553–18560. [CrossRef]
8. Gould, C. A.; McClain, K. R.; Gould, C.A.; McClain, K.R.; Reta, D.; Kragskow, J.G.C.; Marchiori, D.A.; Lachman, E.; Choi, E.-S.; Analytis, J.G.; Britt, R.D.; Chilton, N.F.; et al. Ultrahard magnetism from mixed-valence dilanthanide complexes with metal-metal bonding. *Science* **2022**, *375*, 198–202. [CrossRef]
9. Duan, Y.; Rosaleny, L.E.; Coutinho, J.T.; Giménez-Santamarina, S.; Scheie, A.; Baldoví, J.J.; Cardona-Serra, S.; Gaita-Ariño, A. Data-driven design of molecular nanomagnets. *Nat Commun* **2022** *13*, 7626. [CrossRef]
10. Bader, K.; Dengler, D.; Lenz, S.; Endeward, B.; Jiang, S.-D.; Neugebauer, P.; Van Slageren, J. Room temperature quantum coherence in a potential molecular qubit. *Nat. Commun.* **2014**, *5*, 5304. [CrossRef]
11. Zadrozny, J.M.; Niklas, J.; Poluektov, O.G.; Freedman, D.E. Millisecond coherence time in a tunable molecular electronic spin qubit. *ACS Central Sci.* **2015**, *1*, 488–492. [CrossRef]
12. Atzori, M.; Tesi, L.; Morra, E.; Chiesa, M.; Sorace, L.; Sessoli, R. Room-temperature quantum coherence and rabi oscillations in vanadyl phthalocyanine: toward multifunctional molecular spin qubits. *J. Am. Chem. Soc.* **2016**, *138*, 2154–2157. [CrossRef]
13. Godfrin, C.; Ferhat, A.; Ballou, R.; Klyatskaya, S.; Ruben, M.; Wernsdorfer, W.; Balestro, F. Operating Quantum States in Single Magnetic Molecules: Implementation of Grover's Quantum Algorithm. *Phys. Rev. Lett.* **2017** *119*, 187702. [CrossRef]

14. Gaita-Ariño, A.; Luis, F.; Hill, S.; Coronado, E. Molecular spins for quantum computation. *Nat. Chem.* **2019**, *11*, 301–309. [CrossRef]
15. AlDamen, M.A.; Cardona-Serra, S.; Clemente-Juan, J.M.; Coronado, E.; Gaita-Ariño, A.; Marti-Gastaldo, C.; Luis, F.; Montero, O. Mononuclear lanthanide single molecule magnets based on the polyoxometalates $[Ln(W_5O_{18})_2]^{9-}$ and $[Ln(\beta_2\text{-}SiW_{11}O_{39})_2]^{13-}$, ($Ln^{III}$ = Tb, Dy, Ho, Er, Tm, and Yb). *Inorg. Chem.* **2009**, *48*, 3467–3479. [CrossRef]
16. Guo, F.-S.; Day, B. M.; Chen, Y.-C.; Tong, M.-L.; Mansikkamäki, A.; Layfield, R. A. Magnetic hysteresis up to 80 kelvin in a dysprosium metallocene single-molecule magnet. *Science* **2018**, *362*, 1400-1403. [CrossRef]
17. Goodwin, C.A.; Reta, D.; Ortu, F.; Chilton, N.F.; Mills, D.P. Synthesis and electronic structures of heavy lanthanide metallocenium cations. *J. Am. Chem. Soc.* **2017**, *139*, 18714–18724. [CrossRef]
18. Long, J.; Ivanov, M.S.; Khomchenko, V.A.; Mamontova, E.; Thibaud, J.-M.; Rouquette, J.; Beaudhuin, M.; Granier, D.; Ferreira, R.A.S.; Carlos, L.D.; et al. Room temperature magnetoelectric coupling in a molecular ferroelectric ytterbium(III) complex. *Science* **2020**, *367*, 671–676. [CrossRef]
19. Ghosh, S.; Datta, S.; Friend, L.; Cardona-Serra, S.; Gaita-Ariño, A.; Coronado, E.; Hill, S. Multi-frequency EPR studies of a mononuclear holmium single-molecule magnet based on the polyoxometalate $[Ho^{III}(W_5O_{18})_2]^{9-}$. *Dalton Trans.* **2012**, *41*, 13697–13704. [CrossRef]
20. Shiddiq, M.; Komijani, D.; Duan, Y.; Gaita-Ariño, A.; Coronado, E.; Hill, S. Enhancing coherence in molecular spin qubits via atomic clock transitions. *Nature*, **2016**, *531*, 348–351. [CrossRef]
21. Vonci, M.; Giansiracusa, M.J.; Van den Heuvel, W.; Gable, R.W.; Moubaraki, B.; Murray, K.S.; Yu, D.; Mole, R.A.; Soncini, A.; Boskovic, C. Magnetic excitations in polyoxotungstate-supported lanthanoid single-molecule magnets: an inelastic neutron scattering and ab initio study. *Inorg. Chem.* **2017**, *56*, 378–394. [CrossRef] [PubMed]
22. Giménez-Santamarina, S.; Cardona-Serra, S.; Clemente-Juan, J. M.; Gaita-Ariño, A.; Coronado, E. Exploiting clock transitions for the chemical design of resilient molecular spin qubits. *Chem. Sci.* **2020**, *11*, 10718–10728. [CrossRef] [PubMed]
23. Liu, J.; Mrozek, J.; Duan, Y.; Ullah, A.; Baldoví, J.J.; Coronado, E.; Gaita-Ariño, A.; Ardavan, A. Quantum coherent spin-electric control in molecular nanomagnets. *Nat. Phys.* **2021**, *17*, 1205–1209. [CrossRef]
24. Härtle, R.; Butzin, M.; Rubio-Pons, O.; Thoss, M. Quantum interference and decoherence in single-molecule junctions: how vibrations induce electrical current. *Phys. Rev. Lett.*, **2011**, *107*, 046802. [CrossRef] [PubMed]
25. Ballmann, S.; Härtle, R.; Coto, P.B.; Elbing, M.; Mayor, M.; Bryce, M.R.; Thoss, M.; Weber, H.B. Experimental evidence for quantum interference and vibrationally induced decoherence in single-molecule junctions. *Phys. Rev. Lett.* **2012**, *109*, 056801. [CrossRef]
26. Hackenmuller, L.; Hornberger, K.; Brezger, B.; Zeilinger, A.; Arndt, M. Decoherence of matter waves by thermal emission of radiation. *Nature* **2004**, *427*, 711–714. [CrossRef]
27. Chirolli, L.; Burkard, G. Decoherence in solid-state qubits. *Adv. Phys.* **2008**, *57*, 225–285. [CrossRef]
28. Graham, M.J.; Zadrozny, J.M.; Shiddiq, M.; Anderson, J.S.; Fataftah, M.S.; Hill, S.; Freedman, D.E. Influence of electronic spin and spin–orbit coupling on decoherence in mononuclear transition metal complexes. *J. Am. Chem. Soc.* **2014**, *136*, 7623–7626. [CrossRef]
29. Chen, J.; Hu, C.; Stanton, J.F.; Hill, S.; Cheng, H.-P.; Zhang, X.-G. Decoherence in molecular electron-spin qubits: Insights from quantum many-body simulations. *J. Phys. Chem. Lett.* **2020**, *11*, 2074–2078. [CrossRef]
30. Ullah, A.; Hu, Z.; Cerda, J.; Aragó, J.; Gaita-Ariño, A. Electrical two-qubit gates within a pair of clock-qubit magnetic molecules, *NPJ Quantum Inf.* **2022**, *8*, 133. [CrossRef]
31. Giansiracusa, M.J.; Kostopoulos, A.K.; Collison, D.; Winpenny, R.E.P.; Chilton, N.F. Correlating blocking temperatures with relaxation mechanisms in monometallic single-molecule magnets with high energy barriers (U_{eff} > 600 K). *Chem. Commun.* **2019**, *55*, 7025–7028. [CrossRef]
32. Gu, L.; Wu, R. Origin of the anomalously low Raman exponents in single molecule magnets. *Phys. Rev. B* **2021**, *103*, 014401. [CrossRef]
33. Kragskow, J.G.C.; Marbey, J.; Buch, C.D.; Nehrkorn, J.; Ozerov, M.; Piligkos, S.; Hill, S.; Chilton, N.F. Analysis of vibronic coupling in a 4f molecular magnet with FIRMS. *Nat. Commun.* **2022**, *13*, 825. [CrossRef]
34. Lunghi, A.; Totti, F.; Sessoli, R.; Sanvito, S. The role of anharmonic phonons in under-barrier spin relaxation of single molecule magnets. *Nat. Commun.* **2017**, *8*, 14620. [CrossRef]
35. Lunghi, A. Toward exact predictions of spin-phonon relaxation times: An ab initio implementation of open quantum systems theory. *Sci. Adv.* **2022**, *8*, eabn7880. [CrossRef]
36. Mondal, S.; Lunghi, A. Spin-phonon decoherence in solid-state paramagnetic defects from first principles. *arXiv* **2022**, arXiv: 2212.11705.
37. Santanni, F.; Albino, A.; Atzori, M.; Ranieri, D.; Salvadori, E.; Chiesa, M.; Lunghi, A.; Bencini, A.; Sorace, L.; Totti, F.; et al. Probing vibrational symmetry effects and nuclear spin economy principles in molecular spin qubits *Inorg. Chem.* **2021**, *60*, 140–151.
38. Blockmon, A. L. Ullah, A.; Hughey, K. D.; Duan, Y.; O'Neal, K.R.; Baldoví, J.J.; Aragó, J.; Gaita-Ariño, A.; Coronado, E.; Musfeldt, J.L. Spectroscopic analysis of vibronic relaxation pathways in molecular spin qubit $[Ho(W_5O_{18})_2]^{9-}$: Sparse spectra are key. *Inorg. Chem.* **2021**, *60*, 14096–14104. [CrossRef]
39. Brinzari, T.V.; Haraldsen, J.T.; Chen, P.; Sun, Q.-C.; Kim, Y.; Tung, L.-C.; Litvinchuk, A.P.; Schlueter, J.A.; Smirnov, D.; Manson, J.L.; et al. Electron-phonon and magnetoelastic interactions in ferromagnetic $Co[N(CN)_2]_2$. *Phys. Rev. Lett.* **2013**, *111*, 047202. [CrossRef]

40. Ullah, A.; Cerdá, J.; Baldoví, J.J.; Varganov, S.A.; Aragó, J.; Gaita-Ariño, A. In silico molecular engineering of dysprosocenium-based complexes to decouple spin energy levels from molecular vibrations. *J. Phys. Chem. Lett.* **2019**, *10*, 7678–7683. [CrossRef]
41. Garlatti, E.; Tesi, L.; Lunghi, A.; Atzori, M.; Voneshen, D.; Santini, P.; Sanvito, S.; Guidi, T.; Sessoli, R.; Carretta, S. Unveiling phonons in a molecular qubit with four-dimensional inelastic neutron scattering and density functional theory. *Nat. Comm.* **2020**, *11*, 1751. [CrossRef] [PubMed]
42. Yu, C.-J.; Von Kugelgen, S.; Krzyaniak, M.D.; Ji, W.; Dichtel, W.R.; Wasielewski, M.R.; Freedman, D.E. Spin and phonon design in modular arrays of molecular qubits. *Chem. Mater.* **2020**, *32*, 10200–10206. [CrossRef]
43. Bayliss, S.L.; Deb, P.; Laorenza, D.W.; Onizhuk, M.; Galli, G.; Freedman, D.E.; Awschalom, D.D. Enhancing spin coherence in optically addressable molecular qubits through host-matrix control. *Phys. Rev. X.* **2022**, *12*, 031028. [CrossRef]
44. Zvyagin, S.A.; Graf, D.; Sakurai, T.; Kimura, S.; Nojiri, H.; Wosnitza, J.; Ohta, H.; Ono, T.; Tanaka, H. Pressure-tuning the quantum spin Hamiltonian of the triangular lattice antiferromagnet Cs_2CuCl_4. *Nat. Commun.* **2019**, *10*, 1064. [CrossRef] [PubMed]
45. Escalera-Moreno, L.; Baldoví, J.J.; Gaita-Ariño, A.; Coronado, E. Spin states, vibrations and spin relaxation in molecular nanomagnets and spin qubits: A critical perspective. *Chem. Sci.* **2018**, *9*, 3265–3275. [CrossRef]
46. Mao, H.K.; Bell, P.M.; Shaner, J.W.; Steinberg, D.J. Specific volume measurements of Cu, Mo, Pd, and Ag and calibration of the ruby R_1 fluorescence pressure gauge from 0.06 to 1 Mbar. *J. Appl. Phys.* **1976**, *49*, 3276–3283. [CrossRef]
47. Mao, H.K.; Xu, J.; Bell, P.M. Calibration of the ruby pressure gauge to 800 kbar under quasi-hydrostatic conditions. *J. Geophys. Res.* **1986**, *91*, 4673–4676. [CrossRef]
48. Frisch, M. J.; Trucks, G. W.; Schlegel, H. B.; Scuseria, G. E.; Robb, M. A.; Cheeseman, J. R.; Scalmani, G.; Barone, V.; Mennucci, B.; Petersson, G. A.; et al. *Gaussian09*; Revision D.01; Gaussian, Inc.: Wallingford CT, USA, 2009.
49. Adamo, C.; Barone, V. Toward reliable density functional methods without adjustable parameters: The PBE0 model. *J. Chem. Phys.* **1999**, *110*, 6158–6169. [CrossRef]
50. Cao, X.; Dolg, M. Valence basis sets for relativistic energy-consistent small-core lanthanide pseudopotentials. *J. Chem. Phys.* **2001**, *115*, 7348. [CrossRef]
51. Cao, X.; Dolg, M. Segmented contraction scheme for small-core lanthanide pseudopotential basis sets. *J. Mol. Struc-THEOCHEM* **2002**, *581*, 139–147. [CrossRef]
52. Dolg, M.; Stoll, H.; Preuss, H. Energy-adjusted ab initio pseudopotentials for the rare earth elements. *J. Chem. Phys.* **1989**, *90*, 1730–1734. [CrossRef]
53. Ross, R.; Powers, J.; Atashroo, T.; Ermler, W.; LaJohn, L.; Christiansen, P. Ab initio relativistic effective potentials with spin–orbit operators. IV. Cs through Rn. *J. Chem. Phys.* **1990**, *93*, 6654–6670. [CrossRef]
54. Hariharan, P.; Pople, J. The influence of polarization functions on molecular orbital hydrogenation energies. *Theor. Chim. Acta* **1973**, *28*, 213–222. [CrossRef]
55. Hehre, W.; Ditchfield, R.; Pople, J. Self-Consistent Molecular Orbital Methods. XII. Further Extensions of Gaussian-Type Basis Sets for Use in Molecular Orbital Studies of Organic Molecules. *J. Chem. Phys.* **1972**, *56*, 2257–2261. [CrossRef]
56. Grimme, S.; Ehrlich, S.; Goerigk, L. Effect of the damping function in dispersion corrected density functional theory. *J. Comput. Chem.* **2011**, *32*, 1456–1465. [CrossRef]
57. Baldoví, J.; Borrás-Almenar, J.; Clemente-Juan, J.; Coronado, E.; Gaita-Ariño, A. Modeling the properties of lanthanoid single-ion magnets using an effective point-charge approach. *Dalton Trans.* **2012**, *41*, 13705–13710. [CrossRef]
58. Baldoví, José J.; Cardona-Serra, S.; Baldoví, J.J.; Cardona-Serra, S.; Clemente-Juan, J.M.; Coronado, E.; Gaita-Ariño, A.; Palii, A. SIMPRE: A software package to calculate crystal field parameters, energy levels, and magnetic properties on mononuclear lanthanoid complexes based on charge distributions. *J. Comput. Chem.* **2013**, *34*, 1961–1967. [CrossRef]
59. Musfeldt, J.L.; Brinzari, T.V.; O'Neal, K.R.; Chen, P.; Schlueter, J.A.; Manson, J.L.; Litvinchuk, A.P.; Liu, Z. Pressure-temperature phase diagram reveals spin-lattice interactions in $Co[N(CN)_2]_2$. *Inorg. Chem.* **2017**, *56*, 4950–4955. [CrossRef]
60. Hughey, K.D.; Harms, N.C.; O'Neal, K.R.; Monroe, J.C.; Blockmon, A.L.; Landee, C.P.; Liu, Z.; Ozerov, M.; Musfeldt, J.L. Spin-lattice coupling across the magnetic quantum phase transition in Cu-containing coordination polymers. *Inorg. Chem.* **2020**, *59*, 2127–2135. [CrossRef] [PubMed]
61. Clune, A.; Harms, N. C.; O'Neal, K. R.; Clune, A.; Harms, N.C.; O'Neal, K.R.; Hughey, K.D.; Smith, K.; Obeysekera, D.; Haddock, J.; Dalal, N.; Yang, J.; Liu, Z.; et al. Developing the pressure—Temperature—Magnetic field phase diagram of multiferroic $[(CH_3)_2NH_2]Mn(HCOO)_3$. *Inorg. Chem.* **2020**, *59*, 10083–10090. [CrossRef] [PubMed]

Disclaimer/Publisher's Note: The statements, opinions and data contained in all publications are solely those of the individual author(s) and contributor(s) and not of MDPI and/or the editor(s). MDPI and/or the editor(s) disclaim responsibility for any injury to people or property resulting from any ideas, methods, instructions or products referred to in the content.

Article

Giant Angular Nernst Effect in the Organic Metal α-(BEDT-TTF)₂KHg(SCN)₄

Danica Krstovska [1,*], Eun Sang Choi [2] and Eden Steven [2,3]

1. Faculty of Natural Sciences and Mathematics, Ss. Cyril and Methodius University, Arhimedova 3, 1000 Skopje, North Macedonia
2. National High Magnetic Field Laboratory, Florida State University, 1800 E. Paul Dirac Drive, Tallahassee, FL 32310, USA
3. Emmerich Research Center, Jakarta Utara DKI, Jakarta 14450, Indonesia
* Correspondence: danica@pmf.ukim.mk

Abstract: We have detected a large Nernst effect in the charge density wave state of the multiband organic metal α-(BEDT-TTF)₂KHg(SCN)₄. We find that apart from the phonon drag effect, the energy relaxation processes that govern the electron–phonon interactions and the momentum relaxation processes that determine the mobility of the q1D charge carriers have a significant role in observing the large Nernst signal in the CDW state in this organic metal. The emphasised momentum relaxation dynamics in the low field CDW state (CDW$_0$) is a clear indicator of the presence of a significant carrier mobility that might be the main source for observation of the largest Nernst signal. The momentum relaxation is absent with increasing angle and magnetic field, i.e., in the high-field CDW state (CDW$_x$) as evident from the much smaller Nernst effect amplitude in this state. In this case, only the phonon drag effect and electron–phonon interactions are contributing to the transverse thermoelectric signal. Our findings advance and change previous observations on the complex properties of this organic metal.

Keywords: Nernst effect; organic metal; charge density wave; quantum oscillations; relaxation processes

Citation: Krstovska, D.; Choi, E.S.; Steven, E. Giant Angular Nernst Effect in the Organic Metal α-(BEDT-TTF)₂KHg(SCN)₄. *Magnetochemistry* **2023**, *9*, 27. https://doi.org/10.3390/magnetochemistry9010027

Academic Editors: Laura C. J. Pereira and Dulce Belo

Received: 8 December 2022
Revised: 3 January 2023
Accepted: 7 January 2023
Published: 10 January 2023

Copyright: © 2023 by the authors. Licensee MDPI, Basel, Switzerland. This article is an open access article distributed under the terms and conditions of the Creative Commons Attribution (CC BY) license (https://creativecommons.org/licenses/by/4.0/).

1. Introduction

Thermoelectric effects provide an important information concerning the sign of the transport carriers and may be useful for the study of the relation between the anisotropy of the electronic bands and superconducting state, especially in multiband superconductors. It is well known that measurements of the Seebeck effect (which is a longitudinal voltage induced by a temperature gradient in a magnetic field) as a function of temperature yield important information concerning not only parameters of charge carriers but also the nature of the electron–phonon scattering in the system, which is especially important for a class of materials such as are organic layered conductors, where strong Fermi liquid effects have been anticipated. The Nernst effect, which is an analogue of the Hall effect, is the transverse voltage induced by a temperature gradient in a magnetic field. It is considered as an important probe of strongly correlated electron systems. The Nernst effect was discovered in bismuth, and experimental studies have shown that it is the dominant thermoelectric response in this material [1]. The Nernst effect is usually small in normal metals, but it is large in semimetals. In high-T_c cuprates, a large Nernst effect was observed in the vicinity of the resistive transition temperature T_c [2] and also near the mobility edge [3]. A giant resonant Nernst signal was also discovered in some layered organic conductors such are the quasi one-dimensional (q1D) organic conductors (TMTSF)₂PF₆ [4,5] and (TMTSF)₂ClO₄ [6]. In the former, a giant Nernst effect was detected in tilted magnetic fields near the Lebed magic angles parallel to crystallographic planes. In the latter, a Nernst effect with a greatly enhanced amplitude was found not only in the metallic state but also at high fields in the field-induced spin density wave state. For what concerns the quasi-two dimensional (q2D) organic conductors, the magnetoresistance, Hall resistance and magnetisation have been studied in detail in many compounds [7]. However, the thermoelectric response which yields information about both the thermodynamic and transport properties of charge carriers has been less investigated. The q2D character of the electron energy spectrum of layered organic conductors distinguishes them from both the two-dimensional materials and normal metals. While the magnetoresistance and Hall effect studies were successfully utilized for the Fermi surface (FS) reconstruction of organic conductors and enabled obtaining important information about the charge carriers, the investigation of the thermoelectric effects (especially in strong magnetic fields) allows us to obtain additional insights about the structure of the electron energy spectrum, as these effects are much more sensitive to the changes in the energy spectrum. The thermoelectric effects are also important for revealing the presence of the q1D FS open sheets in multiband organic conductors.

In this work, we investigate the Nernst effect behaviour in a quasi-two dimensional, two-band organic metal α-(BEDT-TTF)$_2$KHg(SCN)$_4$. This organic conductor has been an object of intense experimental studies for many years due to its unusual electronic properties (see Ref. [8] and references therein). The FS consists of both a q2D cylinder and a pair of weakly warped open q1D sheets. The q1D and q2D bands are separated by a substantial gap near the Fermi level. At $T_p = 8$ K, the system undergoes a phase transition from metallic to a charge density wave (CDW) state due to the nesting of the q1D open sheets. At the kink field B_K, the low-field CDW$_0$ state is transformed into the high-field CDW$_x$ state with a field-dependent wavevector. Throughout the years, there have been many reports on the magnetotransport studies at low temperatures in this organic metal aiming to obtain information on the Fermi surface topology as well as to explain the origin of the charge density wave state at temperatures below $T_p = 8$ K [9–13]. On the contrary, the thermoelectric transport has not been studied in much detail. In that regard, our goal is to make a contribution to the investigation of the thermoelectric response in the given compound, especially at low temperatures. Our studies show that at low temperatures, in α-(BEDT-TTF)$_2$KHg(SCN)$_4$, the thermoelectric response is mainly off-diagonal, although both Seebeck and Nernst effect are detected. We have detected a giant angular resonant Nernst effect in the CDW state of α-(BEDT-TTF)$_2$KHg(SCN)$_4$ similar to that reported in some q1D organic compounds. Furthermore, the magnetic field dependence of the Nernst effect reveals a resonant-like behaviour with sign change at certain magnetic fields when the magnetic field is tilted at the magic angles. The temperature dependence reveals that the relaxation processes have a significant role in the observed large Nernst effect in this compound. It also shows and confirms previous observations that in α-(BEDT-TTF)$_2$KHg(SCN)$_4$, there are different phases existing within a given temperature interval.

2. Materials and Methods

The single-crystal sample in this study was grown using conventional electrochemical crystallization techniques and was mounted on a rotating platform. Au wires were attached to the sample along the b-axis and on the edges of the ac-plane of the sample by carbon paste for both the resistance and Nernst effect measurements, respectively (Figure 1). The resistance was measured by a conventional 4-probe technique. The sample was positioned between two quartz blocks, which were heated by sinusoidal heating currents with an oscillation frequency f_0 and phase difference of $\pi/2$ to establish a small temperature gradient along the b-axis. The details about the method used for the thermoelectric measurements are given in Ref. [14]. The direction of the magnetic field is arbitrary, $\mathbf{B} = (B\sin\theta\cos\phi, B\sin\theta\sin\phi, B\cos\theta)$, where θ is the polar angle (measured from the b-axis to ac-plane) and $\phi = 37°$ is the azimuthal angle (measured from the c-axis in the ac-plane).

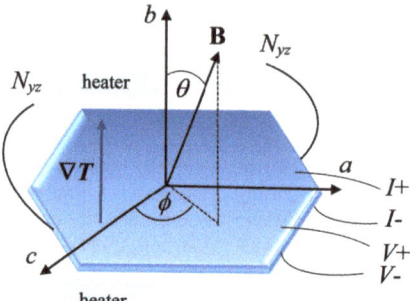

Figure 1. The experimental setup for the Nernst effect measurements as a function of the magnetic field strength and orientation. The temperature gradient ∇T is along the less conducting axis, b-axis of the conductor, perpendicular to the Q2D conducting ac plane. The magnetic field **B** is rotated for angle θ from the b-axis to the ac-plane at a fixed azimuthal angle of $\phi = 37°$ from the c-axis.

3. Results and Discussion

In the following, we present a detailed study of the Nernst effect in the multiband organic conductor α-(BEDT-TTF)$_2$KHg(SCN)$_4$ with the magnetic field magnitude, orientation and temperature. The magnetic field and angular dependencies are obtained in the low-temperature regime in the CDW state of the conductor. The Nernst effect investigation allows us to obtain more insights about the behaviour of different CDW states that develop in this conductor as well as information on the driving mechanism for the CDW instabilities in α-(BEDT-TTF)$_2$KHg(SCN)$_4$. Additionally, it may give information about the role of quasiparticles in the transport processes in multiband systems.

3.1. Angular Hall and Nernst Effect Oscillations

The angular Hall resistance oscillations (AHROs) in the CDW$_0$ state of α-(BEDT-TTF)$_2$KHg(SCN)$_4$ are shown in Figure 2 for several fields: 10 T, 15 T, and 21 T at $T = 0.6$ K. The Hall resistance quantum oscillations, associated with the Landau quantization of the closed FS orbit, are observed superimposed on the angular oscillations for each magnetic field strength. Although in the CDW phase, no q1D FS open sheets should survive, there appear clear angle-dependent Hall resistance oscillations similar to the Lebed resonances, which are characteristic for q1D organic conductors [15]. The AHRO maxima and minima appear at angles $\theta_{\text{max}i}$ and $\theta_{\text{min}i}$ where $i = 1, 2, 3, \ldots$. Interestingly, the dips in the Hall resistance become sharp at angles above $\theta = 60°$, whereas the first minimum occurring at $\theta_{\text{min}1} \sim 45°$ is rather broad especially with increasing magnetic field magnitude. This is in contrast to the angular magnetoresistance oscillations (AMROs) behaviour in α-(BEDT-TTF)$_2$KHg(SCN)$_4$ (see Figure 2) where the maxima in the angular dependence are not as sharply pronounced as those in the AHROs. The Hall resistance is negative, indicating that in α-(BEDT-TTF)$_2$KHg(SCN)$_4$, the in-plane transport is mainly electron-like independently on the field direction with respect to the layers. An important observation is that the AHRO maxima occur at angles at which AMRO minima appear known as Lebed magic angles. Consequently, AHRO minima are observed at angles at which AMRO maxima are detected. However, there is a slight shift between the AMRO (AHRO) maxima (minima) seen only at angles below $\theta = 60°$. This can be more clearly identified from the corresponding curves presented in Figure 3 below. With rotation of the magnetic field toward the plane of the layers, more charge carriers have been involved in the transport processes, which can account for the absence of a shift between the AMRO (AHRO) maxima (minima) at angles above $\theta = 60°$. Interestingly, changes in the angle-dependent Nernst effect also occur around this angle at each magnetic field magnitude. Our results show that in the compound under consideration, the thermoelectric transport is more complex than previously anticipated, as the large Nernst signal arises in total absence of superconducting fluctuations. This opens a new window for investigation of the driving mechanism for the CDW instabilities in α-(BEDT-TTF)$_2$KHg(SCN)$_4$.

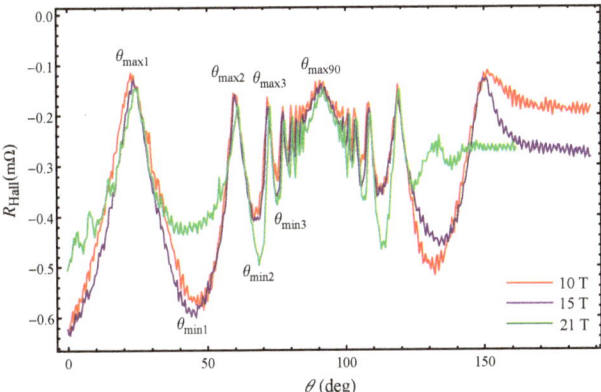

Figure 2. Angular oscillations of the Hall resistance in α-(BEDT-TTF)$_2$KHg(SCN)$_4$ for several fields: $B = 10$ T, 15 T, 21 T at $T = 0.6$ K. The quantum oscillations are visible at each field magnitude in the whole range of angles. The positions of some of the AHROs maxima and minima are indicated.

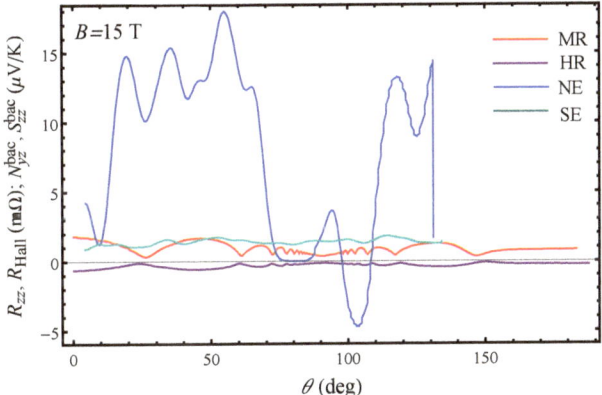

Figure 3. Comparison of the angular dependence of magnetoresistance, Hall resistance, Seebeck and Nernst effect in α-(BEDT-TTF)$_2$KHg(SCN)$_4$ for $B = 15$ T and $T = 0.6$ K. The Nernst effect amplitude is extremely large compared to the amplitude of magnetoresistance, Hall resistance and thermopower in a wide range of angles with exception of the angles near the plane of the layers.

We present in Figure 4 the angular dependence of the total Nernst effect in α-(BEDT-TTF)$_2$KHg(SCN)$_4$ for several fields: $B = 10$ T, 23 T, 26 T and 30 T at $T = 0.6$ K. We have detected an angular Nernst signal with a large amplitude for each magnetic field magnitude. Exceptions from this are certain angles as discussed below. We find that in α-(BEDT-TTF)$_2$KHg(SCN)$_4$, the angular Seebeck effect (thermopower) [16] is much smaller than the Nernst effect, although the thermoelectric measurements are performed for a longitudinal temperature gradient. Both effects show oscillations, but the transverse component is clearly dominant, as it is about 10 times larger than the longitudinal one. This is also a case even for magnetic field orientations close to the less conducting axis (b-axis) where the thermopower is expected to be the dominant thermoelectric response. The rather complex nature of the observed angular Nernst signal in α-(BEDT-TTF)$_2$KHg(SCN)$_4$ is correlated with its multiband character as two kinds of carriers are involved in the thermoelectric transport. The giant Nernst effect previously observed in some of the q1D organic compounds with an undetectable Seebeck effect has been ascribed to the vortex flow [4,5]. On contrary, the superconducting state in α-(BEDT-TTF)$_2$KHg(SCN)$_4$ develops below $T_c = 0.1$ K and for a quasi-hydrostatic pressure of $P_c \sim$2.3–2.5 kbar [17]. This indicates that in the compound under consideration, the quasiparticles are responsible for the observed large amplitude of the Nernst effect.

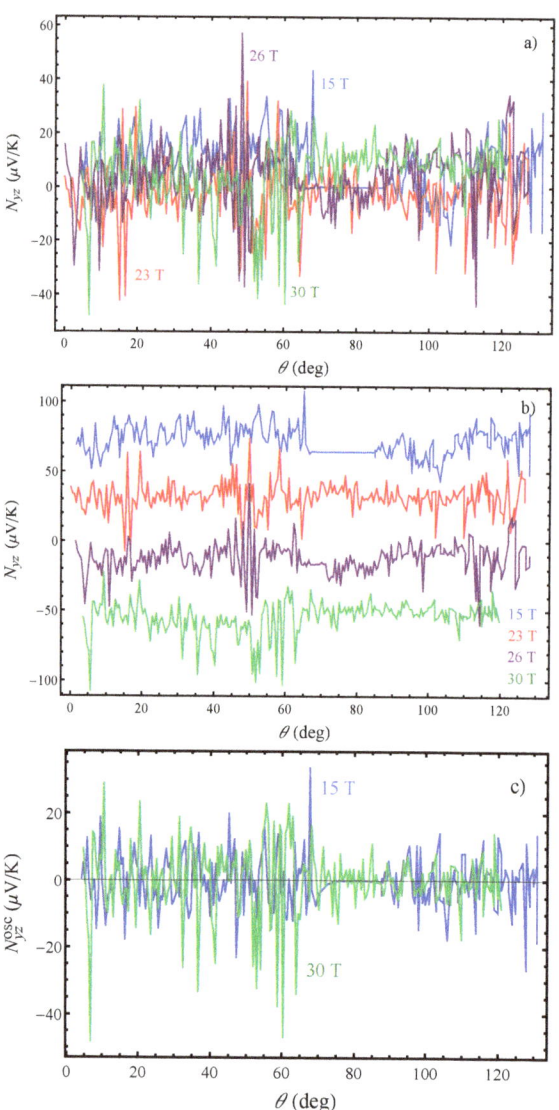

Figure 4. (**a**) Angular oscillations of the total Nernst effect in α-(BEDT-TTF)$_2$KHg(SCN)$_4$ for several fields: $B = 15$ T, 23 T, 26 T and 30 T and $T = 0.6$ K. The Nernst effect quantum oscillations are visible at each magnetic field magnitude in the whole range of angles and are the dominant component in the total Nernst effect. (**b**) The Nernst effect oscillatory curves are shifted for clarity. (**c**) Angular quantum oscillations of the Nernst effect for $T = 0.6$ K at $B = 15$ T and $B = 30$ T differ in amplitude and phase, indicating that the system is driven from one into another CDW state with increasing field and tilt angle.

For obtaining a more detailed picture on the thermoelectric transport in α-(BEDT-TTF)$_2$ KHg(SCN)$_4$, one should study the behaviour of the two components, the quantum oscillating component and the background term. The quantum oscillations are visible at each magnetic field magnitude in the whole range of angles. The exceptions are only the angles close to the layer's plane at 15 T in the CDW$_0$ state (Figure 4c) where the total Nernst signal is vanishing, although it should not be zero because the transverse voltage is the largest along the y direction. The quantum oscillations are mainly due to the fundamental frequency of the closed FS α orbit, $F_α = 671$ T, which is present above $B = 8$ T. The oscillatory part of the transport coefficients is determined by the oscillatory dependence of the relaxation time $τ$ that results from the summation over all the electron states in the incoming term of the collision integral. In the Born approximation, for $\hbar ω_c \ll t_c$, $τ \sim τ_0 τ^{osc}$ where $\hbar$ is the Planck constant divided by $2π$, $ω_c = eB/m^*$ is the cyclotron frequency, m^* is the cyclotron effective mass, t_c is the interlayer transfer integral, and $τ_0$ and $τ_{osc}$ are the non-oscillatory and oscillatory part of the relaxation time. The oscillatory part of the relaxation time is a sum of the contributions from all extremal orbits, $S_{exti} = 2π\hbar eF_i$, on the Fermi surface for a particular orientation in a magnetic field $τ_{osc} \sim \sum_i A_i (B\cos θ)^{1/2} R_T R_S R_D R_{MB} \cos(2π(\frac{F_i}{B\cos θ} - \frac{1}{2}))$, where R_T accounts for finite temperature effects, R_D for impurity scattering, R_S for spin-splitting effects, and R_{MB} for the magnetic breakdown effects. As a result of the oscillatory dependence of $τ$, the thermoelectric coefficient $α_{zz}$ acquires an oscillatory component $α_{zz}^{osc} \sim σ_0 \frac{∂τ_{osc}}{∂ε}|_{ε=μ}$ and so the oscillatory part of the Nernst effect is $N_{yz}^{osc} \sim \frac{α_{zz}^{osc}}{σ_{yz}} \sim \frac{1}{σ_{yz}} \sum_i A_i (B\cos θ)^{-1/2} R_T R_S R_D R_{MB} \sin(2π(\frac{F_i}{B\cos θ} - \frac{1}{2}))$. It is large when a Landau level meets the chemical potential $μ$ and is damped when $μ$ is between two successive Landau levels. The amplitude of the Nernst effect quantum oscillations is much larger than that of the Hall resistance quantum oscillations (about 10^2 times larger) as can be seen by comparing Figures 2 and 4. This is in correlation with the high fundamental frequency of the main oscillations, $F_α$, for $μ \gg \hbar ω_c$ which strongly increases the amplitude of the effect when performing the derivative of the density of states over $μ$.

The background Nernst effect shown in Figure 5 reveals oscillatory features emphasizing the resonant-like character of the effect. The resonance-like behaviour of the background Nernst effect is evident at each magnetic field magnitude similar to that previously observed in the angular dependence of the Seebeck effect [16]. Apart from that, the angular dependence of the Nernst effect changes significantly with increasing magnetic field strength. This is a reflection of the complex electronic properties associated with the existence of different CDW states in this organic metal. It should be specifically emphasized that all of the Nernst effect curves show different behaviour above a certain tilt angle, $θ_c \sim 60°$, which corresponds to the second maximum in the Hall resistance angular dependence. At a lower field ($B = 15$ T), when the system is in the CDW$_0$ state, the Nernst effect decreases above $θ_c$ without changing the sign. A decrease in the Nernst effect amplitude is also seen at $B = 26$ T above $θ_c$ accompanied with a sign change around $θ = 85°$. At $B = 30$ T, the Nernst effect is zero at $θ_c$ above which it is positive with a pronounced increase in the amplitude. For $B = 26$ T and $B = 30$ T, the system is supposed to be in the CDW$_x$ state, but the observed distinct behaviour of the Nernst effect around $θ_c$ implies that this is not the case. The curve obtained for $B = 23$ T demonstrates the most unusual behaviour as the Nernst effect constantly changes sign in the angular range between the first and third AHRO maximum. Most strikingly, the curve is mirroring its angular behaviour exactly at $θ_c$, which is not characteristic for the other curves. The observed different Nernst effect behaviour, especially above $θ_c$, with or without a sign change, indicates that significant changes occur in the density of states.

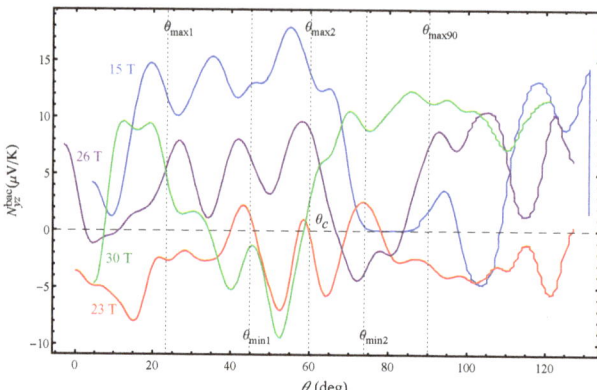

Figure 5. Angular oscillations of the background Nernst effect in α-(BEDT-TTF)$_2$KHg(SCN)$_4$ for several fields: B = 15 T, 23 T, 26 T, 30 T at T = 0.6 K. The positions of some of the AHRO maxima and minima are indicated by the vertical dashed lines.

In order to capture the magnitude of the observed Nernst signal at low temperature, we present in Figure 3 the AMROs, AHROs, Seebeck and Nernst effect angular oscillations at B = 15 T and T = 0.6 K. It is clearly seen that the amplitude of the Nernst effect is exceeding by far the amplitudes of the other effects with an exception of a small interval of field orientations near the layer's plane where the Nernst signal is zero and the Seebeck effect is the main thermoelectric response, although this should not be observed when the temperature gradient is along the normal to layer's plane. In addition to the large amplitude, there are other features that should be addressed concerning the behaviour of the angular Nernst effect in the CDW state. The features observed in the angular Nernst effect for T = 0.6 K neither appear at the magic angles (where the peaks and dips in the AMRO and AHRO are observed) nor are located at the mid angle positions. However, the latter was observed in the Nernst effect angular oscillations for T = 4 K [18]. Another interesting observation is that for T = 4 K, in the CDW$_0$ state, the Nernst effect is also oscillatory up to angles $\theta_c \sim 60°$ but with a 10 times smaller amplitude than that for T = 0.6 K. However, for T = 0.6 K, the Nernst effect is the largest in the CDW$_0$ state up to angles $\theta \sim 70°$ (Figure 5). This implies that at low temperatures, deep down in the CDW$_0$, there are some processes that are dominant leading to an existence of a large transverse thermoelectric response in the current compound, and the Nernst effect is very sensitive to these low-temperature processes.

By comparing Figures 2 and 5, one can see that the transverse thermoelectric transport represents more remarkable signatures of the CDW transitions, as the Nernst effect is very different from the Hall resistance. This is because in α-(BEDT-TTF)$_2$KHg(SCN)$_4$, the Nernst effect is especially sensitive to both the Landau-level spectrum and energy gap in the electronic structure. We find that both the quantum oscillating and background component of the Nernst effect behave differently with increasing angles and fields. At B = 15 T, the amplitude of the quantum oscillations is of the same order as the background up to angles $\theta_c \sim 60°$, while above this angle, there is a sudden decrease in both components, which vanishes for field orientations close to the plane of the layers. This behaviour, however, is not expected, as the Nernst signal is supposed to be the largest when the generated transverse electric field is the largest. With an increasing field, at B = 23 T and B = 26 T, the quantum oscillating component is dominant over the background one in the whole range of angles. For B = 30 T, the quantum oscillating component is larger only below θ_c. Figure 5 shows that above $\theta_c \sim 60°$, the Nernst effect amplitude is large only at high magnetic fields. In the absence of systematic studies of the Nernst effect in other layered q2D organic systems, one may speculate in several directions concerning the origin of the giant Nernst signal. We first note that due to the change of the quantum

Nernst effect component below and above $\theta_c \sim 60°$, there might be a significant change in the quasiparticle density of states at the Fermi energy, which may lead to an enhanced Nernst signal. On the other hand, the background Nernst effect is intimately related to the off-diagonal component of the thermoelectric coefficient tensor α_{yz}, which is the energy derivative of the off-diagonal conductivity component σ_{yz} at the Fermi energy, according to the Mott formula $\alpha_{yz} = \frac{\pi^2 k_B^2 T}{3e} \frac{\partial \sigma_{yz}}{\partial \varepsilon}|_{\varepsilon=\mu}$. Thus, if σ_{yz} changes due to to a small change in the Fermi surface volume, this could also lead to an enhanced α_{yz} and hence to a sizable Nernst signal. Since the Nernst effect is very sensitive to changes in the Fermi surface, this implies that in α-(BEDT-TTF)$_2$KHg(SCN)$_4$, large changes occur in the out-of-plane electronic structure.

With investigation of the magnetic and temperature dependence of the Nernst effect, in addition to the angular dependence, one can obtain information about the possible sources for detection of the large angular Nernst effect in this multiband system. In regard to that, the obtained magnetic field and temperature dependencies of the Nernst effect (discussed below) suggest that the primary source of producing a large Nernst signal is not the change of the quasiparticle density of states but the change of charge carriers relaxation times (i.e., the change of the charge carriers scattering rates). Although the charge relaxation dynamics has not been much considered in studying the thermoelectric effects, in many cases, it can be dominant in producing a sizeable thermoelectric response, especially a transverse one. However, we have detected a large angular Nernst signal in contrast to the observed much smaller magnetic Nernst signal (Figure 6 below). This implies that the observed giant angular Nernst effect is not only due to changes in the density of states and scattering rates but also due to changes in the charge carriers velocity as they move along different parts of the FS. Indeed, with changing the field orientation, the electron trajectories along the corrugated FS also change. For specific field orientations, the velocity is more effectively averaged to zero, and therefore, transport is reduced. For the Fermi cylinder, the velocity vector along the cylinder axis goes to zero when all closed orbits have the same area. For the open Fermi sheets, the velocity vector is reduced when the electrons are not moving along the corrugation axis. This can explain the zero angular Nernst signal observed at certain angles in Figure 5. In addition, the Nernst effect is most clearly manifested when the temperature gradient and the velocity vector of the q1D charge carriers are not perpendicular to each other [19]. This might be the reason for observing the largest angular Nernst effect in the CDW$_0$ state for angles below θ_c and a zero Nernst effect for angles close to the plane of the layers (blue curve in Figure 5). On the other hand, a large Nernst effect is observed above $\theta_c = 60°$ at a high field (green curve for $B = 30$ T in Figure 5). The Nernst signal is positive and large as a result of the formation of a significant number of new closed hole-like orbits obtained as a result of the magnetic breakdown effect. Considering that a field of 30 T is larger than the magnetic breakdown field ($B_{MB} = 20$ T), many new closed orbits with a different area are formed, and thus, the q2D charge carriers velocity vector, along the cylinder axis, significantly differs from zero. For $B = 26$ T (purple curve in Figure 5), the Nernst effect is resembling that for 15 T for angles below $\theta_c = 60°$ but is not zero for field orientations close to the plane of the layers. It seems that at high fields, the CDW$_0$ state is still present for angles below θ_c, although the magnetic field for this measurement is larger than the kink field B_K, and the CDW$_x$ state is expected to be observed instead of the CDW$_0$. However, above θ_c, the system is in the CDW$_x$ state due to the appearance of new electron-like closed orbits resulting from the magnetic breakdown. The peculiar behaviour of the Nernst effect for $B = 23$ T (red curve in Figure 5) is correlated to the proximity of the field (for which this angular dependence is obtained) to the kink field, $B_K \sim 22$ T, at which the CDW$_0 \rightarrow$ CDW$_x$ transition occurs. Therefore, for $B = 23$ T, the Nenst effect curve is most probably a result of the mixing of contributions from several bands, and hence, it reflects the properties of both the CDW$_0$ and CDW$_x$ state.

With rotation of the magnetic field from the direction of the temperature gradient towards the plane of the layers, the average velocity of charge carriers along the z-axis, $\bar{v}_z$, decreases, but the average velocities along the x- and y-axis, $\bar{v}_x = \bar{v}_z \cos\phi \tan\theta$ and

$\bar{v}_y = \bar{v}_z \sin\phi \tan\theta$, are rather large. This can lead to the generation of a substantial in-plane Nernst signal N_{yx}. In other words, the in-plane Nernst effect component can be significant, and its contribution can not be neglected as it can affect the overall behavior of the out-of-plane Nernst signals due to mixing of the components of the thermoelectric tensor α_{ij}. In support of this scenario, Choi et al. [20] have shown that at low temperatures, the in-plane Nernst effect N_{yx} is large enough even when the magnetic field is applied along the b-axis, i.e., along the direction of the temperature gradient.

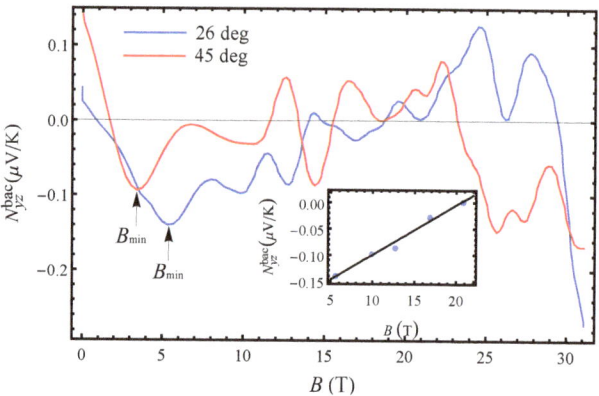

Figure 6. Magnetic field dependence of the background Nernst effect in α-(BEDT-TTF)$_2$KHg(SCN)$_4$, $N_{yz}^{bac}(B)$ after filtering out the quantum oscillation component. The minimum and maximum features that appear in the field dependence at different field orientations are evident. The arrows indicate B_{min}, which is angle-dependent and decreases with the increasing tilt angle. The inset shows the magnetic field dependence of the dip-to-dip amplitude of the Nernst effect at $\theta_{max1} = 26°$.

3.2. Resonant Magnetic Field Nernst Effect

Figure 7a shows the magnetic field dependence of the total Nernst effect in α-(BEDT-TTF)$_2$KHg(SCN)$_4$ obtained at $T = 0.6$ K and $\theta = 26°, 45°$. The angles correspond to the orientations of the magnetic field where the AHROs have a maximum and a minimum (first maximum and minimum in Figure 2), since the Nernst effect behaves differently at these orientations. At low temperatures, the magnetic Nernst effect is small compared to the Seebeck effect [16]. The magnetic Nernst effect has a similar amplitude for orientations corresponding to both the AHRO maximum and minimum. From Figure 7b, which presents the magnetic quantum oscillations of the Nernst effect with the inverse magnetic field, one can see that the total magnetic Nernst effect is dominated by the quantum oscillating component in fields above 8 T. The quantum oscillations are mainly determined by the fundamental α frequency with a small contribution from its second harmonic only for $\theta_{max1} = 26°$. There is no significant change in the amplitude of the magnetic quantum oscillations on crossing the kink field $B_K \sim 24$ T, i.e., with the transition from CDW$_0$ to CDW$_x$ state. This is opposite from what was observed for the angular quantum oscillations where the amplitude of the oscillations changes significantly depending on if the field, for which a given angular dependence is obtained, is below or above the kink field B_K. This implies that if indeed, there are changes happening in the out-of-plane electronic structure in this organic metal, as suggested by the angular Nernst effect behaviour, then these changes are greatly triggered by the magnetic field rotation and are manifested mostly in the Nernst effect and not in the Seebeck effect (as evident from the large angular Nernst effect in this work and the small angular Seebeck effect in Ref. [16]).

The background Nernst effect shown in Figure 6 also manifests minimum and maximum features in the magnetic field dependence similarly to those previously observed in the background Seebeck effect in Ref. [16]. However, the features seen in the Nernst effect are slightly more pronounced (especially with increasing angle), although the Seebeck effect is exceeding the Nernst effect. For angles close to the b-axis ($\theta_{max1} = 26°$, corresponding to the first AHRO maximum), the Nernst effect shows weak oscillatory features and changes sign at $B \sim 18$ T and also around $B \sim 30$ T. The electron–hole asymmetry in the CDW_0 state is evident as the transport changes from electron-like to completely hole-like. In the CDW_0 state, the Nernst signal changes significantly at $B_{min} \sim 5$ T from a relatively linear (below B_{min}) to an approximately linear (increase in slope) with field. For $\theta_{max1} = 26°$, the dip-to-dip Nernst effect amplitude changes linearly with field as shown in the inset in Figure 6. The negative Nernst effect below 18 T, in the CDW_0 state, indicates that the transport is essentially electron-like due to the electrons on the open Fermi sheets. This, on the other hand, confirms that the small closed orbits do not form in the CDW_0 state after the Fermi surface reconstruction; i.e., there are only open electron orbits in this state. However, since magnetic breakdown effects take place around $B = 20$ T, the probability for the formation of closed hole orbits is increasing with increasing fields.

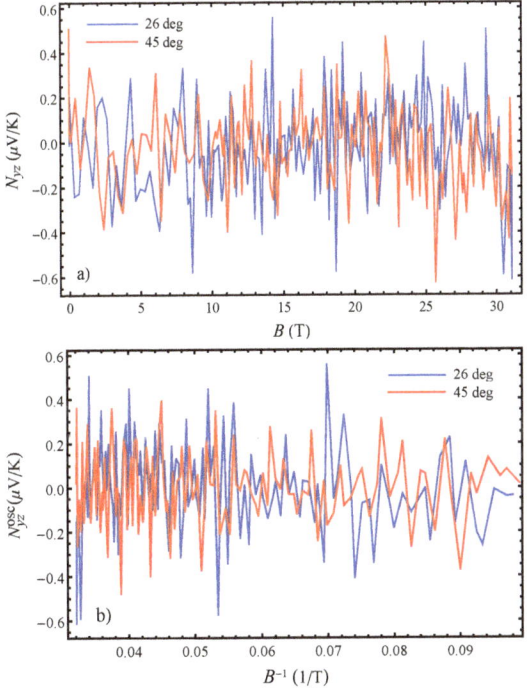

Figure 7. (**a**) Magnetic field dependence of the total Nernst effect in α-(BEDT-TTF)$_2$KHg(SCN)$_4$ obtained at $T = 0.6$ K and different angles: $\theta = 26°, 45°$ that correspond to the first maximum and first minimum in the Hall resistance angular oscillations, respectively. (**b**) Magnetic quantum oscillations of the Nernst effect with the inverse magnetic field at the same angles (B is in the range 10–30 T). The quantum oscillations display a complex structure with multiple periodicities.

With tilting the field at the first AHRO minimum ($\theta_{min1} = 45°$), the oscillatory features become more prominent, but the Nernst effect is in general small. An important observation is that when the field is located at the AHRO minimum, the Nernst effect changes sign first at a low field, then in between 10 and 18 T, and again just before entering the high-field CDW_x state where it becomes negative. Above the kink field for this angle, $B_K = 25$ T, the Nernst signal is weakly field dependent, resembling the behaviour of the Seebeck effect but with a smaller amplitude. Here, the field dependence changes from linear to non-linear at $B_{min} \sim 3.5$ T. The sign change in the field range $B = 10$–18 T is due to the formation of new electron and hole-closed orbits as a result of the magnetic breakdown. In that way, with increasing angle and field, both types of charge carriers become involved in the thermoelectric transport, while their dominance changes with increasing field. For this field orientation, the closed electron/hole orbits are formed much below the magnetic breakdown field. The different Nernst effect behaviour with the magnetic field from that observed for $\theta_{max1} = 26°$ shows that the q1D charge carriers are not the dominant carriers with increasing angle, but the role of the q2D carriers becomes essential in the transport processes at moderate fields of 10 T.

The charge carrier concentration changes depending on the magnetic field direction (maximum or minimum location). Although the carrier density n is difficult to estimate for the present metal with both open and closed bands, we could obtain $n(\theta_{max1}) = 6.9 \times 10^{18}$ cm^{-3} and $n(\theta_{min1}) = 1.6 \times 10^{18}$ cm^{-3} for $B = 15$ T by using the simple relation $n = 1/eR_H$. Since $n(\theta)$ decreases with the rotation of the field at the location of the first minimum, the amplitude of magnetic Nernst effect quantum oscillations slightly grows, and their frequency becomes smaller (Figure 7b). For these carrier concentrations, giant angular Nernst quantum oscillations with a single frequency are observed. Taking into account that these are not very high carrier concentrations and given that the in-plane conductivity is large, it follows that the carrier mobility, $\mu_B = \sigma/en$, will be high. In addition, as $n(\theta)$ changes with the field rotation from the AHRO maximum to the AHRO minimum, the charge mobility changes rapidly with the rotation of the field from the AHRO maxima to the AHRO minima. We further investigate, with the temperature-dependent measurements of the Nernst effect, what processes contribute to the rapidly changing carrier mobility, which can significantly contribute to the Nernst effect enhancement in the given compound.

3.3. Temperature Dependence of the Nernst Effect

The temperature dependence of the out-of-plane Nernst effect presented in Figure 8 shows that the effect is strongly non-linear in the given temperature range, and the temperature profile changes significantly with increasing field and angle. One striking feature of Figure 8a is the presence of a large negative Nernst signal in the CDW$_0$ state and an even larger Nernst effect in the mixed state. The Nernst effect has a magnitude drastically exceeding what is expected for a multiband system. A large Nernst effect has been also discovered in some Bechgaard salts for fields oriented along the magic angles [5] and in some heavy-fermion compounds [21]. More interestingly, our results show that the Nernst effect in the organic metal α-(BEDT-TTF)$_2$KHg(SCN)$_4$ has a very similar behavior and size compared to that in a heavy-fermion superconductor CeCoIn$_5$ [22]. This indicates that quasiparticels can lead to a large Nernst signal in the total absence of superconducting fluctuations.

For a magnetic field of 20 T oriented along the first AHRO maximum ($\theta_{max1} = 26°$), the temperature dependence of the Nernst effect shown in Figure 8a reveals there is a CDW$_0$ state below 8 K with a $N_{yz}(T) = A + B/T + CT^3 + DT^5$ dependence (A, B, C and D are constants). Above 8 K, instead of the expected metallic behaviour, a mixed state with a $N_{yz}(T) = E + FT + GT^3 + HT^5$ dependence (E, F, G and H are constants) is realized. The Nernst signal changes sign from positive to negative around 6 K, indicating a change in the dominant charge carriers in the thermoelectric transport. The observed broad positive and negative maximum in Figure 8a might be an indicator that there is a thermally induced counter-flow of electrons and holes when the field is oriented at the first AHRO maximum. Our results reveal that the temperature profile is complex, as several terms arising from different processes appear in the temperature dependence. The thermal activation over an energy band gap gives a $1/T$ dependence, as expected, since the system is driven into the CDW$_0$ state. The presence of this term indicates that the electron–phonon coupling has an essential role in the transport processes at low temperatures and high fields. However, only the phonon drag term does not provide a reasonable fit to the data. There are two additional terms in the temperature dependence proportional to T^3 and T^5, respectively. This was not observed in the Seebeck effect temperature profile for the same field magnitude and orientation [16]. The presence of these terms in the given temperature dependence is a clear indicator that the relaxation times (on which the components of the kinetic and thermoelectric coefficients depend) have different temperature dependence. The absence of the above terms in the Seebeck effect temperature dependence indicates that the Nernst effect is much more sensitive to the change in the relaxation dynamics in this organic metal. The energy relaxation processes governing the electron–phonon interactions and the momentum relaxation processes governing the charge carriers mobility are described by the temperature-dependent scattering times, $\tau_\varepsilon(T)$ and $\tau_p(T)$. The components of the electrical conductivity tensor σ_{ij} depend on τ_p and the components of the thermoelectric coefficient tensor α_{ij} depend on τ_ε. At temperatures much less than the Debye temperature Θ_D, as it is in the case under consideration, the temperature dependence of τ_ε is different than that of τ_p. With the obtained T^3 and T^5 terms in the Nernst effect temperature profile, our results show that the energy and momentum relaxation processes might have an important role in observing the large Nernst signal, especially in the CDW_0 state. For $T \ll \Theta_D$, electron–phonon scattering processes make a significant contribution to the energy relaxation. In that case, τ_ε is proportional to T^3, and the momentum relaxation time τ_p is proportional to T^5 [19]. The presence of the latter term indicates the presence of a significant momentum relaxation dynamics in the CDW_0 state. A significant gradient of charge relaxation processes can generate a sizeable contribution to the transverse thermoelectric signal. The observed temperature dependence of the Nernst effect arises predominantly from the presence of the q1D group of charge carriers in accordance with the picture of a reconstructed Fermi surface with strongly corrugated open Fermi sheets in the CDW_0 state. Thus, except for the phonon drag, the change of the electron scattering rates can lead to an unusually large Nernst effect signal. As already known, peak structures in the temperature dependence of the Nernst effect appear as a result of the phonon drag effect (since the Nernst effect is more sensitive to the phonon drag effect than thermopower), but in this case, the change in the electron scattering rates additionally contributes to the observed large Nernst effect. It is known that the momentum relaxation time is proportional to the charge carrier mobility, $\tau_p = \mu_B e / m^*$. In that way, any change in τ_p will lead to a change in the charge carrier mobility. If τ_p is large enough, then there is a counter-flow of high mobility charge carriers (thermally induced quasiparticles) that contributes to generating a large transverse voltage. This is in agreement with results previously obtained from the angular and magnetic field dependence of the Nernst effect. In that way, in a material with a changing carrier mobility, resulting from changes in the relaxation dynamics, the Lorenz force acting on the slow and fast carriers of the cold and hot ends of the sample will not be fully compensated, which can lead to a sizeable Nernst effect. Above 8 K, instead of observing only a metallic state, there is a mixed state with a non-metallic behaviour in which the CDW_0 state still exists, indicating that the FS is still reconstructed. The phonon drag effect is not affecting the Nernst effect behaviour, but the relaxation processes are not reduced up to $T \sim 12$ K. In the mixed state, the momentum relaxation is adding more to the Nernst signal than the energy relaxation, leading to a peak around 11.8 K.

For a magnetic field of 25 T oriented along the first AHRO minimum $\theta_{min1} = 45°$ (Figure 8b), the Nernst effect has a significantly smaller amplitude and a $N_{yz}(T) = M + N/T + LT^3$ dependence (M, N and L are constants) below 4.5 K. At this angle and field, the CDW_x state (high field CDW state) is realized in α-(BEDT-TTF)$_2$KHg(SCN)$_4$ below 4.5 K. This shows that in the high field state, the Nernst effect is a result of the phonon drag effect and electron–phonon interactions. The absence of the T^5 term for the CDW state excludes the contribution from the momentum relaxation processes in this state. This refers to the decrease in the carriers mobility with increasing angle and field. Above 4.5 K, a non-metallic behaviour is observed, which is characterised with the following temperature dependence $N_{yz}(T) = P + QT + RT^3$ (P, Q and R are constants). This implies that a mixed state showing properties of a weakened CDW_x state and metallic state is realized. Interestingly, mixed states have been also observed in the thermopower temperature dependence when the magnetic field is located exactly at the AMRO maximum or minimum. No mixed state is found when the field is along the normal to the conducting ac plane [16]. Obviously, the rotation of the magnetic field away from the temperature gradient direction, at the AMRO (AHRO) maxima and minima locations, brings the system into some kind of mixed states, which is an indication of a significant change in the electronic structure. The transitions from a pure CDW state into a mixed state occur at different temperature. The apparent trend is that with increasing angles, the transition is reached at a lower temperature. This shows that, most probably, the FS is not completely restored in the given temperature range, indicating that this organic metal has a very complex electron structure that changes depending on the underlying conditions. It seems that the presence of a temperature gradient contributes greatly to these changes by affecting the dynamics and flow of the quasiparticles. The obtained temperature dependencies clearly show that the relaxation processes change when the system transitions from one into another ordered state. Obviously, they are most pronounced in the CDW_0 state (Figure 8a), while they are strongly reduced in the CDW_x state, which is evident from the much smaller Nernst effect amplitude with increasing field and angle (Figure 8b). The predicted Nernst effect behaviour for temperatures outside of the temperature range used in these measurements is seen from the insets in Figure 8. The presented results greatly contribute not only for revision of the previous findings about the thermoelectric transport but also for obtaining information on the possible mechanisms responsible for the existence of different phases in the given organic metal.

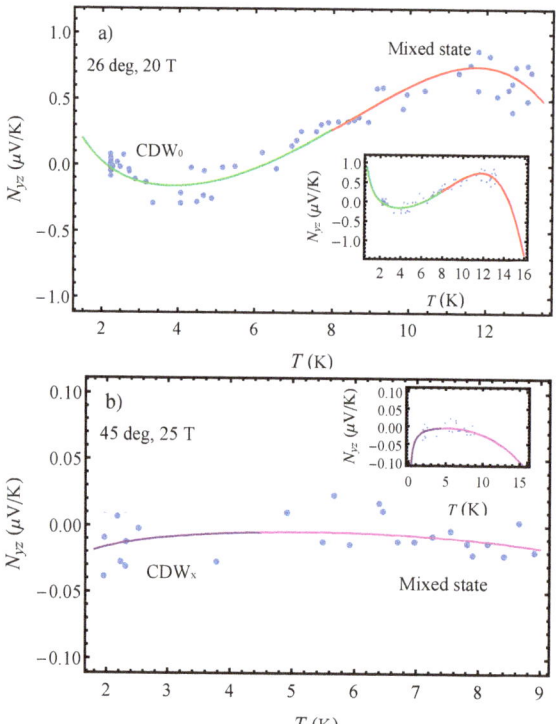

Figure 8. Temperature dependence of the Nernst effect, $N_{yz}(T)$, in α-(BEDT-TTF)$_2$KHg(SCN)$_4$ obtained at: (a) $\theta = 26°, B = 20$ T and (b) $\theta = 45°, B = 25$ T. The prominent change in the temperature profile of Nernst effect is evident with increasing tilt angle and magnetic field strength. The insets show the predicted temperature profile of the Nernst effect outside of the measured temperature range.

4. Conclusions

In conclusion, we report on the study of Nernst effect in the quasi-2D organic metal α-(BEDT-TTF)$_2$KHg(SCN)$_4$, showing charge density wave instabilities. The behaviour of the Nernst effect with the angle, magnetic field and temperature is analyzed. We find that in this multiband system, the transverse thermoelectric response is dominant when the magnetic field is rotated from the least conducting axis towards the plane of the layers. A combination of mechanisms including the phonon drag effect and the relaxation processes can lead to the enhanced Nernst effect with a large amplitude similar to that reported in some quasi-1D organic conductors and heavy-fermion compounds. The energy and momentum relaxation processes have a significant role in observing the largest Nernst signal in the CDW_0 state. The obtained temperature dependence of the Nernst effect when the field is oriented at the location of the first AHRO maximum shows that there is a change in the momentum relaxation dynamics in the CDW_0 state with increasing temperature. The momentum relaxation processes are absent in the CDW_x state for which the main mechanisms are the phonon drag effect and electron–phonon interactions. We suggest that in this organic metal, due to changes in the relaxation dynamics, there is a change in the carrier mobility. When the momentum relaxation time is large enough, then there might be a counter-flow of high mobility quasiparticles that contributes to a generation of a large transverse voltage. Thus, the Lorenz force acting on the slow and fast carriers of the cold and hot ends of the sample will not be fully compensated, leading to a sizeable Nernst effect. The observed giant angular Nernst effect is not only due to the change in the relaxation times and thus in the charge carrier scattering rate but also due to the change in the charge carriers velocity vector (which moves along different parts of the FS) with respect to the temperature gradient direction. Taking into account the change in the direction of the charge carriers' velocity on different FS parts, in relation to the temperature gradient, one can explain the observed specific behaviour of the angular Nernst signal at certain angles.

Author Contributions: Conceptualisation, D.K.; Investigation, D.K., E.S.C. and E.S.; Resources, E.S.C. and E.S.; Writing—original draft, D.K.; Writing—review and editing, D.K. and E.S.C. All authors have read and agreed to the published version of the manuscript.

Funding: This research received no external funding.

Institutional Review Board Statement: Not applicable.

Informed Consent Statement: Not applicable.

Data Availability Statement: Not applicable.

Acknowledgments: This work was performed at the National High Magnetic Field Laboratory, supported by NSF DMR-0654118, by the State of Florida, and by the DOE.

Conflicts of Interest: The authors declare no conflict of interest.

References

1. Behnia, K.; Measson, M.; Kopelevich, Y. Oscillating Nernst-Ettingshausen effect in bismuth across the quantum limit. *Phys. Rev. Lett.* **2007**, *98*, 166602. [CrossRef] [PubMed]
2. Galffy, M.; Freimuth, A.; Murek, U. Unusual Seebeck and Nernst effects in the mixed state of $Bi_{2-x}Pb_xSr_2Ca_2Cu_3O_\delta$. *Phys. Rev. B* **1990**, *41*, 11029. [CrossRef] [PubMed]
3. Alexandrov, A.; Zavaritsky, V. Nernst effect in poor conductors and in the cuprate superconductors. *Phys. Rev. Lett.* **2004**, *93*, 217002. [CrossRef] [PubMed]
4. Wu, W.; Lee, I.J.; Chaikin, P.M. Giant Nernst effect and lock-in currents at magic angles in $(TMTSF)_2PF_6$. *Phys. Rev. Lett.* **2003**, *91*, 056601. [CrossRef] [PubMed]
5. Wu, W.; Ong, N.P.; Chaikin, P.M. Giant angular-dependent Nernst effect in the quasi-one-dimensional organic conductor $(TMTSF)_2PF_6$. *Phys. Rev. Lett.* **2005**, *72*, 235116. [CrossRef]
6. Choi, E.S.; Brooks, J.S.; Kang, H.; Yo, Y.J.; Kang, W. Resonant Nernst effect in the metallic and field-induced spin density wave states of $(TMTSF)_2ClO_4$. *Phys. Rev. Lett.* **2005**, *95*, 187001. [CrossRef] [PubMed]
7. Kartsovnik, M. High magnetic fields: A tool for studying electronic properties of layered organic metals. *Chem. Rev.* **2004**, *104*, 5737–5781. [CrossRef] [PubMed]
8. Kartsovnik, M. Layered Organic Conductors in Strong Magnetic Fields. In *The Physics of Organic Superconductors and Conductors*; Lebed, A., Ed.; Springer: Berlin/Heidelberg, Germany, 2008; pp. 185–246.

9. Osada, T.; Yagi, R.; Kawasumi, A.; Kagoshima, S.; Miura, N.; Oshima, M.; Saito, G. High-field magnetotransport and Fermi-surface topology in the novel quasi-two-dimensional organic conductor bis(ethylenedithiolo)tetrathiafulvalenium mercuric postassium thiocyanate, (BEDT-TTF)$_2$KHg(SCN)$_4$. *Phys. Rev. B* **1990**, *41*, 5428.
10. Sasaki, T.; Toyota, N. Anisotropic galvanomagnetic effect in the quasi-two-dimensional organic conductor α-(BEDT-TTF)$_2$KHg(SCN)$_4$, where BEDT-TTF is bis(ethylenedithio) tetrathiafulvalene. *Phys. Rev. B* **1994**, *49*, 10120.
11. Qualls, J.S.; Balicas, L.; Brooks, J.S.; Harrison, N.; Montgomery, L.K.; Tokumoto, M. Competition between Pauli and orbital effects in a charge-density-wave system. *Phys. Rev. B* **2000**, *62*, 10008.
12. Uchida, K.; Konoike, T.; Osada, T. Angle-dependent magnetoresistance oscillations and magnetic breakdown in α-(BEDT-TTF)$_2$KHg(SCN)$_4$. *Phys. B* **2010**, *405*, S221–S223.
13. Uchida, K.; Yamaguchi, R.; Konoike, T.; Osada, T.; Kang, W. Angle-dependent magnetoresistance oscillations and charge density wave in the organic conductor α-(BEDT-TTF)$_2$KHg(SCN)$_4$. *J. Phys. Soc. Jpn.* **2013**, *82*, 043714.
14. Choi, E.S.; Brooks, J.S.; Qualls, J.S.; Song, Y.S. Low-frequency method for magnetothermopower and Nernst effect measurements on single crystal samples at low temperatures and high magnetic fields. *Rev. Sci. Instrum.* **2001**, *72*, 2392. [CrossRef]
15. Brown, S.E.; Chaikin, P.M.; Naughton, M.J. La Tour des Sels de Bechgaard. In *The Physics of Organic Superconductors and Conductors*; Lebed, A.G., Ed.; Springer: Berlin/Heidelberg, Germany, 2008; pp. 49–88. [CrossRef]
16. Krstovska, D.; Choi, E.S.; Steven, E. Seebeck effect studies in the charge density wave state of organic conductor α-(BEDT-TTF)$_2$KHg(SCN)$_4$. *Phys. Scr.* **2021**, *96*, 125734. [CrossRef]
17. Andres, D.; Kartsovnik, M.V.; Biberacher, W.; Neumaier, K.; Müller, H. Direct evidence for superconductivity in the organic charge density wave compound α-(BEDT-TTF)$_2$KHg(SCN)$_4$ under hydrostatic pressure. *J. Phys. IV* **2002**, *12*, Pr9–Pr87. [CrossRef]
18. Krstovska, D.; Choi, E.S.; Steven, E.; Brooks, J.S. The angular magnetothermoelectric power of a charge density wave system. *J. Phys. Condens. Matter* **2012**, *24*, 265502. [CrossRef]
19. Galbova, O.; Kirichenko, O.V.; Peschansky, V.G. Thermoelectric effect in layered conductors at low temperatures. *Low Temp. Phys.* **2009**, *35*, 810–814. [CrossRef]
20. Choi, E.S.; Brooks, J.S.; Qualls, J.S. Magnetothermopower study of the quasi-two-dimensional organic conductor α-(BEDT-TTF)$_2$KHg(SCN)$_4$. *Phys. Rev. B* **2002**, *65*, 205119.
21. Yang, Y.-F. Universal behavior in the Nernst effect of heavy fermion materials. *Phys. Rev. Res.* **2020**, *2*, 033105. [CrossRef]
22. Bel, R.; Behnia, K.; Nakajima, Y.; Izawa, K.; Matsuda, Y.; Shishido, H.; Settai, R.; Onuki, Y. Giant Nernst Effect in CeCoIn$_5$. *Phys. Rev. Lett.* **2004**, *92*, 217002. [CrossRef]

Disclaimer/Publisher's Note: The statements, opinions and data contained in all publications are solely those of the individual author(s) and contributor(s) and not of MDPI and/or the editor(s). MDPI and/or the editor(s) disclaim responsibility for any injury to people or property resulting from any ideas, methods, instructions or products referred to in the content.

 magnetochemistry

Article

Effect of External Pressure on the Metal–Insulator Transition of the Organic Quasi-Two-Dimensional Metal κ-(BEDT-TTF)$_2$Hg(SCN)$_2$Br †

Sergei I. Pesotskii [1,*], Rustem B. Lyubovskii [1], Gennady V. Shilov [1], Vladimir N. Zverev [2], Svetlana A. Torunova [1], Elena I. Zhilyaeva [1] and Enric Canadell [3,*]

1. Federal Research Center of Problems of Chemical Physics and Medicinal Chemistry, Russian Academy of Sciences, Chernogolovka, Moscow 142432, Russia
2. Institute of Solid State Physics, Russian Academy of Sciences, Chernogolovka 142432, Russia
3. Institut de Ciència de Materials de Barcelona, ICMAB-CSIC, Campus UAB, 08193 Bellaterra, Spain
* Correspondence: pesot@icp.ac.ru (S.I.P.); canadell@icmab.es (E.C.)
† Dedicated to Professor Manuel Almeida on the Occasion of his 70th Birthday.

Abstract: The metal–insulator transition in the organic quasi-two-dimensional metal κ-(BEDT-TTF)$_2$Hg(SCN)$_2$Br at $T_{MI} \approx 90$ K has been investigated. The crystal structure changes during this transition from monoclinic above T_{MI} to triclinic below T_{MI}. A theoretical study suggested that this phase transition should be of the metal-to-metal type and brings about a substantial change of the Fermi surface. Apparently, the electronic system in the triclinic phase is unstable toward a Mott insulating state, leading to the growth of the resistance when the temperature drops below $T_{MI} \approx 90$ K. The application of external pressure suppresses the Mott transition and restores the metallic electronic structure of the triclinic phase. The observed quantum oscillations of the magnetoresistance are in good agreement with the calculated Fermi surface for the triclinic phase, providing a plausible explanation for the puzzling behavior of κ-(BEDT-TTF)$_2$Hg(SCN)$_2$Br as a function of temperature and pressure around 100 K. The present study points out interesting differences in the structural and physical behaviors of the two room temperature isostructural salts of κ-(BEDT-TTF)$_2$Hg(SCN)$_2$X with X = Br, Cl.

Keywords: organic conductors; crystal structure; phase transition; band structures; Fermi surface; conductivity; magnetoresistance

1. Introduction

The organic metal κ-(BEDT-TTF)$_2$Hg(SCN)$_2$Br belongs to the family of quasi-two-dimensional metallic conductors where the cation–radical layers of a molecule such as BEDT-TTF (bis(ethylenedithio)tetrathiafulvalene) alternate with inorganic anion layers. The holes created in the donor layers provide metallic conductivity inside these layers that is several orders of magnitude higher than the conductivity between the layers [1]. Quasi-two-dimensional molecular conductors have aroused a large amount of interest, primarily due to the abundance of exotic quantum states occurring at low-temperatures. In particular, for κ-(BEDT-TTF)$_2$Hg(SCN)$_2$Br, evidence has been found concerning the realization of dipole liquid, spin liquid, and spin glass states; the formation of ferromagnetic polarons; and as a consequence the formation of weak ferromagnetism [2–7].

The main reason for this wealth in physical states resides in the crystalline and electronic structure of the conducting donor layers. In organic metals with the general chemical formula of (BEDT-TTF)$_2$X, the conduction band is one-quarter-filled. However, in metals with κ-type packing, the donor molecules form dimers and the conduction band formally splits into fully filled and half-filled sub-bands [8]; that is, the significant units concerning the electron transfer are the donor dimers. Taking into account that the conduction band

Citation: Pesotskii, S.I.; Lyubovskii, R.B.; Shilov, G.V.; Zverev, V.N.; Torunova, S.A.; Zhilyaeva, E.I.; Canadell, E. Effect of External Pressure on the Metal–Insulator Transition of the Organic Quasi-Two-Dimensional Metal κ-(BEDT-TTF)$_2$Hg(SCN)$_2$Br. *Magnetochemistry* 2022, 8, 152. https://doi.org/10.3390/magnetochemistry8110152

Academic Editors: Laura C. J. Pereira and Dulce Belo

Received: 6 October 2022
Accepted: 4 November 2022
Published: 8 November 2022

Publisher's Note: MDPI stays neutral with regard to jurisdictional claims in published maps and institutional affiliations.

Copyright: © 2022 by the authors. Licensee MDPI, Basel, Switzerland. This article is an open access article distributed under the terms and conditions of the Creative Commons Attribution (CC BY) license (https://creativecommons.org/licenses/by/4.0/).

in organic metals is quite narrow, of the order of 1000 K [9], the possibility arises for such dimerized metals to have an instability toward a Mott insulating state when decreasing the temperature. Such instability is likely when the condition U/W > 1 is fulfilled (U is the Coulomb repulsion energy on one dimer and W is the kinetic energy of an electron in the conduction band). In that case, below the transition temperature the electrons are generally localized within the dimers, forming an ordered antiferromagnetic system.

A metal–insulator transition was detected in κ-(BEDT-TTF)$_2$Hg(SCN)$_2$Br at a temperature of ≈90 K [2]. However, since the dimers form a triangular frustrated lattice, once an electronic localized state below the transition is reached, the long-range spin order is destroyed, leading to the formation at low-temperatures of a spin liquid and other quantum states. The nature of these exotic states was the main object of most of the previous work on this salt. However, the nature of the insulating transition itself, which is the source of the low-temperature exotic states, is not fully understood. In particular, in a previous study [3], a hysteresis was found in the temperature dependence of the susceptibility, suggesting that a first-order transition may take place. At the same time, the absence of changes in the crystal structure during the transition was noted in [2]. In this paper, we provide new information related to the nature of the ≈90 K transition in κ-(BEDT-TTF)$_2$Hg(SCN)$_2$Br. In particular, we present the results of X-ray diffraction studies providing evidence for a structural change at the transition temperature. We also report on the electronic structure above and below the transition calculated on the basis of this new information, as well as an analysis of the magnetoresistance behavior, which is in good agreement with the theoretical results.

2. Crystal Structure of κ-(BEDT-TTF)$_2$Hg(SCN)$_2$Br before and after the Phase Transition

The κ-(BEDT-TTF)$_2$Hg(SCN)$_2$Br crystals were previously studied by our group using an X-ray diffraction analysis [10,11]. The structure was determined and refined in a monoclinic cell with the space group C2/c. Our new conductivity measurements showed that a phase transition occurs below 100 K and that the conductivity regime varies from metallic to semiconducting. In the present work, we determined the crystal structure before and after the ≈90 K transition in order to gain an understanding of the nature of the transition. From 300 K to 100 K, the crystal structure did not experience any change. At temperatures below 100 K, additional weak superstructure reflections appeared in the diffraction field. Above the phase transition temperature, the crystal belongs to the monoclinic system and below it to the triclinic one. The crystal structure of the high-temperature phase can be described in the space group C2/c and the low-temperature phase in P-1. The monoclinic cell at 150 K has parameters of a = 36.7651 (19) Å, b = 8.2240 (5) Å, c = 11.6716 (5) Å, α = 90^0, β = 90.163 (4)^0, γ = 90^0, V = 3529.0 (3) Å^3, Z = 4, and Goof = 1.046, with final R indices of [I > 2σ (I)] R1 = 0.0375, wR2 = 0.0817. The crystal structure of the low-temperature P-1 triclinic cell at 80 K (Figure 1a) can be described with the following parameters: a = 11.6613 (6) Å, b = 16.3444 (6) Å, c = 18.7804 (9) Å, α = 102.592 (4)^0, β = 90.113 (4)^0, γ = 90.005 (3)^0, V = 3493.4 (3) Å^3, Z = 4. The crystal structure below the transition was determined using the direct method and refined using the least squares method in space group P-1, with Goof = 1.011 and final R indices of [I > 2σ (I)] R1 = 0.0336, wR2 = 0.0747. In this case, the number of symmetry-independent BEDT-TTF molecules is four and the unit cell contains eight of them (Figure 1b). The monoclinic cell is related to the triclinic cell of the high-temperature structure by the transition matrix (0 0 -1 0 -2 0 -0.5 0.5 0). Using this transformation, the cell parameters of the 80 K structure lead to the cell parameters a = 36.6575 (17) Å, b = 8.1722 (3) Å, c = 11.6613 (6) Å, α = 90^0, β = 90.117 (6)^0, γ = 90^0 for a monoclinic cell. An attempt was made to determine the low-temperature structure in the space group P1 within the same unit cell. However, this led to a deterioration of the refinement results. In addition, whereas the band structure calculations with the P-1 structure led to results that were easily understandable in terms of the folding of the high-temperature donor layer structure and were in good agreement with the experimental data (see below), those based on the P1 structure led to unreasonable results.

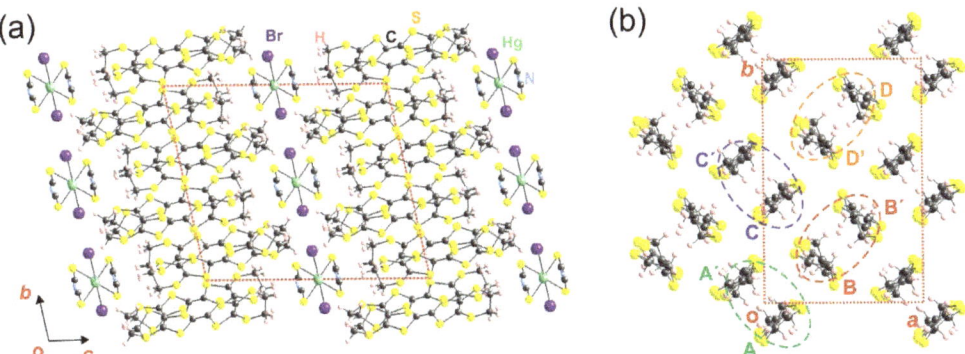

Figure 1. Low-temperature (80 K) crystal structure of κ-(BEDT-TTF)$_2$Hg(SCN)$_2$Br: (**a**) projection view along the *a*-direction; (**b**) BEDT-TTF donor layers showing the different dimers (AA′, BB′, etc.). For clarity, the disorder in half of the donor ethylidenedithio groups (i.e., those not facing the Br atoms) is not shown. The four dimers in the repeat unit of the layer are highlighted with different colors. Note that in the high-temperature structure, the repeat unit contains only two dimers, i.e., the AA′ dimer is identical to CC′ and the BB′ dimer is identical to DD′.

Note that although not shown in Figure 1, half of the ethylidenedithio groups of BEDT-TTF (i.e., those not facing the Br atoms, which strongly interact with the hydrogen atoms of the donor) exhibit disorder in both the high- and low-temperature structures. The disorder degree is not very influenced by the transition if we take into account the effect of the thermal contraction (58/42% at 150 K and 68/31% at 80 K). The internal structure of the anions is not substantially altered by the distortion, except for the decrease in symmetry. A slight change of the Hg-N distance from 2.754 (150 K) to 2.739/2.728 Å (80 K) as well as variations in the C-N (from 1.154 (150 K) to 1.114 Å (80 K) and C-S (from 1.667 (150 K) to 1.701 Å (80 K) bonds of the SCN group are the more significant changes detected.

The main structural features controlling the different transport properties of the high- and low-temperature structures concern the internal structure of the donor layers. The high-temperature crystal structure contains two symmetry-equivalent donor layers [10,11], but as shown in Figure 1a, there is only one after the phase transition. This layer (Figure 1b) is simply the usual one for κ-phases but is doubled along the *b*-direction. The repeat unit of the layer contains eight donors making four dimers (AA′, BB′, CC′, and DD′, shown with different colors in Figure 1b), although only three of them are different, AA′, CC′, and BB′= DD′ (see Table 1). In the usual κ-phases, the repeat unit of the layer contains only two dimers, which can be equivalent or not. The central C = C bond lengths of the four donor molecules (see Figure 1b for the labeling) are reported in Table 1, together with those for the high-temperature structure. Note that in the high-temperature structure, the two dimers are equivalent and contain just one type of donor, while as mentioned above, in the low-temperature structure there are three different types of dimers. One of the dimers in the high-temperature structure becomes a dimer containing two different donors in the low-temperature structure (for instance, BB′). There are two equivalent dimers of this type (i.e., BB′ = DD′). In addition, the other dimer of the high-temperature structure leads to two different symmetrical dimers with equivalent donors (AA′ and CC′) in the low-temperature structure. The short S . . . S intra-dimer contacts for the three different dimers of the 80 K structure are very similar; in all dimers there are two contacts of ~3.68 Å and two of ~3.87 Å. Thus, according to the crystal structure analysis, the structural changes brought about by the transition seem to be mostly located in the donor layers (i.e., the inner C = C bond lengths of the donors). The non-equivalence between symmetrical dimers

(AA′ and CC′) or between the two donors of an unsymmetrical dimer (BB′ = DD′) leads to the opening of gaps in the electronic structure, which are at the origin of the difference transport properties.

Table 1. Central C = C bond lengths for the different BEDT-TTF donors in the low- and high-temperature structures of κ-(BEDT-TTF)$_2$Hg(SCN)$_2$Br (in Å) (see Figure 1b for the labeling).

Donor Molecule	Low-Temperature Structure (80 K)	High-Temperature Structure (150 K)
A	1.357	1.357
A′	1.357	1.357
C	1.386	1.357 (CC′ = AA′)
C′	1.386	1.357 (CC′ = AA′)
B	1.358	1.357
B′	1.367	1.357
D	1.358	1.357 (DD′ = BB′)
D′	1.367	1.357 (DD′ = BB′)

3. Resistance and Magnetoresistance

Figure 2a,b show the temperature dependences of the interlayer resistance in κ-(BEDT-TTF)$_2$Hg(SCN)$_2$Br at different pressures for two different samples. At ambient pressure and $T_{MI} \approx 90$ K, a pronounced metal–insulator transition is observed. It can be clearly seen (Figure 2b) that the transition occurs with a noticeable hysteresis with a width $\Delta T = (4.8 \pm 0.3)$ K, which is almost pressure-independent. The onset of the transition, defined as the temperature of the minimum on the temperature dependence, shifts with increasing pressure toward high temperatures and does not disappear completely up to the maximum available pressure of $p = 8$ kbar. The dependence of the shift of the beginning of the transition with the pressure in the cooling mode is reported in Figure 3. At the same time, the increase in resistance upon cooling below the transition temperature is suppressed by pressure in such a way that even at a pressure of $p = 3$ kbar, the "metallic" behavior of the resistance at low temperatures is quite pronounced. A further increase in pressure enhances this behavior.

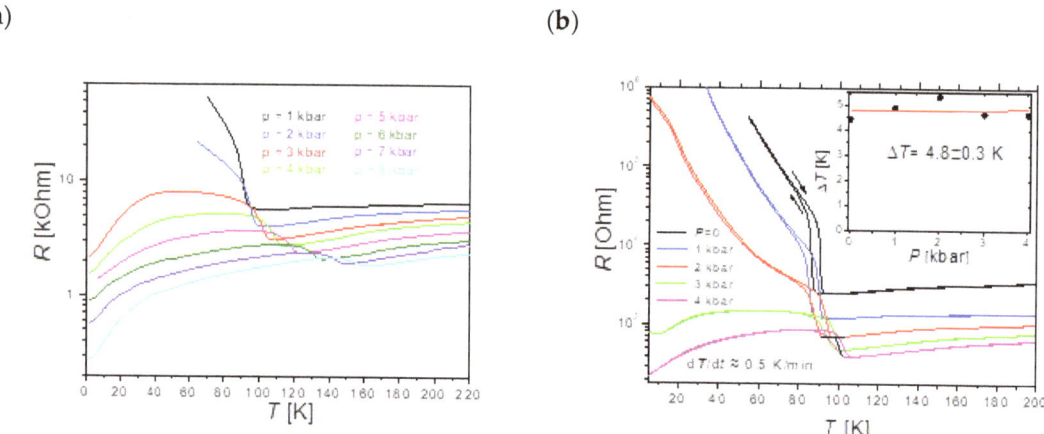

Figure 2. Temperature dependences of the interlayer resistance at various pressures recorded in the sample cooling mode for sample 1 (**a**) and showing the temperature hysteresis for sample 2 (**b**). The pressure dependence of the width of the hysteresis loop is shown in the inset.

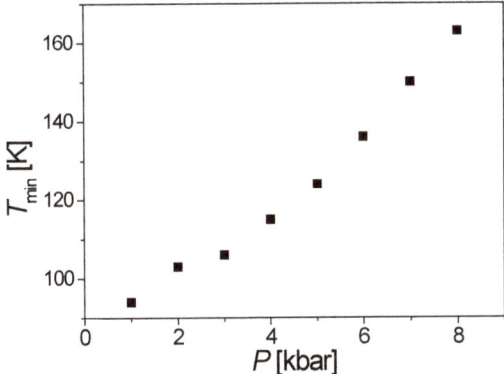

Figure 3. Metal–insulator transition temperature as a function of pressure.

Figure 4a shows the field dependences of the magnetoresistance $R(B)/R(0)$ in κ-(BEDT-TTF)$_2$Hg(SCN)$_2$Br at various pressures. At pressures above 4 kbar, Shubnikov–de Haas oscillations were observed with a frequency $F \approx 240$ T, which hardly change with increasing pressure. At the same time, the value of the non-oscillating part of the magnetoresistance and the amplitude of the oscillations showed noticeable increases with pressure. The pressure dependence of the cyclotron mass associated with the observed oscillations is shown in Figure 4b. The cyclotron mass decreases by about 10% when the pressure changes from 5 to 8 kbar. The inset to the figure shows the dependence of the oscillation amplitude on the magnetic field (Dingle plot) in the coordinates that make it possible to estimate the Dingle temperature ($T_D \approx 1.5$ K at $p = 8$ kbar). It should be noted that the dependence presented does not deviate from the Lifshitz–Kosevich model for fields up to 16.5 T, and the Dingle temperature within good accuracy does not depend on the pressure.

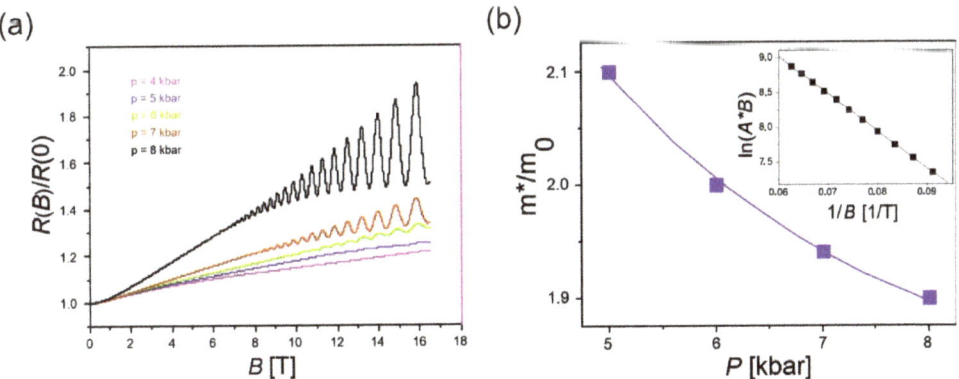

Figure 4. (a) Field dependences of the interlayer magnetoresistance at various pressures for T = 0.5 K. The field is perpendicular to the conducting layers. (b) Dependence of the relative cyclotron mass on the pressure. Inset: Dependence of the oscillation amplitude on the field in Dingle coordinates ($p = 8$ kbar, $T = 0.5$ K).

4. Band Structure Calculations

The calculated band structure for the donor lattice in the low-temperature structure of κ-(BEDT-TTF)$_2$Hg(SCN)$_2$Br is shown in Figure 5a. The eight bands are mostly built from the HOMO (highest occupied molecular orbital) of BEDT-TTF, and because of the stoichiometry, two of these bands should be empty. Since the second and third bands from the top overlap, the salt is predicted to be metallic below the phase transition (see below).

The calculated Fermi surface (FS) is reported in Figure 5b and contains two contributions: first, elongated bone-like electron pockets with a cross-sectional area of 9.4% of the Brillouin zone (BZ); second, two boomerang-like hole pockets with a cross sectional area of 4.6% for each of them, which either touch or very slightly superpose at the M point. Strictly speaking, these two pockets very slightly interact, leading to a pocket with an area of 9.2% and another very small one at M with an area of 0.2%. However, as it is clear from Figure 5b (see lines Γ-M and Γ-S), a minute change in the structure could suppress the overlap and lead to a single hole pocket with an area of 9.4%. This FS can be described as resulting from the superposition and hybridization of a series of ellipses centered at the Y point. This is a surprising result because the FS of the κ-phases usually results from the superposition and hybridization of a series of pseudo-circles centered at the Γ point.

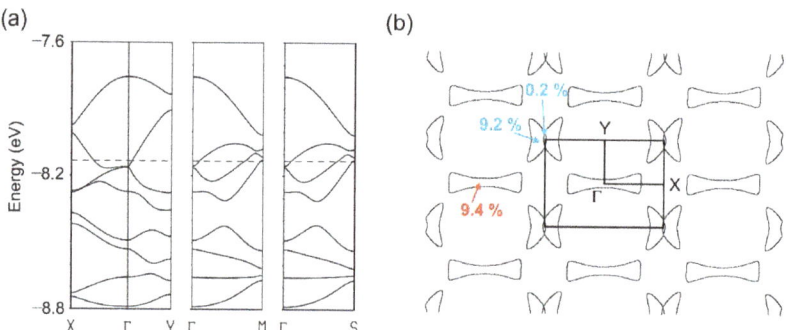

Figure 5. Calculated band structure (**a**) and Fermi surface (**b**) for the donor lattice of the low-temperature crystal structure of κ-(BEDT-TTF)$_2$Hg(SCN)$_2$Br, where Γ = (0, 0), X = (a^*/2, 0), Y = (0, b^*/2), M = (a^*/2, b^*/2), and S = ($-a^*$/2, b^*/2). The dashed line in (**a**) refers to the Fermi level.

In order to understand this unusual shape, we also calculated the FS for the high-temperature structure (see Figure 6a). This FS is the expected one for the κ-phase; it results from the superposition of a series of circles centered at the Γ point with a cross-sectional area of 100% of the BZ. Because of the superposition, closed pockets with a cross-section of 15.7% around the X point are generated. Note that the a- and c-directions are interchanged with respect to those in the low-temperature structure, but we keep the same labelling for the high- and low-temperature structures in order to facilitate the comparison. The FSs for the donor layers of the high- (Figure 6a) and low-temperature (Figure 5b) structures can be easily related if it is taken into account that the b parameter of the crystal structure practically doubles when going from the high- to the low-temperature structures. Thus, the FS of the low-temperature structure should simply result from a folding of the high-temperature one. This process is schematically illustrated in Figure 6b. The FS of the high-temperature structure and the BZ are shown in blue. Because the new BZ of the low-temperature structure is only half the size, half of the high-temperature FS must be translated within the new smaller BZ. Since the b parameter doubles, the folding must be carried out along the b^*-direction and consequently the Y point of the original BZ is translated into the Γ point of the new BZ, as schematized by the red arrow in Figure 6b, and the black lines are generated. In that way, the approximate FS of the low-temperature structure is generated (i.e., the black and blue lines of Figure 6b). Finally, because of the lack of symmetry (except for the trivial symmetry of the inversion center), small gaps are opened at the crossing points and the FS of Figure 5b is generated. Thus, because of the doubling of the repeat unit of the donor layers as a result of the dimer differentiation, the FS changes from a superposition of circles to a superposition of ellipses and finally contains closed orbits differing in area and nature before and after the transition.

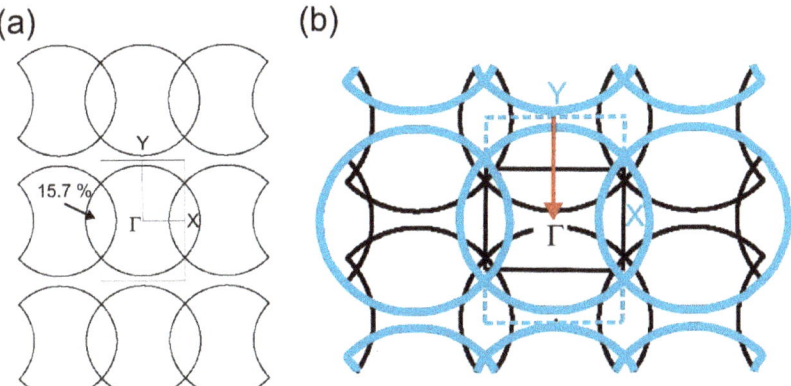

Figure 6. (a) Calculated Fermi surface for the donor lattice of the high-temperature crystal structure of κ-(BEDT-TTF)$_2$Hg(SCN)$_2$Br, where Γ = (0, 0), X = (c^*/2, 0), Y = (0, b^*/2, 0), M = (c^*/2, b^*/2), and S = ($-c^*$/2, b^*/2). Note that the a- and c-directions are interchanged with respect to those in the low-temperature structure. We keep the same labelling for the high- and low-temperature structures in order to facilitate the comparison of the two FSs. (b) Construction of the Fermi surface of the low-temperature donor layer from that of the high-temperature one using a folding process along the b^*-direction (see text).

5. Discussion

The present X-ray diffraction study has shown that when a single crystal of κ-(BEDT-TTF)$_2$Hg(SCN)$_2$Br is cooled at a temperature $T \approx 90$ K, a structural phase transition occurs, with a change in the crystal structure from monoclinic at $T > T_{MI}$ to triclinic at $T < T_{MI}$. According to our theoretical calculations, the electronic structures of the monoclinic and triclinic lattices are different, but in both cases the system is predicted to be metallic. Thus, the structural transition at $T_{MI} \approx 90$ K should not be accompanied by a change toward an activated conductivity regime and a metal-to-metal transition should indeed occur. The electronic structures of the different types of κ-type salts are consistently well described by the types of calculations employed here [12]. In addition, the overlap between the second and third bands from the top in Figure 5a is substantial. We, thus, believe that it is unlikely that an intrinsic band gap separating the top two bands in Figure 5a from the lower ones opens below the transition. Measurements of the temperature dependence of the resistance confirm the occurrence of a transition, and based on the observation of hysteresis, we propose that this is a first-order transition. However, at temperatures below T_{MI} and ambient pressure, κ-(BEDT-TTF)$_2$Hg(SCN)$_2$Br turns out to be an insulator that is at odds with the results of the band structure calculations (but remember that these are one-electron band structure calculations).

Since the BEDT-TTF molecules in the triclinic phase are still combined into dimers, and consequently the band is formally half-filled, it can be assumed that the U/W ratio can subtly change from <1 to >1 around $T < T_{MI}$, so that a Mott transition can interfere with the structural transition. A Mott insulating state is usually suppressed under external pressure, essentially because of a change in W caused by the decrease in the unit cell size. Indeed, as shown in Figure 2a, the behavior of the resistance as a function of temperature below the transition temperature strongly depends on the external pressure. The resistance increase upon cooling at low pressures is gradually suppressed when the pressure increases. At the same time, the transition temperature increases with the pressure. Thus, it seems that at all pressures, at least up to 8 kbar, free and localized electrons may coexist, and the fraction of free electrons increases with increasing pressure.

At 0.5 K and starting from a pressure of $p = 5$ kbar, magnetoresistance oscillations can be observed (see Figure 4a). The oscillation frequency is $F \approx 240$ T, which corresponds to an orbit with a cross-section of approximately 10% of the area of the first BZ in the conducting plane. This is in good agreement with the calculated FS of the low-temperature triclinic cell (see Figure 5b). The increases in magnetoresistance and amplitude of oscillations with increasing pressure (Figure 4a) may indicate an increase in the concentration of free carriers with increasing pressure because of the increase in HOMO–HOMO interactions. Thus, one can probably think of a partial delocalization of electrons and the restoration of the FS of the triclinic phase due to pressure.

The oscillation frequency practically does not change with increasing pressure. An estimate of the cyclotron mass of the electrons associated with quantum oscillations at a pressure of 5 kbar gives a value of $m^* \approx 2.1\, m_0$ (m_0 is the mass of a free electron), which decreases by approximately 10% as the pressure increases to 8 kbar (Figure 4b). Such behavior is typical in the vicinity of the M-I transition and could be due to the mass renormalization by many-body interactions [13]. The decrease could be due to an increase in the concentration of free carriers, and as a consequence a weakening of the electron correlations [13]. The field dependences of the oscillation amplitude in Dingle coordinates for two-dimensional systems (see the example in the inset in Figure 4b) makes it possible to calculate the Dingle temperature. In the present case, $T_D \approx 1.5$ K and does not change with pressure. This fact excludes the influence of a change in scattering under pressure over the amplitude of the magnetoresistance oscillations. In addition, a good agreement between the field dependence of the amplitude and the Lifshitz–Kosevich model (Figure 4b) indicates the absence, at least in our range of fields, of the contribution of the incoherent interlayer transfer to the oscillations [14].

Finally, let us note that the suppression of the dielectric state when localized and delocalized electrons coexist at each pressure looks rather unusual. In particular, in the isostructural metal κ-(BEDT-TTF)$_2$Hg(SCN)$_2$Cl, the Mott-type transition at $T \approx 30$ K shifts under pressure towards lower temperatures and is completely suppressed even at a pressure of 0.7 kbar, where the initial metallic state is completely restored [15,16]. However, in contrast with the present system, the transition is not accompanied by a change in the structure. It is possible that for a more accurate description of the process of stabilization of the metallic state in κ-(BEDT-TTF)$_2$Hg(SCN)$_2$Br, it will be necessary to take into account the low-temperature state of the localized electrons in the absence of pressure, called a dipole liquid, characterized by charge distribution fluctuations inside the dimer [5].

6. Methods

6.1. Synthesis

The crystals of κ-(BEDT-TTF)$_2$Hg(SCN)$_2$Br were prepared using a technique similar to that described in [17] via the electrocrystallization of BEDT-TTF in the presence of Hg(SCN)$_2$, [Me$_4$N]SCN·1.5 KBr, and dibenzo-18-crown-6 in 1,1,2-trichloroethane at a temperature of 40 °C and a constant current of 0.3 µA [17].

6.2. X-ray Structure Determination

The X-ray diffraction studies of κ-(BEDT-TTF)$_2$Hg(SCN)$_2$Br crystals were carried out on an Agilent XCalibur single-crystal diffractometer with an EOS CCD detector in the temperature range of 80–300 K. From 300 to 100 K, the crystal structure did not experience any changes. At temperatures below 100 K, additional weak superstructure reflections appeared in the diffraction field. In the X-ray experiment carried out at 80 K, the diffraction field was limited to a hemisphere. In this way, we avoided the loss of reflections for any choice of unit cell. The hydrogen atoms of ethylene groups were calculated geometrically. All calculations were performed using the SHELXTL software package [18].

6.3. Conductivity Measurements

The resistance and magnetoresistance were measured using the standard four-contact method using a synchronous detector on single-crystal samples with characteristic dimensions of $0.5 \times 0.3 \times 0.03$ mm^3. A measuring current with a frequency of 20 Hz, the value of which did not exceed 10 µA, was directed perpendicularly to the conductive layers of the crystal. The sample was mounted in a high-pressure chamber filled with a silicon oil as a pressure medium, which ensured the operation under quasi-hydrostatic conditions up to a pressure of $p = 8$ kbar at low temperatures. The pressure was controlled through the resistance of the manganin sensor. The measurements were carried out in an insert with ^{3}He pumping, which made it possible to operate in the temperature range from room temperature to 0.5 K. A magnetic field up to 17 T created by a superconducting solenoid was used and was directed normal to the conducting layers along the current direction.

6.4. Band Structure Calculations

The tight-binding band structure calculations were of the extended Hückel type [19]. A modified Wolfsberg–Helmholtz formula was used to calculate the non-diagonal $H_{\mu\nu}$ values [20]. All valence electrons were taken into account in the calculations and the basis set consisted of Slater-type orbitals of double-ζ quality for C 2s and 2p and S 3s and 3p, and of single-ζ quality for H 1s. The ionization potentials, contraction coefficients, and exponents were taken from previous work [21].

7. Concluding Remarks

The crystal structure of the organic quasi-two-dimensional metal κ-(BEDT-TTF)$_2$Hg(SCN)$_2$Br was studied before and after the metal-to-insulator transition at $T_{MI} \approx 90$ K, and the evolution of the FS was calculated based on the new data. The behaviors of the resistance and magnetoresistance as a function of both the magnetic field and temperature under external pressures of up to 8 kbar were studied. The main conclusions of this study are as follows:

(1) The κ-(BEDT-TTF)$_2$Hg(SCN)$_2$Br undergoes a first-order phase transition at $T_{MI} \approx 90$ K, where the crystal structure changes from monoclinic above T_{MI} to triclinic below T_{MI}. The repeat unit of the donor layer below the transition contains twice the number of dimers and the internal structure of these dimers changes;

(2) Band structure calculations have shown that this phase transition can be described on the basis of a folding model that rearranges the FS and suggests a metal–metal transition;

(3) In contrast with the monoclinic high-temperature phase, the metallic state of the triclinic phase is probably unstable to the transition toward a Mott insulating state, which leads to the growth of the resistance when the temperature drops below $T_{MI} \approx 90$ K;

(4) The application of external pressure suppresses the Mott transition and restores the metallic state of the triclinic phase. The frequency of the detected quantum oscillations of the magnetoresistance is in good agreement with the calculated FS for the triclinic phase.

These results provide interesting differences with respect to the physical behavior of the room temperature isostructural κ-(BEDT-TTF)$_2$Hg(SCN)$_2$Cl salt.

Author Contributions: S.I.P., R.B.L. and V.N.Z. carried out the transport property measurements. G.V.S. performed the structural study. E.C. undertook the theoretical study. S.A.T. and E.I.Z. synthesized high-quality crystals, and S.I.P. and E.C. wrote and reviewed the manuscript with contributions from all authors. All authors have read and agreed to the published version of the manuscript.

Funding: The work at the Federal Research Center of Problems of Chemical Physics and Medicinal Chemistry was carried out within the project of state assignment number AAAA-A19-119092390079-8. V.N.Z. acknowledges the support of the Russian Foundation for Basic Research No. 21-52-12027. The work in Spain was supported by the MICIU (Grant PGC2018-096955-B-C44) and Generalitat de Catalunya (2017SGR1506). E.C. acknowledges the support of the Spanish MICIU through the Severo Ochoa FUNFUTURE (CEX2019-000917-S) Excellence Center distinction.

Institutional Review Board Statement: Not applicable.

Informed Consent Statement: Not applicable.

Data Availability Statement: Not applicable.

Conflicts of Interest: The authors declare no conflict of interest.

References

1. Ishiguro, T.; Yamaji, K.; Saito, G. *Organic Superconductors*, 2nd ed.; Springer: Berlin, Germany, 1998.
2. Ivek, T.; Beyer, R.; Badalov, S.; Culo, M.; Tomic, S.; Schlueter, J.A.; Zhilyaeva, E.I.; Lyubovskaya, R.N.; Dressel, M. Metal-insulator transition in the dimerized organic conductor κ−(BEDT-TTF)$_2$Hg(SCN)$_2$Br. *Phys. Rev. B* **2017**, *96*, 085116. [CrossRef]
3. Hemmida, M.; von Nidda, H.A.K.; Miksch, B.; Samoilenko, L.L.; Pustogow, A.; Widmann, S.; Henderson, A.; Siegrist, T.; Schlueter, J.A.; Loidl, A.; et al. Weak ferromagnetism and glassy state in κ-(BEDT-TTF)2Hg(SCN)2Br. *Phys. Rev. B* **2018**, *98*, 241202. [CrossRef]
4. Yamashita, Y.; Sugiura, S.; Ueda, A.; Dekura, S.; Terashima, T.; Uji, S.; Sunairi, Y.; Mori, H.; Zhilyaeva, E.I.; Torunova, S.A.; et al. Ferromagnetism out of charge fluctuation of strongly correlated electrons in κ-(BEDT-TTF)$_2$Hg(SCN)$_2$Br. *NPJ Quantum Mater.* **2021**, *6*, 87. [CrossRef]
5. Hassan, N.M.; Thirunavukkuarasu, K.; Lu, Z.; Smirnov, D.; Zhilyaeva, E.I.; Turunova, S.; Lyubovskaya, R.N.; Drichko, N. Melting of charge order in the low-temperature state of an electronic ferroelectric-like system. *NPJ Quantum Mater.* **2020**, *5*, 15. [CrossRef]
6. Le, T.; Pustogow, A.; Wang, J.; Henderson, A.; Siegrist, T.; Schlueter, J.A.; Brown, S.E. Disorder and slowing magnetic dynamics in κ−(BEDT−TTF)$_2$Hg(SCN)$_2$Br. *Phys. Rev. B* **2020**, *102*, 184417. [CrossRef]
7. Jacko, A.C.; Kenny, E.P.; Powell, B.J. Interplay of dipoles and spins in κ-(BEDT−TTF)$_2$X, where X= Hg(SCN)$_2$Cl, Hg(SCN)$_2$Br, Cu[N(CN)$_2$]Cl, Cu[N(CN)2]Br, and Ag$_2$(CN)$_3$. *Phys. Rev. B* **2020**, *101*, 125110. [CrossRef]
8. Mori, T.; Mori, H.; Tanaka, S. Structural Genealogy of BEDT-TTF-Based Organic Conductors II. Inclined Molecules: θ, α, and κ Phases. *Bull. Chem. Soc. Jpn.* **1999**, *72*, 179–197. [CrossRef]
9. Kartsovnik, M.V. High Magnetic Fields: A Tool for Studying Electronic Properties of Layered Organic Metals. *Chem. Rev.* **2004**, *104*, 5737–5782. [CrossRef] [PubMed]
10. Aldoshina, M.Z.; Lyubovskaya, R.N.; Konovalikhin, S.V.; Dyachenko, O.A.; Shilov, G.V.; Makova, M.K.; Lyubovskii, R.B. A new series of ET-based organic Metals: Synthesis, Crystal structure and properties. *Synth. Met.* **1993**, *56*, 1905–1909. [CrossRef]
11. Yudanova, E.I.; Hoffmann, S.K.; Graja, A.; Konovalikhin, S.V.; Dyachenko, O.A.; Lyubovskii, R.B.; Lyubovskaya, R.N. Crystal structure, ESR and conductivity studies of bis(ethylenedithio)tetrathiafulvalene (ET) organic conductor (H$_8$-ET)$_2$[Hg(SCN)$_2$Br] and its deuterated analogue (D$_8$-ET)$_2$[Hg(SCN)$_2$Br]. *Synth. Met.* **1995**, *73*, 227–237. [CrossRef]
12. Jung, D.; Evain, M.; Novoa, J.J.; Whangbo, M.-H.; Beno, M.A.; Kini, A.M.; Schultz, A.J.; Williams, J.M.; Nigrey, P.J. Similarities and Differences in the Structural and Electronic Properties of κ-Phase Organic Conducting and Superconducting Salts. *Inorg. Chem.* **1989**, *28*, 4516–4522. [CrossRef]
13. Imada, M.; Fujimori, A.; Tokura, Y. Metal—Insulator transition. *Rev. Mod. Phys.* **1998**, *70*, 1041–1263. [CrossRef]
14. Grigoriev, P.M.; Kartsovnik, M.V.; Biberacher, W. Magnetic-field-induced dimensional crossover in the organic metal α-(BEDT-TTF)$_2$KHg(SCN)$_4$. *Phys. Rev. B* **2012**, *86*, 165125. [CrossRef]
15. Lohle, A.; Rose, E.; Singh, S.; Bayer, R.; Tatra, E.; Ivek, T.; Zhylyaeva, E.I.; Lyubovskaya, R.N.; Dressel, M. Pressure dependence of the metal-insulator transition in κ-(BEDT-TTF)$_2$Hg(SCN)$_2$Cl: Optical and transport studies. *J. Phys. Condens. Matter* **2017**, *29*, 055601. [CrossRef] [PubMed]
16. Lyubovskii, R.B.; Pesotskii, S.I.; Zverev, V.N.; Zhilyaeva, E.I.; Torunova, S.A.; Lyubovskaya, R.N. Hydrostatic-Pressure-Induced Reentrance of the Metallic State in the κ-(ET)$_2$Hg(SCN)$_2$Cl Quasi-Two-Dimensional Organic Conductor. *JETP Lett.* **2020**, *112*, 582–584. [CrossRef]
17. Hassan, N.; Cunningham, S.; Mourigal, M.; Zhilyaeva, E.I.; Torunova, S.A.; Lyubovskaya, R.N.; Schlueter, J.A.; Drichko, N. Observation of a quantum dipole liquid state in an organic quasi-two-dimensional material. *Science* **2018**, *360*, 1101–1104. [CrossRef] [PubMed]
18. Sheldrick, G.M. Crystal structure refinement with SHELXL. *Acta Cryst. C* **2015**, *71*, 3–8. [CrossRef] [PubMed]
19. Whangbo, M.-H.; Hoffmann, R. The band structure of the tetracyanoplatinate chain. *J. Am. Chem. Soc.* **1978**, *100*, 6093–6098. [CrossRef]
20. Ammeter, J.A.; Bürgi, H.-B.; Thibeault, J.; Hoffmann, R. Counterintuitive Orbital Mixing in Semiempirical and ab Initio Molecular Orbital Calculations. *J. Am. Chem. Soc.* **1978**, *100*, 3686–3692. [CrossRef]
21. Pénicaud, A.; Boubekeur, K.; Batail, P.; Canadell, E.; Auban-Senzier, P.; Jérome, D. Hydrogen-bond tuning of macroscopic transport properties from the neutral molecular component site along the series of metallic organic-inorganic solvates (BEDT-TTF)$_4$Re$_6$Se$_5$Cl$_9$.[guest], [guest = DMF, THF, dioxane]. *J. Am. Chem. Soc.* **1993**, *115*, 4101–4112. [CrossRef]

Review

Spin-Peierls, Spin-Ladder and Kondo Coupling in Weakly Localized Quasi-1D Molecular Systems: An Overview

Jean-Paul Pouget

Laboratoire de Physique des Solides, Université Paris-Saclay, CNRS, 91405 Orsay, France; jean-paul.pouget@u-psud.fr

Abstract: We review the magneto-structural properties of electron–electron correlated quasi-one-dimensional (1D) molecular organics. These weakly localized quarter-filled metallic-like systems with pronounced spin 1/2 antiferromagnetic (AF) interactions in stack direction exhibit a spin charge decoupling where magnetoelastic coupling picks up spin 1/2 to pair into S = 0 singlet dimers. This is well illustrated by the observation of a spin-Peierls (SP) instability in the (TMTTF)2X Fabre salts and related salts with the o-DMTTF donor. These instabilities are revealed by the formation of a pseudo-gap in the spin degrees of freedom triggered by the development of SP structural correlations. The divergence of these 1D fluctuations, together with the interchain coupling, drive a 3D-SP ground state. More surprisingly, we show that the Per2-M(mnt)2 system, undergoing a Kondo coupling between the metallic Per stack and the dithiolate stack of localized AF coupled spin $\frac{1}{2}$ (for M = Pd, Ni, Pt), enhances the SP instability. Then, we consider the zig-zag spin ladder DTTTF2-M(mnt)2 system, where unusual singlet ground state properties are due to a combination of a $4k_F$ charge localization effect in stack direction and a $2k_F$ SP instability along the zig-zag ladder. Finally, we consider some specific features of correlated 1D systems concerning the coexistence of symmetrically different $4k_F$ BOW and $4k_F$ CDW orders in quarter-filled organics, and the nucleation of solitons in perturbed SP systems.

Keywords: 1D organic conductor; electron–electron correlated system; spin-charge decoupling; magnetoelastic coupling; spin-Peierls transition; kondo coupling; spin ladder

Citation: Pouget, J.-P. Spin-Peierls, Spin-Ladder and Kondo Coupling in Weakly Localized Quasi-1D Molecular Systems: An Overview. *Magnetochemistry* **2023**, *9*, 57. https://doi.org/10.3390/magnetochemistry9020057

Academic Editors: Laura C. J. Pereira and Dulce Belo

Received: 16 January 2023
Revised: 8 February 2023
Accepted: 10 February 2023
Published: 13 February 2023

Copyright: © 2023 by the author. Licensee MDPI, Basel, Switzerland. This article is an open access article distributed under the terms and conditions of the Creative Commons Attribution (CC BY) license (https://creativecommons.org/licenses/by/4.0/).

1. Basic Interactions in Quasi-1D Molecular Systems

Since the discovery of a metal-like conductivity in 1954, when perylene (Per) was exposed to Br, a new field of research opened among organic materials, with the synthesis in 1960 of the molecular acceptor TCNQ (tetra-cyanoquinodimethane), and in 1970 the discovery of the molecular donor TTF (tetra-thiafulvalene), rapidly followed by the combination of acceptor (A) and donor (D) stacks in the same structure. This led to a large family of charge-transfer organic conductors such as Qn(TCNQ)2 in 1960, NMP-TCNQ in 1965, and then TTF-TCNQ in 1972 (the chemical name of organic molecules quoted in the text is given in the annex). These findings were followed by the unexpected discovery in 1977 that, when exposed to AsF5, the conductivity of trans-polyacetylene, (CH)x, can be raised by more than seven orders of magnitude. These pioneering works open an incredibly large and fecund area of research, bolstered by the continuous discovery of new physical phenomena, now forming the so-called domain of "organic electronics". These finding were rapidly followed by potential applications and the realization of technological devices [1].

The originality of organic metals is due to the fact that these materials exhibit low 1D or 2D electronic properties when coupled to a soft molecular structure. The low electronic dimension generally triggers electronic (charge or spin) collective density wave (CDW or SDW) instabilities not found in conventional 3D metals. In addition, because of the presence of a soft underlying lattice, the electronic instability allows the stabilization of unconventional ground states involving coupled charge/spin and lattice modulations, the most popular one being the $2k_F$ Peierls ground state found in 1D systems such as trans-polyacetylene. Incommensurate $2k_F$ CDW/Peierls modulations were initially detected in 1D metallic charge transfer salts such as $TTF^{+\rho}$-$TCNQ^{-\rho}$, formed of segregated stacks of donor (D) and acceptor (A) with a D to A charge transfer of $\rho = 0.59$, and where the Fermi wave vector of each individual 1D electron gas is simply given by $k_F = \rho/4$ in a reciprocal chain unit. CDW physics, conjointly observed in 1D inorganic systems, have been recently summarized in [2]. Another remarkable originality of organic quasi-1D systems is that, due to the large spatial extent of organic molecules, the intra-stack electronic transfer integral, $t_{//}$, is comparable to the intra-site (U) and first neighbor inter-site (V_1, V_2 ...) Coulomb repulsions entering into the 1D extended Hubbard model, which models the properties of the correlated 1D electron gas quite well [3]. This allows the rationalization of the experimental observation of various $2k_F$ density wave instabilities, as well, to account for the unexpected CDW instability (discovered in TTF-TCNQ) at the $4k_F$ wave vector, where the $4k_F$ wave vector is the first Fourier component of a lattice of localized charge of the Wigner type (for more detail, see [4]). Although $4k_F$ collective charge localization waves were initially observed in incommensurately filled electronic systems, a stronger effect occurs in salts with one charge per site or every two sites. Thus, special attention has been devoted in the literature to studying half-filled ($\rho = 1$) and quarter-filled ($\rho = 1/2$) organic systems, leading to a better interplay between charge and antiferromagnetically (AF) coupled spin degrees of freedom. These commensurate systems allow the efficient stabilization of ground states not known in 3D materials, such as the spin-Peierls (SP) ground state typical of 1D systems [4]. SP ground states stabilized in many of them are more specifically considered in this review.

The basic aspect of the electronic phase diagram of half-filled and quarter filled 1D electron gas is described in refs. [5–9], respectively. Then, the modification of phase diagrams, taking explicitly into account the important coupling of electrons with lattice degrees of freedom (electron–phonon and magnetoelastic coupling), was considered in refs. [10,11]. Figures 1 and 2 exhibit a large panel of instabilities and of ground states experimentally observed in various half-filled and quarter-filled 1D molecular systems, respectively. In particular, both Mott insulator and 1D Luttinger liquids stabilized by sizeable electron–electron correlations present either antiferromagnetic or spin-Peierls ground states (Figures 1 and 2). There are, however, subtle differences between half-filled and quarter-filled situations which depend on the degree of charge localization and electronic interactions, such as the magnitude of the gap of charge, $\Delta\rho$, an the antiferromagnetic exchange interaction J between S-$\frac{1}{2}$ (Sections 2 and 3).

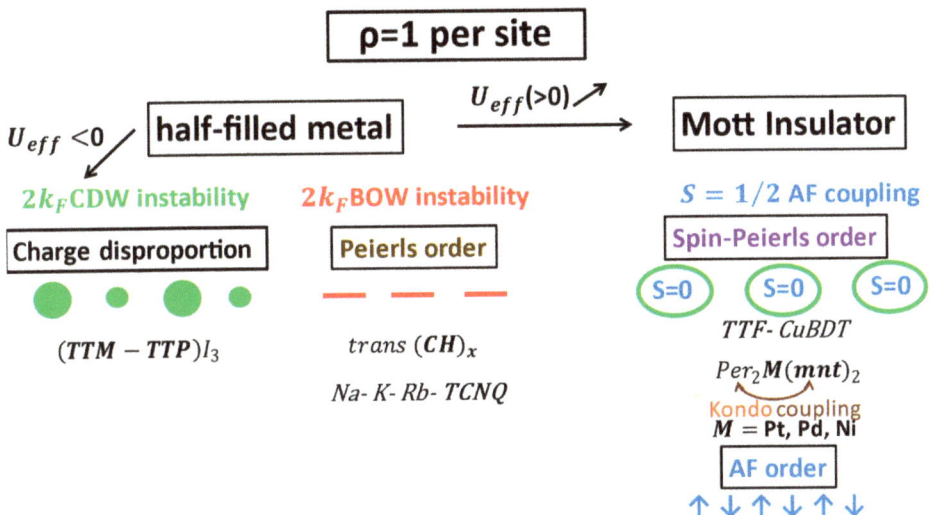

Figure 1. Electronic phase diagram of the half-filled chain. The various electronic instabilities and ground states are schematically described. Typical experimental examples are also indicated. The stack bearing the instability is in bold text.

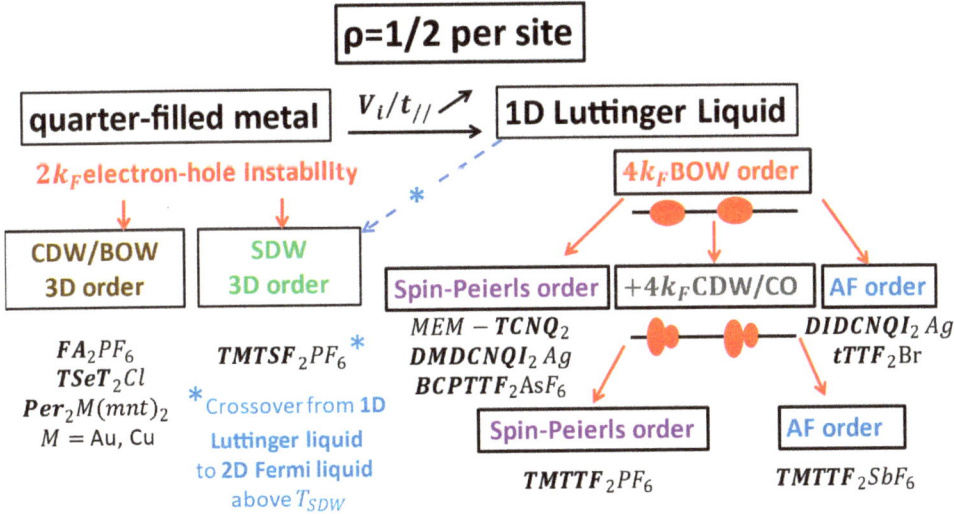

Figure 2. Electronic phase diagram of the quarter-filled chain. The various electronic instabilities and ground states are schematically described. Typical experimental examples are also indicated. The stack bearing the instability is in bold text. The crossover from a 1D Luttinger liquid to a 2D Fermi liquid in TMTSF$_2$PF$_6$ is indicated by the symbol * in the figure.

2. Half-Filled ($\rho = 1$) Versus Quarter-Filled ($\rho = 1/2$) Organic Systems

The dilemma particularly concerns quarter-filled 1D dimerized systems of current interest whose electronic structure is basically that of a quarter-filled HOMO (LUMO) band for the donor (acceptor) stack ($\rho = 1/2$). However, due to lattice dimerization, a band gap, $2\Delta_D$, opens at the Brillouin zone boundary $\pm \pi/a$, where a is the chain parameter of the dimerized stack. For non-interacting electron, this splits the HOMO band by twice the difference of intra- and inter-dimer transfer integrals: $2\Delta_D = 2 | t_{intra} - t_{inter} |$. Thus, if only the upper band of the dimerized chain is relevant, the electronic system could be considered as half-filled ($\rho = 1$). If not, it should be considered as quarter-filled ($\rho = 1/2$). The clarification of the dilemma requires a consideration of all the interactions, knowing that lattice dimerization (or $4k_F$ BOW ordering of the stack- see Figure 2) induces a $4k_F$ lattice potential, allowing second-order electron–electron Umklapp scattering which, in the presence of large electron–electron repulsions, tends to localize the charges [12,13]. By taking into account these effects, the gap of charge, $\Delta\rho$, relevant for the charge localization process, becomes a complex function of Δ_D but also of the coulomb interaction parameters [14].

Thus, if in the T range of interest (let us say below 300 K) $\Delta\rho$ is larger than 1000 K($\sim\pi T_{max}$), the system can be considered as effectively half-filled. In this limit, each electron is localized in the bonding state (antibonding state) of the acceptor (donor) dimer. Charges are frozen, so that there is no charge degree of freedom available in the dimer. This means that there is no possibility for a charge disproportion (or charge ordering, CO) inside the dimer. As in the case for a half-filled system (Figure 1), the only low T instabilities are those driven by Heisenberg AF coupling between localized S-$\frac{1}{2}$ in each dimer, eventually coupled to a phonon field in order to achieve an SP ground state. This situation occurs in quarter-filled strongly localized A$_2$Y compounds, where Y is a monovalent cation, such as MEM(TCNQ)$_2$, which exhibits at 335K a first order transition to a $4k_F$ BOW phase (with dimerization of the acceptor TCNQ stack accompanied by a modification of the cation Y sublattice), opening a very large energy gap of $\Delta\rho = 0.64$ eV, then an SP ground state at 18 K. To a lesser extent, this also occurs in quarter-filled D$_2$X salts, where D is a derivative of the TTF molecule and X a monovalent anion, adopting the prototypal structure of the (TMTSF)$_2$X/(TMTTF)$_2$X (i.e., Bechgaard/Fabre salts) [15]. In this category, organics (DIMET)$_2$SbF$_6$ and (t-TTF)$_2$Br (where $\Delta\rho \sim 3400$ K) exhibit an AF ground state at 12 K and 35 K, respectively, while (BCP-TTF)$_2$PF$_6$ and AsF$_6$ (where $\Delta\rho \sim 2000$ K) exhibit an SP ground state at 32.5 K and 36 K, respectively.

In the opposite situation, where the gap of charge is smaller, $\Delta\rho \sim \pi T_\rho < 1000$ K, the system is really quarter-filled with non-frozen intra-dimer charge degrees of freedom. In these systems, T_ρ is the temperature of minimum of conductivity, corresponding to the $4k_F$ BOW charge localization temperature. This situation is relevant for the Fabre salts (TMTTF)$_2$X where $T_\rho \sim 230$ K, with an activation energy of conductivity of below $\frac{\Delta\rho}{2} \sim 350$ K. At intermediate temperature, around 100 K, those salts undergo a CO/ferroelectric phase transition causing a polar charge rearrangement inside the dimers (Figure 2), accompanied by an anion X displacement. This transition is followed at lower temperature by another phase transition to either AF or SP ground states (for a recent review on quarter-filled organic systems, see [16]). Note that, for a true quarter band-filling, fourth-order Umklapp scattering processes are relevant to interpret the physical properties, as in the Fabre salts [13].

3. Antiferromagnetic Coupling in Spin-1/2 Weakly Localized Organic Stacks

When charges are localized, their S-$\frac{1}{2}$ interact via an antiferromagnetic Heisenberg coupling. In the case of first neighbor interaction, J, the Hamiltonian simply reads:

$$H_{spin} = J \sum_i S_i S_{i+1} \quad (1)$$

No magnetic transition occurs for an isolated S = 1/2 AF chain due to thermal and quantum fluctuations. In fact, in the presence of these fluctuations, the spin susceptibility, χ_{spin}, can be exactly calculated [17]. From both the magnitude and the thermal dependance of χ_{spin}, the first neighbor AF J interaction defined by (1) can be obtained. The fit of the experiment data of organics using expression (1) leads to a large panel of J values. J increases along the sequence J = 77 K for the half-filled ($\rho = 1$) TTF donor chain of the 1:1 TTF—CuBDT compound [CuBDT is CuS$_4$C$_4$(CH)$_4$] and, for various A and D quarter-filled chain ($\rho = 1/2$) 2:1 compounds, J = 138 K in MEM-$(TCNQ)_2$ [4], J = 270 K (see Figure 4 in Section 4) in $(BCP-TTF)_2AsF_6$ [18], J = 460 K in $(TMTTF)_2PF_6$-D_{12} [19](see Figure 3), and similar J values are found for other Fabre salts with different anion X [20]; somewhat larger J are found for the substituted donor o-DMTTF: J = 490(30)K–530(30)K in (o-DMTTF)$_2$X with X = Br and I, respectively, [21] and J~520(50)K with X = NO_3 [22]. The increase of J seems to be correlated with an enhanced charge delocalization in the stack direction. Note that a satisfactory fit of J is obtained if the experimental spin susceptibility (measured at 1 bar) is corrected by the thermal volume expansion. This correction is shown in Figure 3 for deuterated $TMTTF_2PF_6$. In these Fabre salts, J is quite large, which means that the spins (charge) are weakly localized. Additionally, for a significantly delocalized spin system, χ_{spin}, can be equally well accounted from a first neighbor 1D extended Hubbard model. Such a calculation, detailed in note [23], is also shown in Figure 3 which provides Coulomb interaction parameters U and V_1, scaled by the DFT transfer integral $t \approx 0.18$–0.2 eV on the TMTTF stack. These values are twice as small as those calculated for the TTF stack of TTF-TCNQ [24]. Note that deviation between the fit and experimental data of Figure 3 could be reduced at high temperatures by using a larger U (~6t), and at low temperatures by including a non-negligeable second neighbor interaction $V_2 \approx V_1/2$.

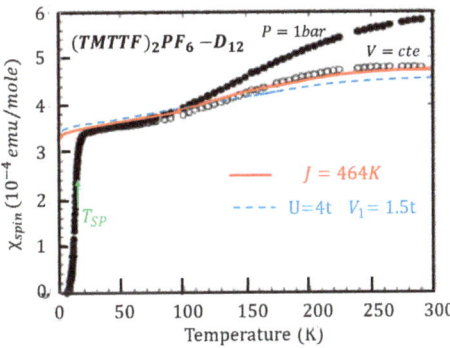

Figure 3. Temperature dependence of the spin susceptibility, χ_{spin}, of deuterated $TMTTF_2PF_6$. The filled circles give the ambient pressure SQUID measurements, and the empty circles give the spin susceptibility at constant volume. The continuous red line is the fit of χ_{spin} with the expression of the spin susceptibility of the localized S—1/2 AF Heisenberg chain [17]. The adjustment shown in the figure, which takes into account both the absolute value and the thermal dependence of χ_{spin}, gives J ≈ 464 K [19]. The dotted blue line is the fit of χ_{spin} for the 1D extended Hubbard model, as detailed in note [23].

A large range of first neighbor AF coupling J_1 values is also obtained from the fit of the high temperature spin susceptibility, χ_{spin}, in the half-filled dithiolate chain ($\rho = 1$) of $Per_2M(mnt)_2$ [25]: $J_1 = 38$ K for M = Pt, from 100 K to 240 K in M = Ni and 270 K for M = Pd (see Figure 7 in Section 5). χ_{spin} is additionally modified upon cooling, firstly by the SP fluctuations, and then by the Kondo interaction with the electronic degrees of freedom on the Per stack for M = Ni and Pd. Kondo interactions add a frustrated second neighbor AF coupling J_2 on the dithiolate stack (see Section 5).

4. Spin-Peierls Fluctuations on Donor Stacks in D_2X Weakly Localized Conductors

All AF organics considered in the last section are subject to a spin-Peierls (SP) instability when the S = 1/2 AF donor stack is coupled to a phonon field. This leads to a 3D phase transition at finite T_{SP} toward an S = 0 non-magnetic ground state accompanied by a dimerization of the S = 1/2 AF chain. Additionally, this second order phase transition is announced by an important regime of SP pre-transitional structural fluctuations which develop below the so-called mean-field SP transition temperature T_{SP}^{MF} of the chain, which is significantly higher than T_{SP}. In most of the SP compounds, magneto-structural coupling picks up singlet fluctuations of the AF chain and correlates them into an SP short-range order. This local order is revealed by a decrease of χ_{spin}, due to the formation of a pseudo-gap in the spin degrees of freedom, which develops in the T range of structural fluctuations below T_{SP}^{MF} [26]. The observation of a pseudo-gap in χ_{spin} is the signature of a classical (adiabatic) regime of instability. It is found in many SP materials [2,4,18,27], some of which are considered below. However, there are other compounds such as TTF—CuBDT, MEM-(TCNQ)$_2$ [4,18] and Per$_2$Pt(mnt)$_2$ [25] where SP pre-transitional lattice effects do not affect the spin susceptibility. There, the SP instability occurs in the anti-adiabatic (quantum) regime of fluctuation. Anti-adiabatic effects are observed in materials exhibiting the smallest AF exchange interaction J, or where charge/spin are strongly localized. Non-adiabaticity is due to the fact that the phonon energy $\hbar\Omega_0$ of the critical lattice mode is larger than the magnetic energy ~J [28]. More precisely, the classical–quantum crossover for an SP transition in a Heisenberg chain occurs for $\hbar\Omega_0 \approx \Delta_{MF}/2$, where the mean-field SP gap, Δ_{MF}, is related to T_{SP}^{MF} by $\Delta_{MF} \approx 2.47 k_B T_{SP}^{MF}$ [29]. Additionally, when non-adiabaticity effects are important, the Hamiltonian (1) is modified by the introduction of next near neighbor exchange interactions. This is the case for the SP inorganic compound CuGeO$_3$ [30].

In the adiabatic limit, SP fluctuations develop a pseudo-gap in the magnetic excitation spectrum of the Heisenberg chain. The energy dependence of the pseudo-gap is controlled by the spatial extension of the SP correlation length, quantitatively calculated in ref. [26]. Figure 4 shows the associated drop of spin susceptibility below about $T_{SP}^{MF} \approx 120$ K in (BCP-TTF)$_2$AsF$_6$. It occurs in the temperature range where quasi-1D SP fluctuations are developing (Figure 4). Below $T_{SP} = 35$ K, these short-range fluctuations condensed into 3D superlattice reflections, which implies a long-range dimerized SP order [18,26,31]. The development of the SP pseudo-gap in deuterated $TMTTF_2PF_6$ has been precisely studied by inelastic neutron scattering measurements of the energy dependance of magnetic excitations. It reveals a drop in the magnetic density of states in an energy range of Δ_{MF} upon cooling below $T_{SP}^{MF} \approx 40$ K [32]. Figure 5 shows both the drop in spin susceptibility and, in inset the decrease in amplitude of magnetic excitation energy. The pseudo-gap, a precursor at an SP transition, is also found in other D$_2$X weakly localized conductors with large J such as (o-DMTTF)$_2$NO$_3$ [22].

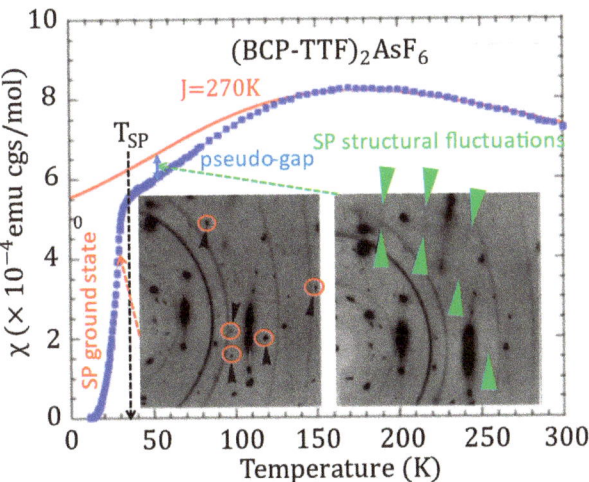

Figure 4. Spin susceptibility of $(BCP\text{-}TTF)_2AsF_6$, showing the development of a pseudo-gap below $T_{SP}^{MF} \approx 120$ K when quasi-1D SP fluctuations of chain dimerization develop. Spin susceptibility abruptly drops at the $T_{SP} = 35$ K SP phase transition, below which satellite reflections due to the dimerization are detected. The inset presents the PF_6 salt X-ray diffuse scattering patterns taken (right part) in the regime of quasi-1D SP fluctuations (green arrows), and (left part) in the 3D-SP dimerized phase (superlattice reflections surrounded by red circles). The figure combines results taken from refs. [18,31].

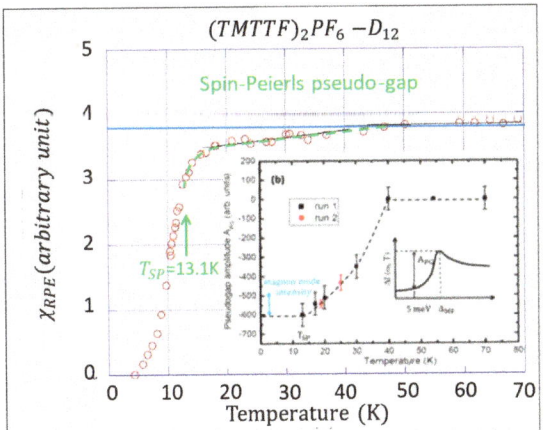

Figure 5. Spin susceptibility of deuterated $TMTTF_2PF_6$ near the SP transition, showing the development of a pseudo-gap below 40 K (χ_{RPE} has been kindly provided by C. Coulon). The inset gives the thermal dependance of the amplitude of the pseudo-gap at 5 meV in the same T range. The energy dependence of the pseudo-gap, obtained from neutron measurements [32], is also schematically represented.

5. Kondo Coupling between Localized and Delocalized Stacks in the $Per_2\text{-}M(mnt)_2$ Series

A very original family of 2:1 charge transfer salt is the $\alpha\text{-}Per_2\text{-}M(mnt)_2$ series, whose array projected perpendicularly onto the chain direction is shown in Figure 6a. It contains a metallic quarter-filled Per stack subject to a Peierls instability and dithiolate chains. For M = Au and Cu close shell dithiolate molecules are non-magnetic, while for M = Pt, Ni and Pd dithiolate derivatives, localized S = 1/2 chains with AF interactions are formed (Figure 6b). Such a coupled stack array presents very original physical properties which are reviewed in [1,33]. It exhibits, in particular, three types of substantial interchain exchange interactions $J_\perp$ (named J′, J″ and J‴ in Figure 6a), whose importance is assessed by the observation of a single EPR line at a g value intermediate between those of dithiolate and Per molecules in the Pd [34] and Pt [35] derivatives. This coupling allows a subtle interplay between Peierls and spin-Peierls instabilities, respectively, located in the Per and M = Pt, Ni and Pd dithiolate stacks [36]. In this respect, and based on the finding of substantial $J_\perp$ values, a Kondo-type of inter-chain magnetic coupling (Figure 6b) was proposed to be a relevant interaction in the Pt salt [35]. However, it was recently realized [25] that most visible effects arise from Per-dithiolate Kondo coupling in the less studied Ni and Pd salts. In them, the SP instability occurs in the adiabatic regime, at the difference of the non-adiabatic SP regime of the Pt salt [25].

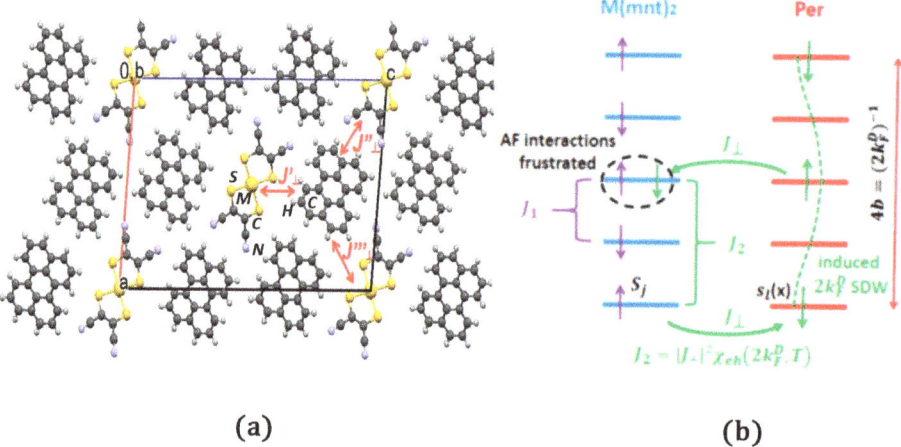

Figure 6. (a) Crystal structure of α-Per$_2$-M(mnt)$_2$ projected along the stack direction b, which reveals the presence of segregated Per and dithiolate stacks. The atoms are labelled. First, neighbor inter-stack [M(mnt)$_2$]—Per AF exchange coupling $J_\perp$ are schematically indicated (note that there are three different types of interactions per Per). (b) Schematic representation of competing S = 1/2 AF exchange couplings in α-Per$_2$-M(mnt)$_2$ derivatives for M = Ni, Pd and Pt. J_1 is the first neighbor direct AF exchange interaction on dithiolate stack. J_2 is the second neighbor indirect RKKY exchange interaction mediated by the induced $2k_F^D$ SDW on the Per stack.

Here, we consider specifically the Pd salt, where the effects of SP instability and of Kondo coupling are very remarkable. Figure 7 shows the thermal dependance of the spin susceptibility of the Pd salt obtained from ref. [36]. At high temperature, χ_{spin} follows the thermal dependance of the spin susceptibility of the S = 1/2 AF Heisenberg Hamiltonian (1) with J = 270 K (J_1 in Figure 6b). Then, below ~100 K χ_{spin} drops strongly when 1D SP structural correlations on the Pd(mnt)$_2$ stack develop [37] (see insert of Figure 7). This behavior bears some resemblance to that exhibited by (BCP-TTF)$_2$AsF$_6$ below T_{SP}^{MF} (Figure 4), with, however, a larger rate of decrease of χ_{spin}. The rate of decrease is enhanced below about 40 K, as revealed by a change of slope (brown dotted lines in Figure 7). This behavior is ascribed to the relevance of the Per-dithiolate exchange Kondo interaction, which sets a J_2 AF interaction of the same sign as J_1 (see Figure 6b and ref [25] for more detail). This induces a drop of χ_{spin} which is so rapid that, at the SP transition of 28 K, only 1/3 of χ_{spin} remains (this is different to (BCP-TTF)$_2$AsF$_6$, where, without AF frustration, 90% of χ_{spin} remains at T_{SP}—see Figure 4).

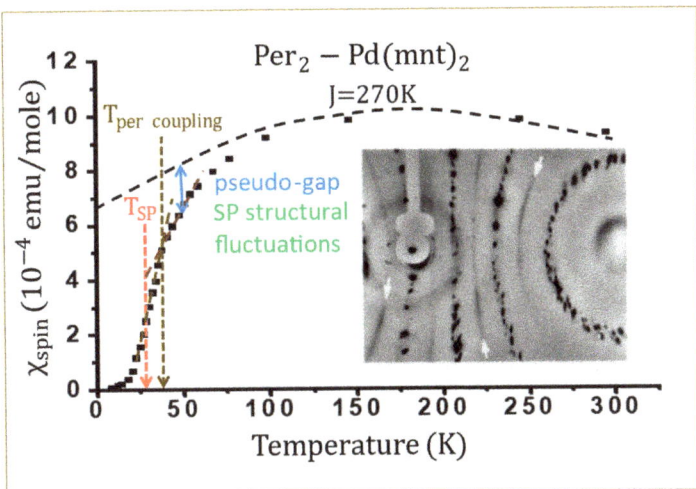

Figure 7. Thermal dependence of the spin susceptibility of Per$_2$-Pd(mnt)$_2$ (data taken from ref. [36] and kindly provided by M. Almeida). At high temperature, χ_{spin} behaves as the spin susceptibility of the S-1/2 AF Heisenberg Hamiltonian (1) with J = 270 K on the dithiolate stack. Below 100 K, χ_{spin} deviates from this dependence due to the onset of SP fluctuations forming a pseudo-gap. Below 40 K ($T_{per\ coupling}$), the deepest rate of decrease is due to the Kondo coupling with the Per stack. The X-ray pattern in inset reveals the presence of 1D-SP structural fluctuations on the Pd(mnt)$_2$ stack at 40 K which are responsible for the pseudo-gap regime (see [37] for more details). The 3D SP transition achieving the condensation of diffuse lines into superstructure spots occurs at T_{SP} = 28 K.

6. Spin-Ladder Behavior in the (DT-TTF)$_2$M(mnt)$_2$ Series

Spin-ladders are a class of low dimensional magnetic system built with a finite number of magnetically (generally AF) coupled spins chains. These spin systems are intermediate between 1D isolated magnetic chains and the 2D magnetic layer. However, their unexpected magnetic properties have attracted a lot of attention relevant for the study of low dimensional and topological quantum systems [38]. Earlier studies show that, depending on the number of interacting spin chains (legs in the spin-ladder), quite different magnetic behaviors are observed. AF coupled S = 1/2 spin-ladders with an odd number of legs behave as isolated chains with a finite spin susceptibility upon approaching zero temperature and spin–spin correlations with a power-law dependance. This contrasts with spin-ladders with an even number of legs, which present a gap in the magnetic excitation spectrum so that the spin susceptibility drops exponentially towards zero upon cooling into a spin-liquid ground state lacking long-range order. The even-leg ladders can be seen as spin singlet pairs, with spin–spin correlation decaying exponentially due to the presence of a finite spin gap.

Two-leg spin ladders are found both in inorganic and organic states. In the inorganics, such as the transition metal oxide α'-NaV$_2$O$_5$ and the family of 1D incommensurate composite crystals M$_{14}$Cu$_{24}$O$_{41}$ with M = La, Y, Sr, Ca, ladders exhibit a rectangular structure with transverse AF exchange interactions along the rung of the ladder. In the organics considered below, the two legs are linked by AF zig-zag interactions.

Organic ladders were first reported in the DTTTF$_2$-M(mnt)$_2$ family of 2:1 charge transfer salt whose structure is represented in Figure 8. These salts are composed of layers of alternating dithiolate stacks and double stacks of DTTTF donors. Double stacks are related by screw axis symmetry, so that donors form a zig-zag chain. (DT-TTF)$_2$M(mnt)$_2$ salts with M = Au, Cu, Pt and Ni are the first organic materials found to exhibit spin-ladder physics [39]. Since then, many other organic compounds, reviewed in [40], have exhibited similar spin-ladder behavior.

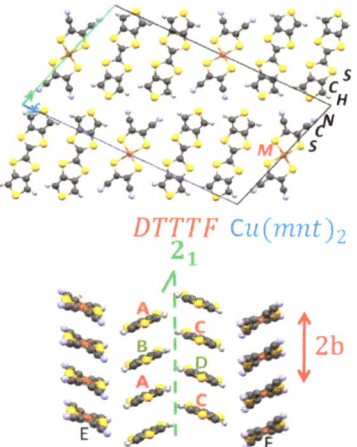

Figure 8. Crystal structure of DTTTF$_2$-Cu(mnt)$_2$ perpendicular to the stack direction (top) and along the stack b direction (bottom), showing the 2$_1$-screw axis symmetry between donor stacks in the metallic state. The atoms are labelled in the top figure. The break of 2$_1$ symmetry below the 4k$_F$ transition doubles the b periodicity and leads to a charge disproportion with alternating hole rich (red) and hole poor (grey) donors.

The most interesting compound, DTTTF$_2$-Cu(mnt)$_2$, incorporates the non-magnetic close shell Cu(mnt)$_2$ dithiolate molecule. It is a quarter-filled D$_2$X 1D metal at ambient conditions with the P2$_1$/c monoclinic space group. It undergoes a second-order electronic transition at 235 K (T_{4k_F}) [41], from a weakly localized regime to gapped insulator, together with the development of a 3D superstructure doubling the b periodicity [41] (Figure 9a). As this transition leaves the spin susceptibility unaffected [42] (Figure 9b), only the charge degrees of freedom are involved. The metal insulator transition thus corresponds, for a 1D system, to a 4k$_F$ charge localization (Figure 2). One should discard a 2k$_F$ Peierls transition, which should also open a (non-observed) gap in the spin degrees of freedom. In a single chain compound, a 4k$_F$ charge localization occurs either on one bond out of two (4k$_F$ BOW) or one site out of two (4k$_F$ CDW). These two types of 4k$_F$ modulation exhibit an inversion symmetry either on the bonds or on the sites, respectively. In the case of the DTTTF zig zag ladder, where the 2$_1$-screw axis is removed (Figure 8) the inversion symmetry elements located on the left and right chains forming the ladder in its metallic state are both removed. Thus, the 4k$_F$ modulation appears to be a mixture of BOW and CDW (or CO) modulations (see Section 7). However, the structural refinement performed at 120K in the Cu derivative [43] shows, basically, the formation of CDW stack alternating "neutral" and "ionic" nearly equidistant donors along b (lower part of Figure 8). The same type of 4k$_F$ transition was previously reported in DTTTF$_2$-Au(mnt)$_2$ [39], with, however, the observation of a short-range order below ~220 K instead of the long-range order below 235 K in the Cu salt. In the Au derivative, infra-red spectroscopy indicates an incomplete charge disproportion (0.5 $\pm$ δ) of about 2δ~0.35–0.4 between donors [44]. Finally, note that inversion symmetry, located in the middle of the ladder, and relating its two peripheric chains (see Figure 11 at the end of the section), could be absent since the structural refinement of the Cu derivative [43] indicates that, with different charges, the inversion symmetry is weakly broken between stacks (lower part of Figure 8).

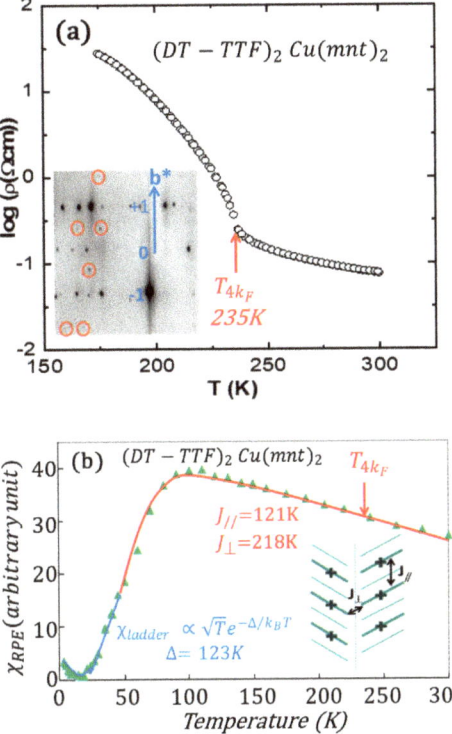

Figure 9. Thermal dependance of the logarithm of the resistivity (**a**) and of the RPE spin susceptibility (**b**) of DTTTF$_2$-Cu(mnt) (data originally published in ref. [41,42], respectively, are kindly provided in a modified form by M. Almeida). The b*/2 superlattice reflections appearing below T_{4k_F} are shown (red circles) in the X-ray pattern of the insert of figure (**a**). The thermal dependance of χ_{RPE} if fitted at high temperature by χ_{spin} of the frustrated ladder (red curve) with $J_\perp$ and $J_{//}$ is represented in the figure, and at low temperature by χ_{ladder} (blue curve), whose analytical expression is given. In (**a**), the weak increase of resistivity on approaching T_{4k_F} from higher T could be due to the development of a pseudo-gap, due to 4k$_F$ fluctuations in the metallic state.

The most interesting aspect of DTTTF$_2$-Cu(mnt)$_2$ concerns the thermal behavior of the spin susceptibility, which reveals the formation of a spin gap, Δ. Above 50 K, the fit of χ_{spin} allows the determination of frustrated AF exchange interactions; they are nearly twice as large along the zig-zag (rung) direction ($J_\perp$ = 218 K) as along the stack direction ($J_{//}$ = 121 K)—see the insert of Figure 9b [42]. In this temperature range, the spin gap result from frustrated AF interactions between $J_\perp$ and $J_{//}$. In this respect, the zig-zag ladder chain bears some resemblance with the J_1-J_2 frustrated AF chain on the Pd(mnt)$_2$ dithiolate stack considered in Section 5. The spin gap is more visible at a low temperature, where the spin susceptibility exhibits activated behavior below 50 K (Figure 9b). Here, χ_{spin} clearly behaves as the spin susceptibility of an SP compound (see Section 4). From the high and low temperature fits, the same spin gap value Δ ≈ 123 K is obtained [42]. However, since the high temperature and low temperature gap values are nearly identical, there is no apparent magnetic symmetry breaking in the temperature, from which one deduces that there is no real enhancement of a low temperature 2k$_F$-SP lattice modulation. However, optical measurements indicate a modification of the vibrational spectrum below 70 K in the Au derivative [45], which has been attributed to the formation of the spin gap. Therefore, the problem of the coexistence of 2k$_F$ SP and of 4k$_F$ charge modulations in the whole temperature range remains ambiguous. Below, we provide an argument for the presence of a high temperature SP modulation.

The electronic structure of the zig-zag ladder in its quarter-filled metallic state is quite subtle. A simple tight binding analysis of the ladder structure shown in Figure 10a, which reveals the presence of two bands. Depending on the ratio of zig-zag (t_d) and intra-chain (t_s) transfer integrals, two different conduction band fillings can be found. In a free electron representation, one obtains the following: (1) for dominant t_s (>1.7t_d), two partially-filled 1D conduction bands which could exhibit a combined inter-band 2(k_{F1}+ k_{F2}) nesting process at $b^*/2$ (Figure 10b); and (2) for dominant t_d (>0.58t_s), an upper partly filled 1D conduction band with a simple 2k_F nesting process could occur at the same $b^*/2$ wave vector (Figure 10c). A more accurate electronic structure determination, with a dispersion resembling the ones shown in Figure 10, is calculated in ref. [46] with an extended Hückel Hamiltonian. With a t_s/t_d~1.7 ratio, the electronic structure of the Au derivative is at the borderline between dispersions given in Figure 10b,c. However, with a larger t_s/t_d~3 ratio, the Ni and Pt derivatives have a two-conduction band structure, as shown in Figure 10b.

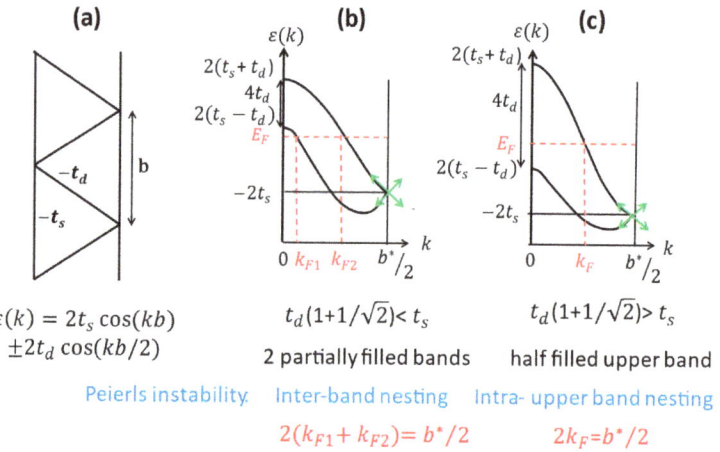

Figure 10. Band structure of the quarter fill zig-zag ladder shown in (**a**). With one-half hole per donor and for decreasing ratio of transfer integrals t_s/t_d one shifts from a two partially filled band structure (**b**) to a half-filled upper band structure (**c**). The Peierls instability is achieved by an inter-band nesting process in (**b**) and by an intra-upper band nesting process in (**c**). For half a hole per donor, both FS nesting processes always occur at $b^*/2$.

At first glance, the stabilization of a 2k_F Peierls ground state, as predicted by the free electron band structures of Figure 10, does not correspond to physical reality because χ_{spin} does not drop at T_{4k_F} (Figure 9b). However, a 2k_F SP instability of the zig-zag lattice of localized spin, which exhibits a modulation of the same symmetry as a 2k_F Peierls instability on the same zig-zag chain, could be stabilized without affecting χ_{spin}. Here, the quadrupling of the zig-zag periodicity d in a diagonal direction (2k_Fd = d*/4) due to an SP instability induces a doubling of the chain periodicity along b, as does the 4k_Fb = $b^*/2$ charge localization in the stack direction. Thus, as shown in Figure 11, both types of instabilities of the same periodicity could coexist. This argument is assessed from theoretical calculations of the instability of a quarter-filled ladder chain treated with the extended Hubbard Hamiltonian coupled to a phonon field [47]. The CDW and BOW patterns thus calculated, assuming an important diagonal coupling, are shown in Figure 11. In this representation it is easy to understand that the spin gap should open from spin singlet pairing along the diagonal bond exhibiting the strongest AF exchange coupling,

$J_\perp > J_{//}$. However, this interpretation should be validated by a precise determination of the SP modulated structure.

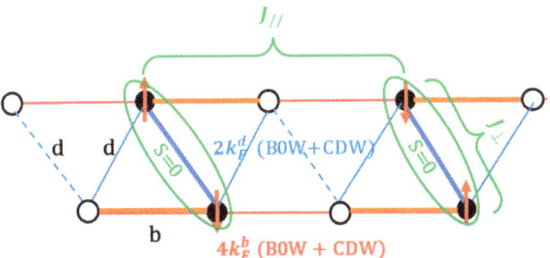

Figure 11. Schematic structure of the BOW and CDW in the quarter-filled zig-zag ladder combining $2k_F^d$ modulation in zig-zag (blue) and $4k_F^b$ modulation in stack (red) directions. Frustrated AF exchange interactions are indicated, as well as the location (in green) of spin singlets. Note that, in this figure, the centrosymmetric bond and charge patterns are kept for simplicity.

More complex behaviors, not yet completely understood, are found in the DTTTF$_2$-M(mnt)$_2$ series with M = Pt and Ni [46,48]. As for the Per$_2$-M(mnt)$_2$ materials with the same M (Section 5), the complexity arises from the additional presence of dithiolate paramagnetic ions forming an $S = 1/2$ AF chain, which compete with the frustrated AF interactions previously found in the donor ladder. DTTTF$_2$-M(mnt)$_2$ with M = Pt and Ni is a better conductor than the Au and Cu analogues. However, they still exhibit a metal–insulator transition, but at a twice smaller T_{4k_F} [46,48]. In the Pt derivative the transition stabilizes the 2b periodicity as for the Cu and Au derivatives [46]. However, the Pt and Ni salts exhibit two interacting magnetic subsystems, as shown by EPR measurements for which the width of the single line is observed to increase dramatically as the conductivity increases. Both salts show two different types of AF interaction, and at a lower temperature exhibit a transition not yet fully understood: it is possibly of the AF type in the Ni salt and of the SP type in the Pt salt. The SP instability of the ladder found in the Au and Cu derivatives is apparently less stable in Pt and Ni derivatives. However, it could be restored by the dimerization of the dithiolate stack, setting a 2b SP lattice periodicity, as for Per$_2$-M(mnt)$_2$. However, the Pt derivative exhibits upon cooling an enhanced disorder between neighboring stacks, which is certainly caused by frustrated coupling between dimerized dithiolate and DTTTF stacks.

7. Coexistence of $4k_F$ BOW and $4k_F$ CDW Orders in Quarter-Filled Systems

The spin ladder DTTTF$_2$-M(mnt)$_2$ with M = Au or Cu superimposes, in a single phase, three modulations which generally occur through different transitions in the Fabre salts (TMTTF)$_2$PF$_6$: namely a $4k_F$ BOW on the dimerized stack in as grown material, which is enhanced below $T_\rho \sim 230$ K, and then a $4k_F$ CDW around $T_{CO} \sim 60$ K followed by a low temperature SP transition at 18 K. The $4k_F$ modulations are interesting because the BOW destroyed the inversion centers on the site and the CDW on the bonds (Figure 2). As a result, the combined $4k_F$ (BOW + CDW) modulation destroys all the inversion centers so that each stack becomes polar. This property is conserved in the 3D stack array of (TMTTF)$_2$X with octahedral anion X, so that these Fabre salts exhibit electronic ferroelectricity below T_{CO} [49]. Note that, in addition, the 3D ferroelectric ground state should be stabilized by a cooperative displacement of the anions X towards the hole rich donors [16,50].

In spite of the presence of giant dielectric anomalies at T_{CO}, structural modifications are weakly apparent in the ferroelectric phase. Charge disproportion is revealed, in addition to optical means [51], by X-ray determination of the modulation of electronic density (and the loss of inversion symmetry in the structure) [52]. Atomic displacements below T_{CO} were evidenced by neutron diffraction experiments [53], which probably reveal the deformation of the H-bond network, which couple methyl terminal groups of the TMTTF with the anions [16]. The T_{CO} transition is also revealed by small lattice anomalies probed by high resolution thermal expansion measurements in the c* direction, where donor layers and anions alternate [54,55]. However, the most surprising finding is the observation of a splitting of the ESR spectrum in the CO ground state, with magnetic axes rotation indicating a broken magnetic symmetry [56–58]. Such measurements reveal, in addition to a modification of the thermal dependence on the ESR linewidth ascribed to an enhanced anion displacement below T_{CO} [57], a superlattice modulation between the spin chains along c [58]. This unexpected result has been interpreted as linked to a space modulation of the CDW, whose origin remains elusive because there is no structural evidence of a doubling of the c periodicity while keeping the ferroelectric phase.

Finally, note that defects of charge ordering, inducing phase shift in the $4k_F$ CDW, strongly perturb the ferroelectric order. This is, in particular, the case for irradiation defects in the Fabre salts [50]. Here, defects nucleate local polarization and induce a relaxor type of dynamics.

8. Soliton Nucleation in Perturbed Spin-Peierls Compounds

The low energy magnetic excitations of the Heisenberg chain are a continuum of non-interacting spinons [59,60]. In this continuum, in the presence of a magnetoelastic coupling, "$2k_F$" pre-transitional SP lattice fluctuations progressively develop, in the adiabatic limit, a pseudo-gap below T_{SP}^{MF}, as previously discussed in Section 4. The energy dispersion of the pseudo-gap has been measured by inelastic neutron scattering in the Fabre salts [32]. Issued from the magnetic excitation spectrum, a fraction of spinons is pinned by defects under the form of solitons. This is revealed by the detection of Rabi oscillations of pinned solitons in AF magnetic chains of $(TMTTF)_2X$ (X = PF_6, AsF_6) [61] and o-$(DMTTF)_2X$ (X = Cl, Br) [62]. The formation of static solitons breaks the longitudinal coherence of the SP dimerized order, as schematically represented in Figure 12. In this situation, and in order to recover the optimal lateral coupling between dimers located on neighboring SP chains, each defect should nucleate a pair of a soliton and anti-soliton, where each dimerization defect bears an unpaired S = 1/2. Defects thus reduce the intrachain SP correlation length, on $\xi_{//}$, but also break the inter-chain SP phase–phase correlations more drastically (Figure 12). In this situation, the superstructure reflections of the SP ground state are transformed into diffuse lines (the Fourier transform of the local SP order) whose width allows the determination of $\xi_{//}^{-1}$. Diffuse lines thus remain the signature of a local (basically 1D) SP order spatially limited by the presence of defects and various disorders. This effect is well documented in irradiated Fabre salts, where point defects due to irradiation centers break the SP lattice coherence and create unpaired S = 1/2, which manifest with a low temperature Curie-type paramagnetic response revealed by ESR [63,64]. Detailed ESR studies of defects have been performed in as-grown $(TMTTF)_2X$ [61], o-$(DMTTF)_2X$ [62,65,66]. One-dimensional local SP correlations are also revealed in as-grown $Per_2M(mnt)_2$ systems with M = Ni [67] and Pt [25] compounds. Unpaired spins 1/2 have been identified by low temperature NMR measurements in the Pt compound [68]. Conversely, for low amounts of substituent, local 1D–SP structural correlations coexist with 3D–SP dimerization superstructure reflections in Si doped $CuGeO_3$ [69]. This proves that dimerization in SP organics is more sensitive to defects than in SP inorganics such as $CuGeO_3$. The SP pattern of $CuGeO_3$ doped with various elements has been quantitively analyzed in ref. [69].

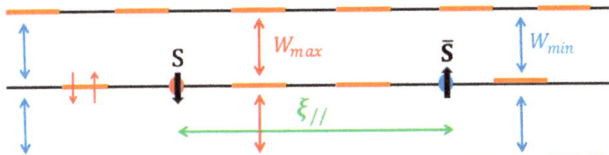

Figure 12. Schematic representation of a pair of a spin 1/2 soliton (S) and spin+1/2 anti-soliton ($\bar{S}$) shifting the phase of the SP intra-chain dimerization on $\xi_{//}$. In addition, between the S and $\bar{S}$ entities the phase shift between neighboring dimers changes from π to 0. This enhances the inter-dimer coupling energy from W_{min} to W_{max}, as schematically indicated.

Note that coherent spin dynamics of solitons have been detected by ESR in the organic spin chain compounds (o-DMTTF)$_2$X [62] and (TMTTF)$_2$X [70] with natural defects. More precisely, defects polarize the AF coupled spins in their vicinity. This leads to a finite local alternating magnetization around the site of the defect, which can be described in terms of a soliton, i.e., a S-1/2 quasiparticle built of many correlated spins, pinned to the defect. Thus, in the presence of a large number of defects, the S-1/2 quasiparticles can interact AF so that the overlap between these magnetic regions is able to achieve a 3D AF order. Of course, this order coexists with local SP correlated 1D regions previously considered. The coexistence of both types of order are observed [71,72] in the quarter-filled organic salt (TMDMTSF)$_2$PF$_6$, where the TMDMTSF non-centrosymmetric donor, built with one half of the TMTTF molecule and one half of the TMTSF, is randomly oriented in a stack direction. Similar phase diagrams revealing a coexistence between SP and AF orders are determined in the SP system CuGeO$_3$ substituted by various elements [73].

Finally, note that under a high magnetic field of the magnitude of the SP gap, the SP dimerized ground state is destabilized towards a lattice of periodic solitons exhibiting an incommensurate periodicity in chain direction, which varies with the amplitude of the applied magnetic field. The resulting phase diagram is observed in many SP compounds [74], including the organic Fabre salts [75] under a high magnetic field. The structure of the high field soliton lattice of CuGeO$_3$ has been determined both by magnetic and structural measurements [76,77].

9. Concluding Remarks

The organic conductors where electronic kinetic energy is comparable to Coulomb repulsion provide an exceptional panel of correlated systems in which unexpected ground states are stabilized. These effects are more clearly revealed in systems exhibiting a 1D electronic anisotropy, and in compounds with a commensurate charge transfer such as the quarter-filled "conductors". This leads, in particular, to spin-charge decoupled physics and the stabilization of new ground states such as the spin-Peierls one. Such ground states are especially stabilized in the organic state, presenting an important lattice response due to the presence of an underlying soft molecular lattice, which thus provides a significant magneto-elastic coupling. In this review, we summarize such features in three families of quarter filled donor organics: (1) the single stack D$_2$X salts which provide clear-cut examples of spin-charge decoupling and low temperature SP instabilities; (2) the segregated two-stack Per$_2$-M(mnt)$_2$ compounds where the Kondo coupling between the AF dithiolate chain and the metallic Per chain unexpectedly enhance the SP instability; and (3) the spin ladder DTTTF$_2$-M(mnt)$_2$ compounds where the double donor stack forming a zig-zag ladder exhibits, simultaneously, a charge localization and an SP gapped ground state. In this review, we have especially presented a unified description of the electronic and structural instabilities in these families of materials.

Funding: This research received no external funding.

Institutional Review Board Statement: Not applicable.

Informed Consent Statement: Not applicable.

Data Availability Statement: Not applicable.

Acknowledgments: This review is a tribute to Manuel Almeida for his 70th birthday. Its content is in great part constructed on scientific results obtained in collaboration with him and his research team at the Universidade de Lisboa. A collaboration between Lisboa and the Laboratoire de Physique des Solides (LPS) of Université Paris-Sud was initiated at the beginning of 1980s by Luis Alcácer, then by Manuel Almeida with long stays at LPS of R.T. Henriques and V. Gama during their PhD period in my group and the one of D. Jérome. With time, the initial research on Per$_2$-M(mnt)$_2$ was extended to the study of the parented spin ladder organics DTTTF$_2$-M(mnt)$_2$ in collaboration with C. Rovira in Barcelona and M. Almeida in Lisboa. The other half of the review is devoted to more classical SP organics, with a special emphasis on recent results obtained in Fabre and analogous salts incorporating substituted donors in collaboration with M. Fourmigué, P. Foury-Leylekian and S. Petit. For this review, fruitful discussions with P. Alemany, S. Bertaina, C. Bourbonnais and E. Canadell are acknowledged.

Conflicts of Interest: The author declares no conflict of interest.

Annex: Chemical Names of Organic Molecules Quoted in the Main Text

Per	perylene
(CH)$_x$	polyacetylene
TCNQ	tetracyanoquinodimethane
Qn	quinolinium
NMP	N-methylphenazinium
MEM	N-methyl-N-ethylmorpholinium
TTF	tetrathiafulvalene
DIMET	dimethyl-ethylene-tetrathiafulvalene
t-TTF	trimethylene-tetrathiafulvalene
TMTTF	tetramethyl-tetrathiafulvalene
TMTSF	tetramethyl-tetraselenafulvalene
BCP-TTF	benzo-cyclopentyl-tetrathiafulvalene
o-DMTTF	*ortho*-dimethyl-tetrathiafuvalene
DTTTF	dithiophen-tetrathiafulvalene
mnt	maleonotril-dithiolate

References

1. Alcácer, L. *The Physics of Organic Electronics: From Molecules to Crystals and Polymers*; IOP Publishing: Bristol, UK, 2022.
2. Pouget, J.-P. The Peierls instability and charge density wave in one-dimensional electronic conductors. *Comptes Rendus Phys.* **2016**, *17*, 332–356. [CrossRef]
3. Hubbard, J. Generalized Wigner lattices in one dimension and some applications to tetracyanoquinodimethane (TCNQ) salts. *Phys. Rev. B* **1978**, *17*, 494–505. [CrossRef]
4. Pouget, J.-P. Bond and charge ordering in low-dimensional organic conductors. *Phys. B Condens. Matter.* **2012**, *407*, 1762–1770. [CrossRef]
5. Nakamura, M. Mechanism of CDW-SDW Transition in One Dimension. *J. Phys. Soc. Jpn.* **1999**, *68*, 3123–3126. [CrossRef]
6. Tsuchiizu, M.; Furusaki, A. Ground-state phase diagram of the one-dimensional half-filled extended Hubbard model. *Phys. Rev. B* **2004**, *69*, 035103. [CrossRef]
7. Emery, V.J.; Noguera, C. Critical Properties of a Spin-1/2 Chain with Competing Interactions. *Phys. Rev. Lett.* **1988**, *60*, 631–634. [CrossRef]
8. Mila, F.; Zotos, X. Phase Diagram of the One-Dimensional Extended Hubbard Model at Quarter-Filling. *Europhys. Lett.* **1993**, *24*, 133–138. [CrossRef]
9. Schmitteckert, P.; Werner, R. Charge-density-wave instabilities driven by multiple umklapp scattering. *Phys. Rev. B* **2004**, *69*, 195115. [CrossRef]
10. Bakrim, H.; Bourbonnais, C. Nature of ground states in one-dimensional electron-phonon Hubbard models at half filling. *Phys. Rev. B* **2015**, *91*, 085114. [CrossRef]
11. Clay, R.T.; Mazumdar, S.; Campbell, D.K. Pattern of charge ordering in quasi-one-dimensional organic charge-transfer solids. *Phys. Rev. B* **2003**, *67*, 115121. [CrossRef]
12. Emery, V.J.; Bruinsma, R.; Barišić, S. Electron-Electron Umklapp Scattering in Organic Superconductors *Phys. Rev. Lett.* **1982**, *48*, 1039–1043. [CrossRef]

13. Giamarchi, T. Theoretical Framework for Quasi-One-Dimensional Systems. *Chem. Rev.* **2004**, *104*, 5037–5055. [CrossRef]
14. Penc, K.; Mila, F. Charge gap in the one-dimensional dimerized Hubbard model at quarter-filling. *Phys. Rev. B* **1994**, *50*, 11429–11445. [CrossRef] [PubMed]
15. Pouget, J.-P.; Ravy, S. Structural Aspects of the Bechgaard Salts and Related Compounds. *J. Phys. I Fr.* **1996**, *6*, 1501–1525. [CrossRef]
16. Pouget, J.-P.; Alemany, P.; Canadell, E. Donor–anion interactions in quarter-filled low-dimensional organic conductors. *Mater. Horiz.* **2018**, *5*, 590–640. [CrossRef]
17. Eggert, S.; Affleck, I.; Takahash, M. Susceptibility of the Spin 1/2 Heisenberg Antiferromagnetic Chain. *Phys. Rev. Lett.* **1994**, *73*, 332–335. [CrossRef] [PubMed]
18. Liu, Q.; Ravy, S.; Pouget, J.P.; Coulon, C.; Bourbonnais, C. Structural fluctuations and spin-Peierls transitions revisited. *Synth. Met.* **1993**, *56*, 1840–1845. [CrossRef]
19. Foury-Leylekian, P.; Petit, S.; Coulon, C.; Hennion, B.; Moradpour, A.; Pouget, J.P. Inelastic neutron scattering investigation of magnetic excitations in the spin-Peierls ground state of $(TMTTF)_2PF_6$. *Phys. B Condens. Matter* **2009**, *404*, 537–540. [CrossRef]
20. Dumm, M.; Loidl, A.; Fravel, B.W.; Starkey, K.P.; Montgomery, L.K.; Dressel, M. Electron spin resonance studies on the organic linear-chain compounds $(TMTCF)2X$ (C = S, Se; X = PF_6, AsF_6, ClO_4, Br). *Phys. Rev. B* **2000**, *61*, 511–521. [CrossRef]
21. Soriano, L.; Pilone, O.; Kuz'Min, M.D.; Vezin, V.; Jeannin, O.; Fourmigué, M.; Orio, M.; Bertaina, S. Electron-spin interaction in the spin-Peierls phase of the organic spin chain $(o\text{-}DMTTF)_2X$ (X = Cl, Br, I). *Phys. Rev. B* **2022**, *105*, 064434. [CrossRef]
22. Jeannin, O.; Reinheimer, E.W.; Foury-Leylekian, P.; Pouget, J.P.; Auban-Senzier, P.; Trzop, E.; Collet, E.; Fourmigué, M. Decoupling anion-ordering and spin-Peierls transitions in a strongly one-dimensional organic conductor with a chessboard structure, $(o\text{-}Me_2TTF)_2NO_3$. *IUCrJ* **2018**, *5*, 361–372. [CrossRef] [PubMed]
23. Fuseya, Y.; Tsuchiizu, M.; Suzumura, Y.; Bourbonnais, C. Role of Interchain Hopping in the Magnetic Susceptibility of Quasi-One-Dimensional Electron Systems. *J. Phys. Soc. Jpn.* **2007**, *76*, 014709. [CrossRef]
24. Cano-Cortés, L.; Dolfen, A.; Merino, J.; Behler, J.; Delley, B.; Reuter, K.; Koch, E. Coulomb parameters and photoemission for the molecular metal TTF-TCNQ. *Eur. Phys. J. B* **2007**, *56*, 173–176. [CrossRef]
25. Pouget, J.P.; Foury-Leylekian, P.; Almeida, M. Peierls and Spin-Peierls Instabilities in the $Per2[M(mnt)2]$ Series of One-Dimensional Organic Conductors; Experimental Realization of a 1D Kondo Lattice for M = Pd, Ni and Pt. *Magnetochemistry* **2017**, *3*, 13. [CrossRef]
26. Dumoulin, B.; Bourbonnais, C.; Ravy, S.; Pouget, J.P.; Coulon, C. Fluctuation effects in low-dimensional spin-Peierls systems: Theory and experiment. *Phys. Rev. Lett.* **1996**, *76*, 1360–1363. [CrossRef]
27. Pouget, J.P. Microscopic interactions in $CuGeO3$ and organic Spin-Peierls systems deduced from their pretransitional lattice fluctuations. *Eur. Phys. J. B* **2001**, *20*, 321–333, Erratum in *Eur. Phys. J. B* **2001**, *24*, 415. [CrossRef]
28. Caron, L.G.; Moukouri, S. Density Matrix Renormalization Group Applied to the Ground State of the XY Spin-Peierls System. *Phys. Rev. Lett.* **1996**, *76*, 4050–4053. [CrossRef]
29. Citro, R.; Orignac, E.; Giamarchi, T. Adiabatic-antiadiabatic crossover in a spin-Peierls chain. *Phys. Rev. B* **2005**, *72*, 024434. [CrossRef]
30. Uhrig, G.S. Nonadiabatic Approach to Spin-Peierls Transitions via Flow Equations. *Phys. Rev. B* **1998**, *57*, R14004. [CrossRef]
31. Liu, Q.; Ravy, S.; Moret, R.; Pouget, J.P.; Coulon, C.; Bechgaard, K. Structural investigation of new 2:1 series of organic conductors. *Synth. Met.* **1991**, *42*, 1879–1883. [CrossRef]
32. Pouget, J.P.; Foury-Leylekian, P.; Petit, S.; Hennion, B.; Coulon, C.; Bourbonnais, C. Inelastic neutron scattering investigation of magnetostructural excitations in the spin-Peierls organic system $(TMTTF)_2PF_6$. *Phys. Rev. B* **2017**, *96*, 035127. [CrossRef]
33. Almeida, M.; Henriques, R.T. Perylene Based Conductors. Chapter 2. In *Handbook of Organic Conductive Molecules and Polymers Volume 1 "Charge Transfer Salts, Fullerenes and Photoconductors"*; Nalva, H.S., Ed.; John Wiley & Sons Ltd.: Chichester, UK, 1997; pp. 87–149.
34. Alcacer, L.; Maki, A.H. Magnetic Properties of Some Electrically Conducting Perylene-Metal Dithiolate Complexes. *J. Phys. Chem.* **1976**, *80*, 1912–1916. [CrossRef]
35. Bourbonnais, C.; Henriques, R.T.; Wzietek, P.; Kongeter, D.; Voiron, J.; Jérom, D. Nuclear and electronic resonance approaches to magnetic and lattice fluctuations in the two-chain family of organic compounds $(perylene_2[M(S_2C_2(CN)_2)_2]$ (M = Pt, Au). *Phys. Rev. B* **1991**, *44*, 641–651. [CrossRef] [PubMed]
36. Gama, V.; Henriques, R.T.; Bonfait, G.; Almeida, M.; Ravy, S.; Pouget, J.P.; Alcacer, L. The interplay between conduction electrons and chains of localized spins in the molecular metals $(Per)_2M(mnt)_2$, M = Au, Pt, Pd, Ni, Cu, Co and Fe. *Mol. Cryst. Liq. Cryst.* **1993**, *234*, 171–178. [CrossRef]
37. Henriques, R.T.; Alcacer, L.; Pouget, J.P.; Jérome, D. Electrical conductivity and X-ray diffuse scattering study of the family of organic conductors $(perylene)_2M(mnt)_2$, (M = Pt, Pd, Au). *J. Phys. C Solid State Phys.* **1984**, *17*, 5197–5208. [CrossRef]
38. Dagotto, E.; Rice, T.M. Surprises on the way from one-to two-dimensional quantum magnets: The ladder materials. *Science* **1996**, *271*, 618–623. [CrossRef]
39. Rovira, C.; Veciana, J.; Ribera, E.; Tarrks, J.; Canadell, E.; Rousseau, R.; Mas, M.; Molins, E.; Almeida, M.; Henriques, R.T.; et al. An Organic Spin-Ladder Molecular Material. *Angew. Chem. Int. Ed. Engl.* **1997**, *36*, 2323–2326. [CrossRef]
40. Silva, R.A.L.; Almeida, M. Spin-ladder behaviour in molecular materials. *J. Mater. Chem. C* **2021**, *9*, 10573–10590. [CrossRef]

41. Dias, J.C.; Lopes, E.B.; Santos, I.C.; Duarte, M.T.; Henriques, R.T.; Almeida, M.; Ribas, X.; Rovira, C.; Veciana, J.; Foury-Leylekian, P.; et al. Structural and Electrical Properties of (DT-TTF)$_2$[Cu(mnt)$_2$]. *J. Phys. IV Fr.* **2004**, *114*, 497–499. [CrossRef]
42. Ribas, X.; Mas-Torrent, M.; Pérez-Benítez, A.; Dias, J.C.; Alves, H.; Lopes, E.B.; Henriques, R.T.; Molins, E.; Santos, I.C.; Wurst, K.; et al. Organic Spin-Ladder from Tetrathiafulvalene (TTF) Derivatives. *Adv. Funct. Mater.* **2005**, *15*, 1023–1035. [CrossRef]
43. Silva, R.A.L.; Santos, I.C.; Wright, J.; Coutinho, J.T.; Pereira, L.C.J.; Lopes, E.B.; Rabaça, S.; Vidal-Gancedo, J.; Rovira, C.; Almeida, M.; et al. Dithiophene-TTF Salts; New Ladder Structures and Spin-Ladder Behavior. *Inorg. Chem.* **2015**, *54*, 7000–7006. [CrossRef] [PubMed]
44. Musfeldt, J.L.; Brown, S.; Mazumdar, S.; Clay, R.T.; Mas-Torrent, M.; Rovira, C.; Dias, J.C.; Henriques, R.T.; Almeida, M. Infrared investigation of the charge ordering pattern in the organic spin ladder candidate (DTTTF)$_2$Cu(mnt)$_2$. *Solid State Sci.* **2008**, *10*, 1740–1744. [CrossRef]
45. Wesołowski, R.; Haraldsen, J.T.; Musfeldt, J.L.; Barnes, T.; Mas-Torrent, M.; Rovira, C.; Henriques, R.T.; Almeida, M. Infrared investigation of the low-temperature structural and magnetic transitions in the spin-ladder candidate (DT-TTF)$_2$Au(mnt)$_2$. *Phys. Rev. B* **2003**, *68*, 134405. [CrossRef]
46. Ribera, E.; Rovira, C.; Veciana, J.; Tarés, J.; Canadell, E.; Rousseau, R.; Molins, E.; Mas, M.; Schoeffel, J.P.; Pouget, J.-P.; et al. The [(DT-TTF)$_2$M(mnt)$_2$] Family of Radical Ion Salts: From a Spin Ladder to Delocalised Conduction Electrons That Interact with Localised Magnetic Moments. *Chem. Eur. J.* **1999**, *5*, 2025–2039. [CrossRef]
47. Clay, R.T.; Mazumdar, S. Cooperative DensityWave and Giant Spin Gap in the Quarter-Filled Zigzag Electron Ladder. *Phys. Rev. Lett.* **2005**, *94*, 207206. [CrossRef] [PubMed]
48. Ribera, E.; Rovira, C.; Veciana, J.; Tarés, J.; Canadell, E.; Rousseau, R.; Molins, E.; Mas, M.; Schoeffel, J.P.; Pouget, J.-P.; et al. The (DT-TTF)-M(mnt)$_2$ Family of Compounds. *Synth. Met.* **1999**, *102*, 1743–1746. [CrossRef]
49. Monceau, P.; Nad, F.Y.; Brazovskii, S. Ferroelectric Mott-Hubbard Phase of Organic (TMTTF)$_2$X Conductors. *Phys. Rev. Lett.* **2001**, *86*, 4080–4083. [CrossRef]
50. de Souza, M.; Squillante, L.; Sônego, C.; Menegasso, P.; Foury-Leylekian, P.; Pouget, J.P. Probing the ionic dielectric constant contribution in the ferroelectric phase of the Fabre salts. *Phys. Rev. B* **2018**, *97*, 045122. [CrossRef]
51. Dressel, M.; Dumm, M.; Knoblauch, T.; Masino, M. Comprehensive Optical Investigations of Charge Order in Organic Chain Compounds (TMTTF)$_2$X. *Crystals* **2012**, *2*, 528–578. [CrossRef]
52. Kitou, S.; Fujii, T.; Kawamoto, T.; Katayama, N.; Maki, S.; Nishibori, E.; Sugimoto, K.; Takata, M.; Nakamura, T.; Sawa, H. Successive Dimensional Transition in (TMTTF)$_2$PF$_6$ Revealed by Synchrotron X-ray Diffraction. *Phys. Rev. Lett.* **2017**, *119*, 065701. [CrossRef]
53. Foury-Leylekian, P.; Petit, S.; Andre, G.; Moradpour, A.; Pouget, J.-P. Neutron scattering evidence for a lattice displacement at the charge ordering transition of (TMTTF)$_2$PF$_6$. *Phys. B Condens. Matter* **2010**, *405*, S95–S97. [CrossRef]
54. de Souza, M.; Foury-Leylekian, P.; Moradpour, A.; Pouget, J.-P.; Lang, M. Evidence for Lattice Effects at the Charge-Ordering Transition in (TMTTF)$_2$X. *Phys. Rev. Lett.* **2008**, *101*, 216403. [CrossRef]
55. de Souza, M.; Pouget, J.P. Charge-ordering transition in (TMTTF)$_2$X explored via dilatometry. *J. Phys. Condens. Matter* **2013**, *25*, 343201. [CrossRef]
56. Yasin, S.; Salameh, B.; Rose, E.; Dumm, M.; von Nidda, H.A.K.; Loidl, A.; Ozerov, M.; Untereiner, G.; Montgomery, L.; Dressel, M. Broken magnetic symmetry due to charge-order ferroelectricity discovered in (TMTTF)$_2$X salts by multifrequency ESR. *Phys. Rev. B* **2012**, *85*, 144428. [CrossRef]
57. Dutoit, C.-E.; Bertaina, S.; Orio, M.; Dressel, M.; Stepanov, A. Charge-ordering induces magnetic axes rotation in organic materials (TMTTF)$_2$X (with X = SbF$_6$, AsF$_6$, and PF$_6$). *Low Temp. Phys.* **2015**, *41*, 942–944. [CrossRef]
58. Dutoit, C.E.; Stepanov, A.; Van Tol, J.; Orio, M.; Bertaina, S. Superlattice Induced by Charge order in the organic Spin Chain (TMTTF)$_2$X (X = SbF$_6$, AsF$_6$, and PF$_6$) Revealed by High-Field Electron Paramagnetic Resonance. *J. Phys. Chem. Lett.* **2018**, *9*, 5598–5603. [CrossRef] [PubMed]
59. Tennant, D.A.; Perring, T.G.; Cowley, R.A.; Nagle, S.E. Unbound spinons in the $S = 1/2$ antiferromagnetic chain KCuF$_3$. *Phys. Rev. Lett.* **1993**, *70*, 4003–4006. [CrossRef]
60. Arai, M.; Fujita, M.; Motokawa, M.; Akimitsu, J.; Bennington, S.M. Quantum Spin Excitations in the Spin-Peierls System CuGeO$_3$. *Phys. Rev. Lett.* **1996**, *77*, 3647–3652. [CrossRef]
61. Bertaina, S.; Dutoit, C.-E.; Van Tol, J.; Dressel, M.; Barbara, B.; Stepanov, A. Rabi oscillations of pinned solitons in spin chains: A route to quantum computation and communication. *Phys. Rev. B* **2014**, *90*, 060404(R). [CrossRef]
62. Zeisner, J.; Pilone, O.; Soriano, L.; Gerbaud, G.; Vezin, H.; Jeannin, O.; Fourmigué, M.; Büchner, B.; Kataev, V.; Bertaina, S. Coherent spin dynamics of solitons in the organic spin chain compounds (o-DMTTF)$_2$X (X = Cl, Br). *Phys. Rev. B* **2019**, *100*, 224414. [CrossRef]
63. Coulon, C.; Lalet, G.; Pouget, J.-P.; Foury-Leylekian, P.; Moradpour, A.; Fabre, J.-M. Anisotropic conductivity and charge ordering in (TMTTF)$_2$X salts probed by ESR. *Phys. Rev. B* **2007**, *76*, 085126. [CrossRef]
64. Coulon, C.; Foury-Leylekian, P.; Fabre, J.-M.; Pouget, J.-P. Electronic instabilities and Irradiation effects in the (TMTTF)$_2$X series. *Eur. Phys. J. B* **2015**, *88*, 85. [CrossRef]
65. Soriano, L.; Zeisner, J.; Kataev, V.; Pilone, O.; Fourmigué, M.; Jeannin, O.; Vezin, H.; Orio, M.; Bertaina, S. Electron Spin Resonance of Defects in Spin Chains o-(DMTTF)$_2$X: A versatile system behaving like molecular magnet. *Appl. Mag. Res.* **2020**, *51*, 1307–1320. [CrossRef]

66. Soriano, L.; Orio, M.; Pilone, O.; Jeannin, O.; Reinheimer, E.; Quéméré, N.; Auban-Senzier, P.; Fourmigué, M.; Bertaina, S. A tetrathiafulvalene salt of the nitrite (NO^{2-}) anion: Investigations of the spin-Peierls phase. *J. Mater. Chem. C* **2023**. [CrossRef]
67. Gama, V.; Henriques, R.T.; Almeida, M.; Pouget, J.P. Diffuse X-ray Scattering Evidence for Peierls and "Spin Peierls" Like Transitions in the Organic Conductors (Perylene)$_2$ [M(mnt)$_2$] (M = Cu, Ni, Co and Fe). *Synth. Met.* **1993**, *56*, 1677–1682. [CrossRef]
68. Green, E.L.; Lumata, L.L.; Brooks, J.S.; Kuhns, P.; Reyes, A.; Brown, S.E.; Almeida, M. ^{1}H and ^{195}Pt NMR Study of the Two-Chain Compound Per$_2$[Pt(mnt)$_2$]. *Crystals* **2012**, *2*, 1116–1135. [CrossRef]
69. Pouget, J.P.; Ravy, S.; Schoeffel, J.P.; Dhalenne, G.; Revcolevschi, A. Spin-Peierls lattice fluctuations and disorders in CuGeO$_3$ and its solid solutions. *Eur. Phys. J. B* **2004**, *38*, 581–598. [CrossRef]
70. Bertaina, S.; Dutoit, C.E.; Van Tol, J.; Dressel, M.; Barbara, B.; Stepanov, A. Quantum coherence of strongly correlated defects in spin chains. *Phys. Procedia* **2015**, *75*, 23–28. [CrossRef]
71. Gotschy, B.; Auban-Senzier, P.; Farrall, A.; Bourbonnais, C.; Jérome, D.; Canadell, E.; Henriques, R.T.; Johannsen, I.; Bechgaard, K. One-dimensional physics in organic conductors (TMDTDSF)$_2$X, X = PF$_6$, ReO$_4$: ^{77}Se-NMR experiments. *J. Phys. I* **1992**, *2*, 677–694. [CrossRef]
72. Liu, Q.; Ravy, S.; Pouget, J.P.; Johannsen, I.; Bechgaard, K. X-Ray Investigation of the Tetramethyldithiadiselenafulvalene (TMDTDSF)$_2$X Series of Organic Conductors: II. Influence of the Orientational Disorder on the Structural Instabilities. *J. Phys. I Fr.* **1993**, *3*, 821–837.
73. Grenier, B.; Renard, J.P.; Veillet, P.; Paulsen, C.; Dhalenne, G.; Revcolevschi, A. Scaling in dimer breaking by impurities in CuGeO$_3$: A comparative experimental study of Zn-, Mg-, Ni-, and Si-doped single crystals. *Phys. Rev. B* **1998**, *58*, 8202. [CrossRef]
74. Zeman, J.; Martinez, G.; van Loosdrecht, P.H.M.; Dhalenne, G.; Revcolevschi, A. Scaling of the *H-T* Phase Diagram of CuGeO$_3$. *Phys. Rev. Lett.* **1999**, *83*, 2648–2651. [CrossRef]
75. Brown, S.E.; Clark, W.G.; Zamborszky, F.; Klemme, B.J.; Kriza, G.; Alavi, B.; Merlic, C.; Kuhns, P.; Moulton, W. ^{13}C NMR Measurements of the High-Magnetic-Field, Low-Temperature Phases of (TMTTF)$_2$PF$_6$. *Phys. Rev. Lett.* **1998**, *80*, 5429–5432. [CrossRef]
76. Horvatíc, M.; Fagot-Revurat, Y.; Berthier, C.; Dhalenne, G.; Revcolevschi, A. NMR Imaging of the Soliton Lattice Profile in the Spin-Peierls Compound CuGeO$_3$. *Phys. Rev. Lett.* **1999**, *83*, 420–423. [CrossRef]
77. Rønnow, H.M.; Enderle, M.; McMorrow, D.F.; Regnault, L.-P.; Dhalenne, G.; Revcolevschi, A.; Hoser, A.; Prokes, K.; Vorderwisch, P.; Schneider, H. Neutron Scattering Study of the Field-Induced Soliton Lattice in CuGeO$_3$. *Phys. Rev. Lett.* **2000**, *84*, 4469–4472. [CrossRef] [PubMed]

Disclaimer/Publisher's Note: The statements, opinions and data contained in all publications are solely those of the individual author(s) and contributor(s) and not of MDPI and/or the editor(s). MDPI and/or the editor(s) disclaim responsibility for any injury to people or property resulting from any ideas, methods, instructions or products referred to in the content.

Perspective

Lanthanide-Based Metal–Organic Frameworks with Single-Molecule Magnet Properties †

Fabio Manna [1,2,3], Mariangela Oggianu [1,2], Narcis Avarvari [3,*] and Maria Laura Mercuri [1,2,*]

1. Department of Chemical and Geological Sciences, University of Cagliari, Highway 554, Crossroads for Sestu, I-09042 Monserrato, Italy; f.manna@etud.univ-angers.fr (F.M.); mariangela.oggianu@unica.it (M.O.)
2. National Interuniversity Consortium of Materials Science and Technology, INSTM, Street Giuseppe Giusti, 9, I-50121 Florence, Italy
3. CNRS, MOLTECH-Anjou, SFR MATRIX University of Angers, F-49000 Angers, France
* Correspondence: narcis.avarvari@univ-angers.fr (N.A.); mercuri@unica.it (M.L.M.); Tel.: +33-241-735-084 (N.A.); +39-070-6754-474 (M.L.M.)
† Dedicated to Professor Manuel Almeida on the Occasion of His 70th Birthday.

Abstract: Lanthanide metal–organic frameworks (Ln-MOFs) showing single-molecule magnet (SMM) properties are an ever-growing family of materials where the magnetic properties can be tuned by various interrelated parameters, such as the coordinated solvent, temperature, organic linkers, lanthanide ions and their coordination environment. An overview of the general synthetic methodologies to access MOFs/Ln-MOFs and the peculiarities and parameters to control and/or fine-tune their SMM behavior is herein presented. Additionally, diverse challenging strategies for inducing SMM/SIM behavior in an Ln-MOF are discussed, involving redox activity and chirality. Furthermore, intriguing physical phenomena such as the CISS effect and CPL are also highlighted.

Keywords: lanthanide metal–organic frameworks; single-molecule magnets; single-ion magnets

Citation: Manna, F.; Oggianu, M.; Avarvari, N.; Mercuri, M.L. Lanthanide-Based Metal–Organic Frameworks with Single-Molecule Magnet Properties. *Magnetochemistry* **2023**, *9*, 190. https://doi.org/ 10.3390/magnetochemistry9070190

Academic Editors: Laura C. J. Pereira and Dulce Belo

Received: 27 June 2023
Revised: 17 July 2023
Accepted: 19 July 2023
Published: 22 July 2023

Copyright: © 2023 by the authors. Licensee MDPI, Basel, Switzerland. This article is an open access article distributed under the terms and conditions of the Creative Commons Attribution (CC BY) license (https:// creativecommons.org/licenses/by/ 4.0/).

1. Introduction

Metal–organic frameworks (MOFs), crystalline porous materials formed by the self-assembling of metal ions (nodes) with organic ligands (linkers) [1–3], are attracting considerable interest in material science given their fascinating architectures, which can be tailored by a proper chemical design of linkers and nodes, and the richness of their chemical and physical properties [4–6], which make them promising platforms in a *plethora* of applications such as fuel storage, CO_2 capture, drug delivery, sensing and catalysis [7–9]. As far as magnetic MOFs are concerned, those exhibiting molecular properties, i.e., single-molecule magnets (SMMs), have attracted extensive attention in the field of chemistry, physics and material science due to their potential applications in cutting-edge molecular spintronics and quantum computing devices [10]. SMMs are fascinating materials [11] that have been continuously investigated since the first discovery of $Mn_{12}O_{12}(OAc)_{16}(H_2O)_4$ (known as Mn_{12}) by Sessoli et al. and the synthesis of the first Ln^{III}–SMM complex based on Dy^{III} [12–14]. The SMM efficiency is evaluated by three different parameters: (i) the blocking temperature (T_B), i.e., the highest temperature at which the magnetization is preserved during a given period of time [15]; (ii) the coercive magnetic field (H_c), described as the magnetic field needed to turn to zero the magnetization of an SMM; and (iii) the effective energy barrier (U_{eff}), the most used parameter, which indicates the energy required to convert an SMM into a paramagnet.

Although reported in the literature are many examples of magnetic MOFs, very few show SMM behavior [16]. On the other hand, porosity, the intrinsic property of MOFs, is typically favored by the use of long linkers, i.e., linkers bearing bulky pendant arms, whereas magnetic interactions require short distances between the metal nodes [17]. Thus, the chemical design of MOF-based SMMs is a challenging aspect, and, among the synthetic strategies envisaged to develop MOF-based SMMs with efficient T_B, H_c and U_{eff} values, the use of Ln^{III} ions with large magnetic anisotropy occupies a prominent place.

Indeed, lanthanide-based MOFs (Ln–MOFs) are a class of porous materials that have attracted great interest during the last few decades thanks to their intrinsic advantages such as lanthanide coordination versatility and a wide range of applications due to their unique properties based on f-electrons offering the possibility to combine both luminescent centers and magnetic properties in the same crystal lattice, building up multifunctionality [18]. Lanthanides' valence electrons are located in 4f orbitals, which are shielded from the ligand field by fully filled $5s^2$ and $5p^6$ orbitals. For this reason, the 4f orbitals do not split due to weak interactions between the ligands and the orbitals, and the coordination environment around the 4f ion remains almost unaltered, giving rise to strong spin–orbit coupling interactions. The most used Ln^{III} ions to fabricate SMMs are Tb^{III}, Dy^{III}, Er^{III} and Ho^{III}. Particularly, Tb^{III} and Dy^{III} are the best candidates, since their electronic structure exhibits large magnetic anisotropy due to the strong angular dependence of 4f orbitals [19].

Recently a second generation of SMM-based MOFs, known as single-ion magnet (SIM) MOFs, started to rapidly develop since the first report of the Ln(bipyNO)$_4$(TfO)$_3$ (bipyNO = 4,40-bypyridyl-N,N0-dioxide, TfO = triflate) 3D coordination framework by Espallargas et al. [20] (*vide infra*). SIMs consist of single centers that exhibit slow magnetic relaxation [10], and, consequently, SIM-MOFs are formed by ordered assemblies of lanthanide ions and different bridging linkers, featuring extended and porous high-dimensional frameworks with tunable magnetic properties.

The present work aims to provide an overview, through selected examples, of the different strategies currently employed to integrate SMM/SIM functionality into an MOF. The MOF framework is then used as a suitable scaffold to constrain or tune the local geometries of lanthanide nodes (ions or clusters) and arrange diverse organic linkers into ordered assemblies, featuring a combination of structural diversity and SMM/SIM behavior. In this context, conventional and/or unconventional synthetic approaches toward SMMs/SIMs are also exploited.

2. Synthetic Methodologies

MOFs in general and Ln-MOFs in particular are conventionally obtained by combining organic linkers and metal ions via solvothermal and slow diffusion reactions (layering). Recently, other synthetic strategies have been explored as microwave-assisted, electrochemical, mechanochemical or sonochemical methods, as reported in Scheme 1. The final product is often obtained in the form of single crystals, suitable for X-ray structural studies, microcrystalline powders, nanocrystals or films [21]. Solvothermal synthesis is the most used technique to develop crystalline MOFs and consists of a combination of linkers and metal salts in a closed autoclave, heated under autogenous pressure. The crystal size and morphology are strongly influenced by several parameters, such as the solvent, temperature, concentration, pH, reaction time, etc. The major advantages are the control of the reaction conditions and the obtainment of single crystals suitable for structural characterization. The diffusion method consists of a slow mixing of metal salts and linker solutions by layering one on top of the other, taking advantage of their different densities. This technique is particularly useful for growing single crystals suitable for structural analysis. In the microwave-assisted method, the reactor is heated by electromagnetic waves. The parameters that influence the size and morphologies of the final product are similar to the solvothermal one, but the major advantages are low reaction time, energy saving and microcrystalline materials. In the mechanochemical methods, reactions between metal salts and linkers occur under high-speed grinding in a ball mill at ambient pressure. Metal oxides are frequently employed as starting precursors, and the crystallinity of the final product is generally low, but it could be improved by using different precursors or by adding a small amount of solvent. This method is economical and eco-friendly and very few or no solvents are used, but it affords poor crystalline final products. The electrochemical method is used for the synthesis of MOF thin films (from nano to microscale) whose morphologies could be finely tuned by changing the electrosynthesis parameters (e.g., electrolytes, current density, additives and reaction time). This technique has in principle several advantages over the aforementioned methodologies, such as high-purity products, mild conditions and the facile control of film features, but in the case of Ln-MOFs, it may suffer from a reduction of Ln^{III} ions on the cathode during the process. In the sonochemical method, MOF synthesis occurs at room temperature by the application of high-energy ultrasound to a reaction mixture; this method is easy, rapid, environmentally friendly and allows for the obtainment of nanoscale MOFs with high crystallinity and different morphologies [22,23].

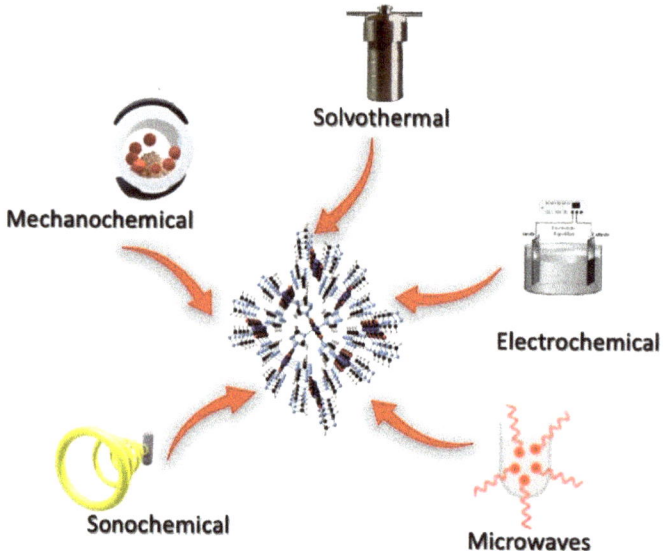

Scheme 1. Survey of synthetic methods.

The strategies to obtain high-temperature SMM-based Ln-MOFs are almost the same as for their molecular counterpart, and they could be summarized as follows: (i) make a proper choice of the crystal field for the Ln^{III} metal ion, taking into account its axial anisotropy; (ii) conduct large crystal field splitting by choosing the proper ligand field, whose shape should match the electronic density of the free metal ions [11,24,25]; (iii) commence strong magnetic interaction between the Ln^{III} ions, induced by the proper choice of diamagnetic or radical paramagnetic bridging ligands; and (iv) use isotopic Ln^{III} ions to suppress the hyperfine coupling, which favors the quantum tunnel of the magnetization relaxation process [26].

Furthermore, Ln-MOFs/coordination polymers (CPs) could be post-modified by external stimuli, such as (i) solvato-switching, principally due to the change in the coordination environment after solvent exchange or removal; (ii) photo-switching, through a mechanism where a change in the ligand with photochromic moieties is observed after irradiation, which could affect Ln-Ln distances and/or the coordination environment; (iii) redox-switching, through post-reduction/oxidation of the linker affording open-shell species, which could affect the exchange interactions of Ln^{III} ions, and also through a change in the open-shell species shape; and (iv) metal-ion exchange, where the incorporation of Ln^{III} ions in a preformed MOF framework occurs via exchange with a suitable cation, as reported by Pardo et al. [27]. Here, Ca^{II} ions could be exchanged by different Ln^{III} ions (Ln^{III} = Dy, Ho, Er) and NO_3^-, for charge compensation, inside the preformed MOF (*vide infra*). Recently, Long et al. reported on the use of mixed-valence $Dy^{II/III}$ ions in a dinuclear complex affording strong magnetic exchange interactions due to M–M bond formation and, consequently, high U_{eff} [28]; moreover, it has been reported that magnetic exchange between Dy^{III} ions could be enhanced through the presence of soft donors, leading to excellent SMM behavior with high U$_{eff}$ [14,29].

The post-modification (*vide supra*) of Ln-SMM-MOFs/CPs [27,30] occurred, in some cases, via an interesting on/off mechanism, particularly suitable for data manipulation applications, while in other cases, it led to a change in the SMM behavior, but in both cases, particular attention should be devoted to the reversibility of the process [31–33].

As highlighted by Powell et al. [34], another promising synthetic strategy is the use of a secondary building unit (SBU) acting as an SMM center, which can govern coordination numbers and the local geometry of the metal nodes of the assemblies in 2D/3D frameworks. Indeed, MOFs have ordered arrangement of SMMs units, one of the most significant advantages with respect to their molecular counterpart [18,34]. For example, Murugesu et al. reported on a binuclear complex [Dy$_2$(hmi)$_2$(NO$_3$)$_2$(MeOH)$_2$] (H$_2$hmi = 2-hydroxy-3-methoxyphenyl methylene (isonicotino)hydrazine) exhibiting SMMs, which could form a 2D network due to the coordination of Dy^{III} ions to the peripheral pyridyl N atoms of the linker. The 2D framework [Dy$_2$(hmi)$_2$(NO$_3$)$_2$(MeOH)$_2$]n$_\infty$ MeCN shows remarkably higher U$_{eff}$ than the binuclear complex [35,36].

3. Single-Molecule Magnet/Ion-Based Ln-MOFs (SMM/SIM-MOFs)

The use of MOFs as a rigid platform (*vide supra*) to constrain and chemically tune the coordination geometries of lanthanide nodes, featuring high coordination numbers, has been shown to be a straightforward strategy for the rational control of SMM behavior, as demonstrated by a proper change in the synthetic conditions (nature of the solvents, temperature, etc., *vide infra*).

The solvent can play a crucial role in determining the supramolecular architecture, as reported by Cheng et al. [37]. Two different MOFs, formulated as {[Dy$_2$(FDA)$_3$(DMF)$_2$]·1.5DMF}$_n$ (**1**) and [Dy$_2$(FDA)$_3$(DMF)$_2$(CH$_3$OH)]$_n$ (**2**) (H$_2$FDA=furan-dicarboxylic acid, see Chart 1), have been obtained, where the use of the DMF/CH$_3$OH solvent mixture leads to a moderate distortion of the Dy^{III} ions coordination sphere in **2** from an ideal D$_{4d}$ with respect to **1**, resulting in a considerable improvement of the SMM behavior (Figure 1). Theoretical calculations demonstrated that Dy^{III} ions in **1** and **2** MOFs exhibit biaugmented trigonal-prismatic geometries (C$_{2v}$), and the estimated (calculated) deviation parameters are 1.579 (1.076) and 0.671 (1.786) for **1** and **2**, respectively, supporting the observed moderate deviation from an ideal D$_{4d}$ symmetry in the latter than in the former. Because of the occupancy of the two coordinated solvents, the FDA^{2-} ligands in **2** adopt a μ_2-η_1:η_1-bridging coordination mode, resulting in a less distorted DyO$_8$ polyhedron of D$_{4d}$ symmetry. When methanol was removed from **2**, a different MOF, formulated as {[Dy$_2$(FDA)$_3$(DMF)$_2$]·2.2DMF} (**1′**), was obtained, where a seven-coordinate Dy^{III} ion is present, surrounded by six carboxylic oxygen atoms and one DMF molecule. The Dy^{III} ion shows a monocapped triangular-prismatic C$_{2v}$ geometry, and a three-dimensional *stp* topological framework is constructed by connecting, via FDA^{2-} ligands, Dy^{III} ions into one-dimensional chains. Remarkably **1′** does not show slow relaxation of magnetization, highlighting the role played by the lanthanide coordination geometry in tuning the SMM behavior. Furthermore, the corresponding **1@Y** and **2@Y**, diluted by diamagnetic Y(III) ions, which should reduce or eliminate magnetic interactions between dysprosium(III) ions, show similar SMM behavior as the pristine **1** and **2** MOFs, confirming that magnetization relaxation depends on the Dy(III) ions' crystal field induced by the observed moderate deviation from the ideal D$_{4d}$ geometry.

The temperature can also induce a fine-tuning of the coordination geometry of lanthanide ions in Ln-MOFs, which can significantly influence the SMM properties, also providing insights into the magneto-structural relationship. Cheng et al. reported on [Dy$_3$(OBA)$_4$(HCOO)(H$_2$O)(DMF)]$_n$ (**3**) and [Dy(OBA)(HOBA)(H$_2$O)$_2$]·3DMF}$_n$ (**4**) (H$_2$OBA = 4,4′-oxybis(benzoate)acid) Ln-MOFs, obtained by reacting Dy(NO$_3$)$_3$·6H$_2$O with H$_2$OBA for 3 days at 160 °C (**3**, orthorhombic, C$_{mc2_1}$ space group) and 80 °C (**4**, triclinic space group P1). Interestingly, **3** undergoes a reversible structural transformation to **4** by standing in its mother liquor for two days at room temperature and vice versa by heating **4** in its mother liquor up to 120 °C for 3 days. This process, thermally induced, provokes a dramatic change in SMM properties, leading to switchable "ON/OFF" SMM behavior due to a significant change in the coordination geometries of the Dy^{III} ion from the C_{2v}/C_{4v} in **3** (paramagnet) to D_{4d} symmetry in **4** (SMM) because of a variation of its coordination mode.

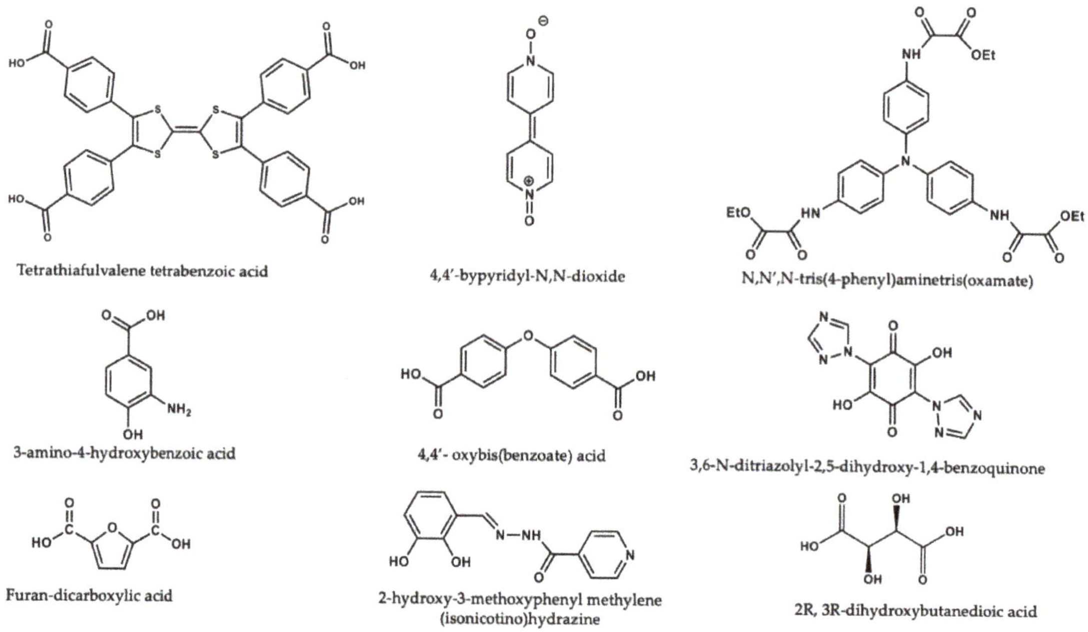

Tetrathiafulvalene tetrabenzoic acid

4,4'-bypyridyl-N,N-dioxide

N,N',N-tris(4-phenyl)aminetris(oxamate)

3-amino-4-hydroxybenzoic acid

4,4'- oxybis(benzoate) acid

3,6-N-ditriazolyl-2,5-dihydroxy-1,4-benzoquinone

Furan-dicarboxylic acid

2-hydroxy-3-methoxyphenyl methylene (isonicotino)hydrazine

2R, 3R-dihydroxybutanedioic acid

Chart 1. Organic linkers reported in Section 3.

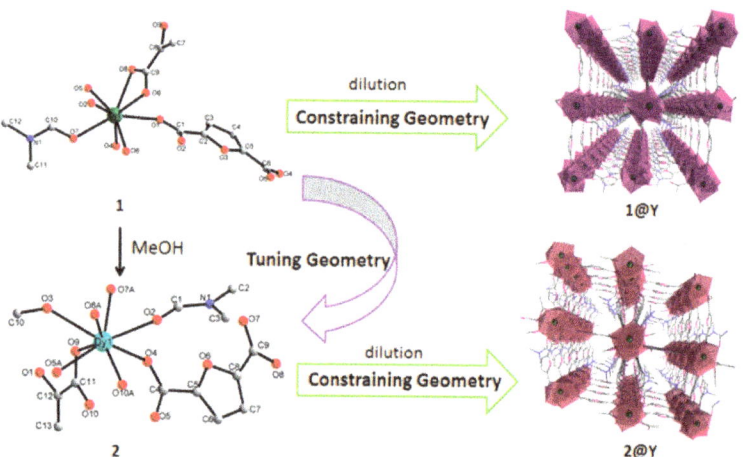

Figure 1. Constraining and Tuning the Local Geometries of Lanthanide Ions in Dy MOFs **1** and **2**. Reprinted with permission from [37].

The study of the mechanism of the structural conversion process from **4** to **3** evidenced the crucial role of temperature in constructing the 3D framework, breaking inter-layer π···π stacking interactions and hydrogen bonds and leading to the coordination environment around the Dy^{III} ion of C_{2v}/C_{4v} symmetry, observed in **3**, as shown in Figure 2 [38].

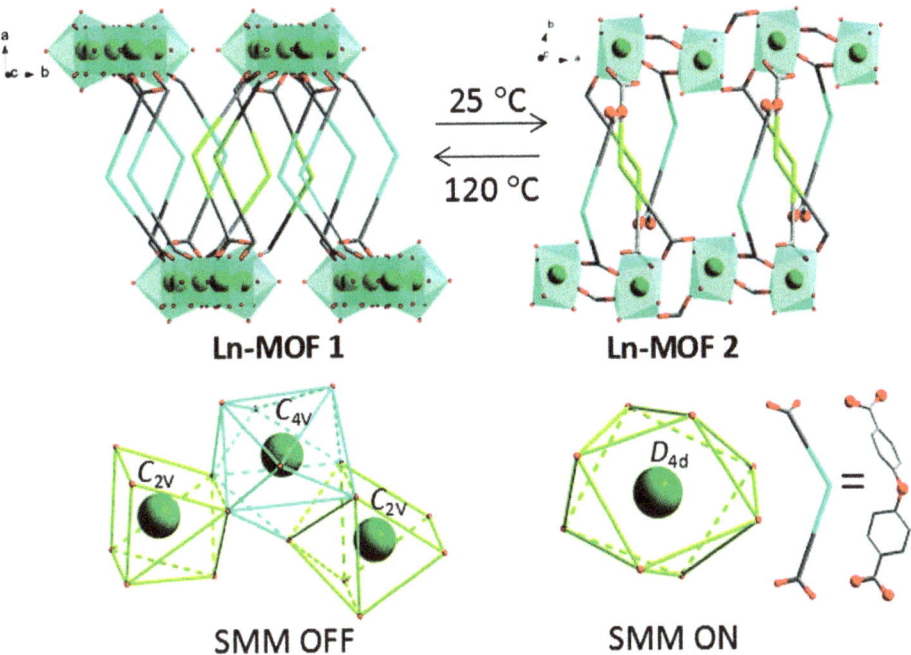

Figure 2. Schematic representation of reversible 3→4 structural transformation (**top**). Dy(III) ions' coordination symmetries and their related "ON/OFF" switching of magnetization relaxation in a zero *dc* field (**bottom**). Reprinted with permission from Ref. [38].

The careful choice of lanthanide nodes with a proper coordination environment is also a challenging strategy to construct an MOF with SMM properties. Particularly, mononuclear lanthanide complexes, where the first coordination sphere consists of the Ln^{III} ion surrounded by hard atoms (O, N) in a square-antiprismatic geometry, which is shown to favor SIM behavior, are promising candidates, as reported by Coronado et al. [20] through the novel family of Ln-MOFs formulated as [Ln(bipyNO)$_4$](TfO)$_3$·xsolvent (Ln=Tb (**5**); Dy (**6**); Ho (**7**); Er (**8**); and bipyNO= 4,4′-bypyridyl-N,N-dioxide, TfO = triflate]. The choice of the linker is very important as well, as long linkers (*vide supra*), such as bipyNO, are preferred since they are capable of keeping lanthanide centers isolated and reducing dipole interactions.

These 3D Ln-MOFs exhibit slow magnetic relaxation, behaving as SIM-MOFs, as confirmed by magnetic measurements (Figure 3).

Interestingly, the incorporation of large anions as polyoxometalates (POMs), particularly [Mo$_6$O$_{19}$]$^{2-}$ polyanions, reported in Figure 4, does not affect the magnetic behavior, and retainment of slow magnetic relaxation is observed, despite the exchange with triflate anions. This is due to the preservation of the structural features of the 3D framework, despite the sterically hindered POM anions, which are located in the cavity of MOFs but far from the lanthanide centers. These findings further demonstrate that SIM behavior is strictly related to the lanthanide coordination geometry and also pave the way to introduce non-innocent POM anions into multifunctional SIM-MOFs or to use POM-MOFs, which are materials with very large surface areas and thus potential candidates for applications in heterogeneous catalysis, where large porosity is required [20].

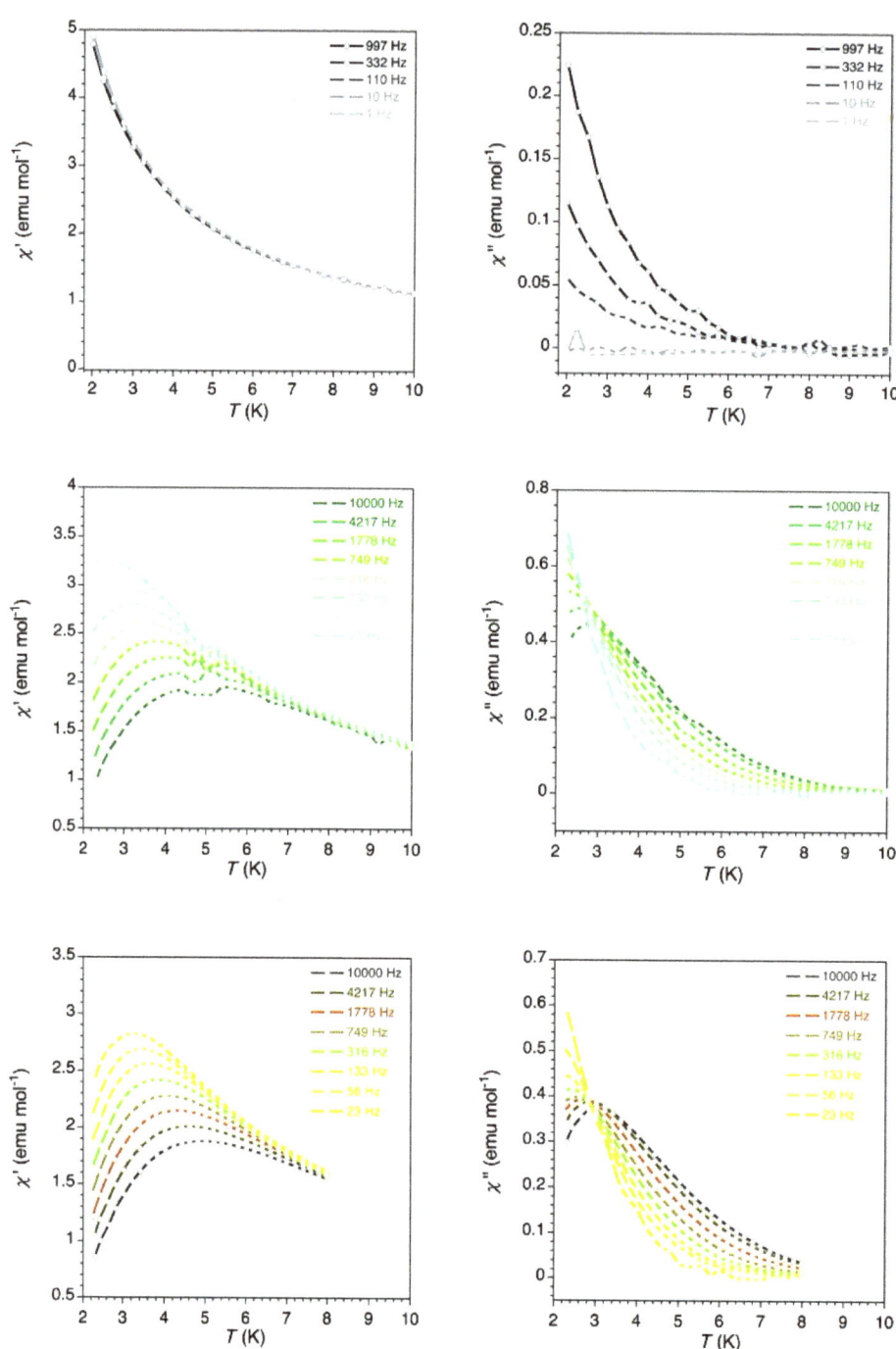

Figure 3. Dynamic susceptibility of **6**: left, in-phase; right, out-of-phase, in applied dc magnetic fields; and 0 G (top) and 1000 G (bottom). Adapted with permission from [20].

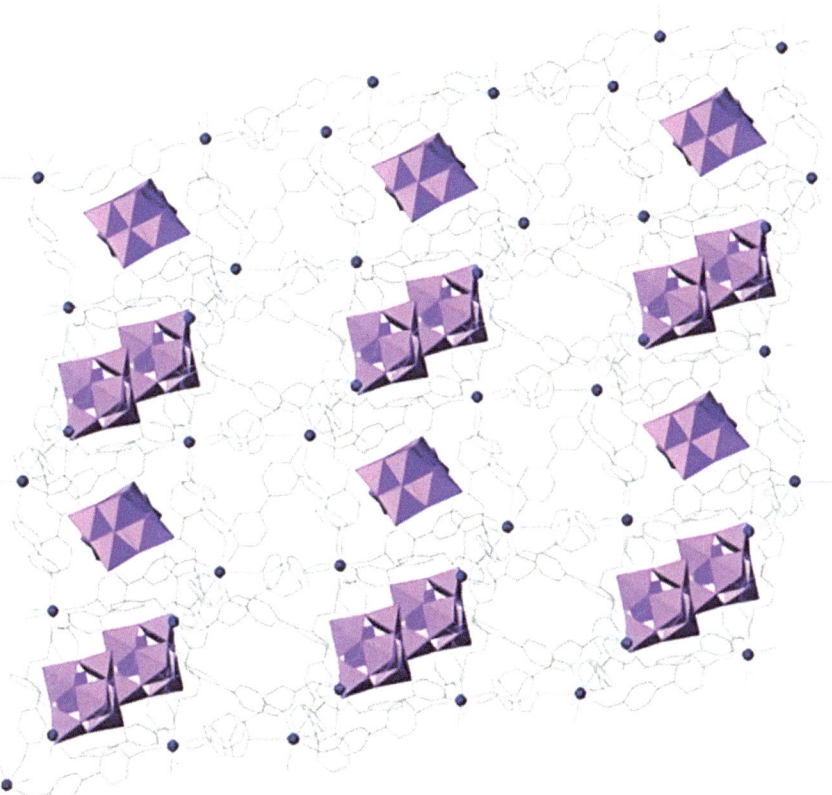

Figure 4. Crystal structure of 6-POM, with POMs (purple) in the cavities. Dy atoms and bipyNO linkers are in blue and grey, respectively. Reprinted with permission from [20].

The combination of post-synthetic methodologies (PSMs) with the molecular approach, consisting of the use of metal complexes as suitable linkers toward selected metal nodes, is a powerful strategy for governing the structure of SMM-MOFs and tuning their magnetic properties through a fine modulation of their structural characteristics. In fact, it has been shown that small structural changes in the coordination geometry of the lanthanide ion in an SMM-MOF provoke significant changes in its magnetic behavior. Pardo et al. [27] reported on the use of an oxamato-based tetranuclear Co^{III} complex as a suitable building block toward Ca^{II} metal ions to construct a diamagnetic Ca^{II}-Co^{III} 3D MOF, formulated as $\{Ca^{II}_6(H_2O)_{24}[Co^{III}_4(tpatox)_4]\}\cdot 44H_2O$ (**9**), (tpatox^{6-} = N,N',N-tris(4-phenyl)aminetris(oxamate)). Interestingly, the oxamate-based complex is formed by tetrahedral cages obtained by combining a multitopic linker, the bulky tripodal tpatox ligand (N,N,N-tris(4-phenyl)aminetris(oxamate), with Co^{III} ions. When the metal-ion exchange post-synthetic method is used, by replacing Ca^{II} ions with Tb^{III}, Dy^{III}, Ho^{III} and Er^{III} ions, isostructural Ln^{III}–Co^{III} 3D MOFs formulated as $\{[Tb^{III}_6(H_2O)_{24}[Co^{III}_4(tpatox)_4]](NO_3)_6\}\cdot 49H_2O$ (**10**), $\{[Dy^{III}_6(H_2O)_{24}[Co^{III}_4(tpatox)_4]](NO_3)_6\}\cdot 53H_2O$ (**11**), $\{[Ho^{III}_6(H_2O)_{30}[Co^{III}_4(tpatox)_4]](NO_3)_6\}\cdot 44H_2O$ (**12**) and $\{[Er^{III}_6(H_2O)_{24}[Co^{III}_4(tpatox)_4]](NO_3)_6\}\cdot 58H_2O$ (**13**) are obtained through a single-crystal-to-single-crystal solid-state process (Figure 5). It is noteworthy that **11** and **13** MOFs, containing Kramers' ions as metal nodes, exhibit slow relaxation of magnetization, thus behaving as SIM-MOFs, while the one-pot reaction between the related $Na_{12}[Co^{III}_4(tpatox)_4]\cdot 6H_2O$ and $Ln(NO_3)_3\cdot 5H_2O$ salts did not afford **10–13**. The crucial role of post-synthetic and molecular approaches to construct MOFs with SIM properties is therefore evidenced.

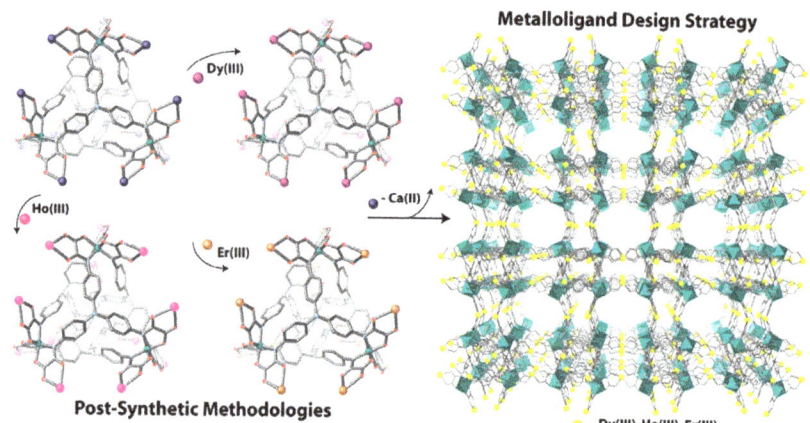

Figure 5. Perspective views of $([Co_4(tpatox)_4])^{12-}$ tetrahedral cages containing Ca^{II} or Dy^{III}, Ho^{III} and Er^{III} exchanged by post-synthetic methods, as building blocks of isostructural **10–13** MOFs (**left**); **9** MOF, along the *a* crystallographic axis (**right**). Co^{III} and Ca^{II} metal ions and linkers are in gray, cyano and blue, respectively. Reprinted with permission from [27].

It has been shown that the coordination geometry of lanthanide ions, in a rigid scaffold as an MOF, can significantly influence its SIM/SMM properties. It is noteworthy that in the case of flexible MOFs [39,40], flexibility may be due to a reversible dehydration/hydration process, when coordinated water molecules are present in the first coordination sphere of the Ln^{III} ion. As a consequence, a change in the coordination environment is observed, resulting in a significant modulation of the relaxation of the magnetization barrier. Very recently, some of us [40] reported on two multifunctional 3D polymorphic frameworks (**14a,b**), formulated as $[Er_2(trz_2An)_3(H_2O)_4]_n\cdot xH_2O$ ($x = 10$, **a**; $x = 7$, **b**), showing a combination of luminescent and magnetic properties, where **14a** is a 3D flexible MOF. These two polymorphs were obtained by combining the NIR-emitting Er^{III} ion with the 3,6-N-ditriazolyl-2,5-dihydroxy-1,4-benzoquinone (H_2trz_2An) linker, bearing bulky triazole-based pendant arms. Moreover, **14a** shows a reversible structural phase transition to a partially dehydrated 3D structure, formulated as $[Er_2(trz_2An)_3(H_2O)_2]_n\cdot 2H_2O$ (**14a_des**; des = desolvated), from ennea- to octa-coordination of the Er^{III} coordination geometry through a reversible loss of a water molecule, which occurred under vacuum or heating up 360 K. Remarkably, structural flexibility provokes the fine-tuning of both luminescent and SIM properties, involving a moderate change in NIR emission and a slight improvement of the magnetic blocking temperature in the dehydrated structural phase. The dehydration/hydration mechanism has been investigated by theoretical calculations, which evidenced that the observed improvement of magnetic properties is mainly due to the variation in the ligand field induced by the loss of the water molecule (Figure 6).

Redox activity has been demonstrated to be an efficient strategy for tuning the conducting and magnetic properties in MOFs due to the generation of organic radicals, which can improve magnetic exchange between metal nodes [4,40–42]. Redox-active linkers can affect also the SMM behavior; therefore, they can act as switchable nodes for constructing redox-controlled SMM-MOFs. Linkers based on a tetrathiafulvalene (TTF) *core*, which possess different oxidation states, such as radical $TTF^{\cdot+}$ and diamagnetic TTF^{2+} species, are efficient ON/OFF redox switches. Zhou et al. [32] have combined, via a solvothermal reaction, ennea-nuclear lanthanide clusters showing SMM behavior, as inorganic nodes, and

TTF-based linkers, as redox switches, to construct 3D SMM-MOFs. Interestingly, crystals of Ln$_9$(μ_3-OH)$_{13}$(μ_3-O)(H$_2$O)$_9$(TTFTB)$_3$] (H$_2$TTFB = tetrathiafulvalene tetrabenzoic acid)] were obtained with Ln^{III} = Dy (**15**), Tb (**16**) and Er (**17**). As an example, in **15**, each Dy$_9$ cluster is linked to TTFTB^{4-} linkers through 12 carboxylate groups, which coordinate in a *syn-syn* bidentate mode, leading to a framework formed by 12-connected hexagonal prisms combined with 4-connected TTFTB^{4-} linkers, which coordinate in a square planar manner, resulting in a shp-a topology (shp = square hexagonal prism) as reported in Figure 7a,b.

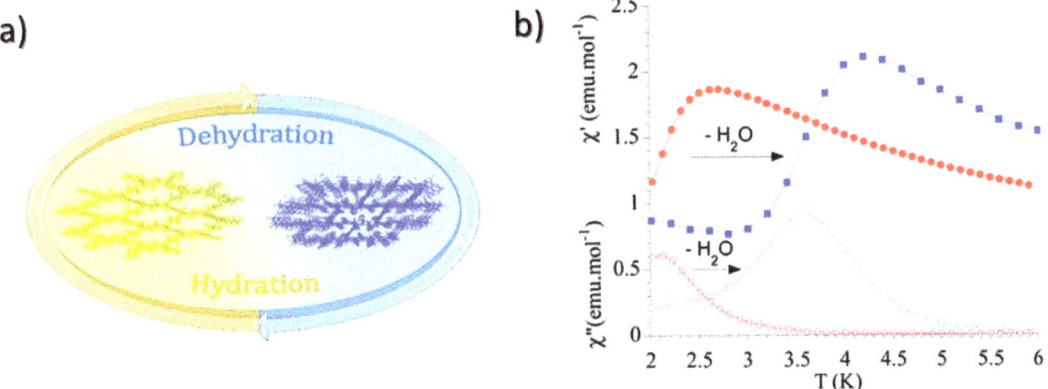

Figure 6. (**a**) Reversible expansion/shrinkage from hexagonal distorted cavities in **14a** to distorted 3,6-brickwall rectangular cavities in **14a_des** (left); (**b**) plot of χ' and χ'' vs T of 14a(full and empty red circles, respectively and 14a_desfull and empty blue circles)) under an applied dc field of 0.09 T at 10,000 Hz. Adapted with permission from [40].

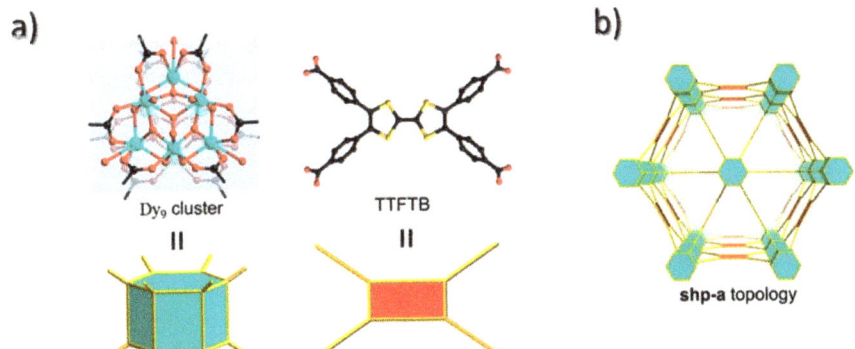

Figure 7. (**a**) The 12-connected Dy9 cluster and 4-connected TTFTB linker in a hexagonal prism and square planar coordination modes, respectively. (**b**) shp-a topology. Adapted with permission from [32].

Remarkably, these multifunctional MOFs, which are isostructural, possess, other than high porosity (Brunauer–Emmett–Teller surface areas of 1515, 1635 and 1800 m$^2 \cdot$g^{-1} for **15**, **16** and **17**, respectively), redox activity and slow relaxation of magnetization, which can be switched OFF upon post-synthetic oxidation with I$_2$ of the TTFTB linker to its corresponding radical S = 1/2 cation and I$_3^-$ anion or by solvent exchange of DMF (N, N-dimethylformamide) guest molecules with cyclohexane or I$_3^-$ anions inside the cavities. Most importantly, this process is reversible, and the SMM behavior can be switched ON through a reduction in the DMF solution. Furthermore, structural features are also retained in the corresponding oxidized MOFs. These promising results show that the fine-tuning of the SMM behavior via redox activity of the organic building blocks can provide a valuable way to control, through external stimuli and/or solvent exchange, the SMM properties of lanthanide clusters within a rigid platform as an MOF for future applications in functional devices based on redox-switchable and multi-stimuli responsive SMM-MOFs.

Very recently, Seco et al. [43] reported on a novel family of 3D magneto-luminescent Ln-MOFs, formulated as a $\{[Ln_5L_6(OH)_3(DMF)_3]\cdot 5H_2O\}_n$-based 3-amino-4-hydroxybenzoic acid linker (H_2L) and Ln^{III} ions (Nd, Sm, Eu, Gd, Tb, Dy, Ho, Er, Tm and Yb), including NIR emitters (Nd, Er and Yb). They are isostructural, and, most importantly, they are multifunctional, showing a combination of porosity, photoluminescence and SMM behavior. In particular, the latter is observed in Dy (**18**), Er (**19**) and Yb (**20**) MOFs due to the oblate or prolate character of lanthanide ions, which depend on the electronic distribution of the coordination geometry. Specifically, in oblate-type ions (Dy^{III}), magnetic anisotropy is enhanced by axially coordinated donor atoms of the linker, showing the highest electron density, while in prolate-type ions (Er^{III}, Yb^{III}), coordination at equatorial positions is preferred to induce greater magnetic anisotropy. Magnetic measurements on **18–20**, reported in Figure 8, evidence that no frequency dependence is present without applying an external field due to quantum tunneling of magnetization (QTM). Under a static field (1000 Oe), QTM is suppressed, and, interestingly, frequency dependence is observed only in **20**, which represents, *according to the authors*, the first 3D SMM-MOF containing NIR-emitting Yb^{III} ions, since Yb-MOFs are rare. Magnetic diluted yttrium-based MOFs were prepared to obtain insights on the magnetization relaxation process, and the magnetic studies evidence the crucial role of the electron density distribution, the prolate vs. oblate nature of lanthanide ions in tuning the magnetic anisotropy and thus the SMM properties. Photoluminescence studies on Eu/Tb-MOFs and their corresponding heterobimetallic MOFs (stoichiometric ratio 50:40:10/5) evidence the potential of the doped materials as ratiometric thermometers. Additionally, the microporous structure of these MOFs makes them potential candidates for CO_2 capture and separation, particularly Dy-MOF (**18**), which is the material showing the best performance in terms of sorption capacity.

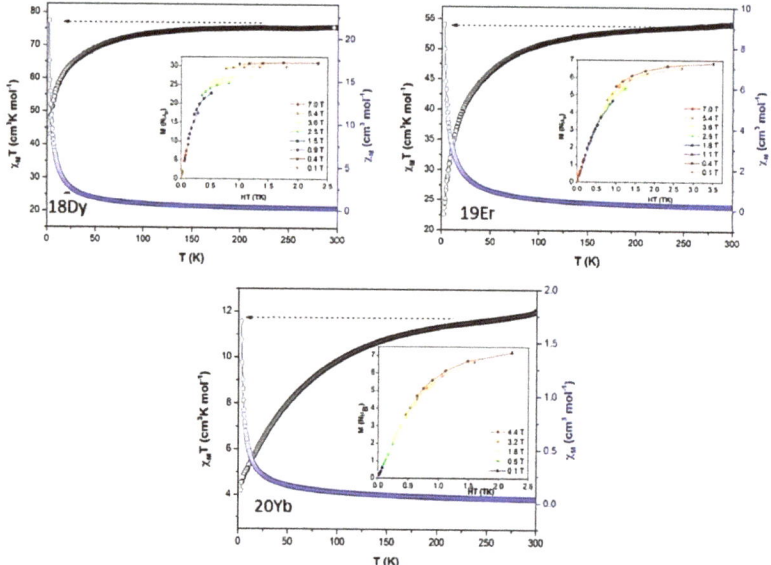

Figure 8. Thermal variation in $\chi_M T$ vs. T, under an applied field of 1000 Oe, for **18–20**. Adapted with permission from Ref. [43].

The use of chiral linkers in a hybrid platform as an MOF is a challenging strategy to study physical phenomena related to the interaction of chiral materials with a polarized electromagnetic field, such as circularly polarized luminescence (CPL) [44], the emission of circularly polarized light upon irradiation, which can have potential applications in optical data storage and spintronics [45]. Ln-MOFs are then promising candidates, since lanthanide(III) ions can boost the CPL performances [46], even though the studies on chiral Ln-MOFs, including those showing SMM behavior, are still scarce due to the difficulties of governing chirality in a hybrid framework. Cepeda et al. [47] reported on a new family of 3D chiral magneto-luminescent Ln-MOFs of formula $\{[Ln_2(\mu_4\text{-tar})_2(\mu\text{-tar})(H_2O)_2]\cdot 3H_2O\}_n)$, with [$Ln^{III}$ = Tb, Dy, Ho, Er and Tm and x = 3 or 4, depending on the counteranion with the *D*- or *L*- tartrate (tar = 2R, 3R-dihydroxybutanedioate) chiral linker. These enantiomerically pure MOFs are all isostructural, thermally stable and flexible, showing different phases depending either on the ion size or the temperature variation, ranging from partially (Ln-L' and Ln-L'') to fully (Ln-L''') dehydrated structures. Magnetic studies show that Tb-, Dy- and Er-L-MOFs (**21**, **22** and **23**) exhibit SMM behavior upon application of an external field and discrete values of an energy barrier (U_{eff}) due to a residual QTM (in particular for Dy-MOF, **22**), which is suppressed through a magnetic dilution with yttrium(III) in the related yttrium-based MOFs, as reported in Figure 9. The dehydration/hydration process is reversible and, remarkably, modulates both the magnetic and photoluminescence responses.

Number **22**, which shows tunable SMM and luminescent properties, which depend on the dehydration/hydration process, has the potential to be used in humidity-sensing applications. Number **21** represents the second example of a chiral MOF showing the CPL property. In the CPL spectrum of **21**, reported in Figure 10, the emission assigned to the $^7F_5 \leftarrow {}^5D_4$ transition is split mainly into two bands, centered at 540 and 555 nm, due to the degeneration removal of sublevels induced by the environment chirality.

These two bands show opposite signs and luminescence dissymmetry factor values around 4×10^{-3}, which, although modest, are similar to those reported in the literature for terbium coordination polymers. Most importantly, **22** (Dy-L-MOF, *vide supra*) represents the first example, to the best of our knowledge, of a multifunctional chiral Ln MOF showing chirality-induced spin selectivity (CISS) [48,49], the generation of a spin-polarized current by injecting electrons through a chiral material without the presence of a magnetic field. The first report on the performance of Dy-L-MOF as an almost ideal spin filter is made by San Sebastian et al. [50]. The impressive spin selectivity found is due to the 3D helicoidal packing pattern, which may boost the electronic conduction along the crystallographic axes through the helices. The CISS effect is also influenced by the large spin-orbit coupling of dysprosium (III) ions, which, along with the 3D helicity exhibited by this MOF, may be responsible for the observed remarkably high spin polarization values. In Figure 11, a schematic representation of the CISS effect is reported.

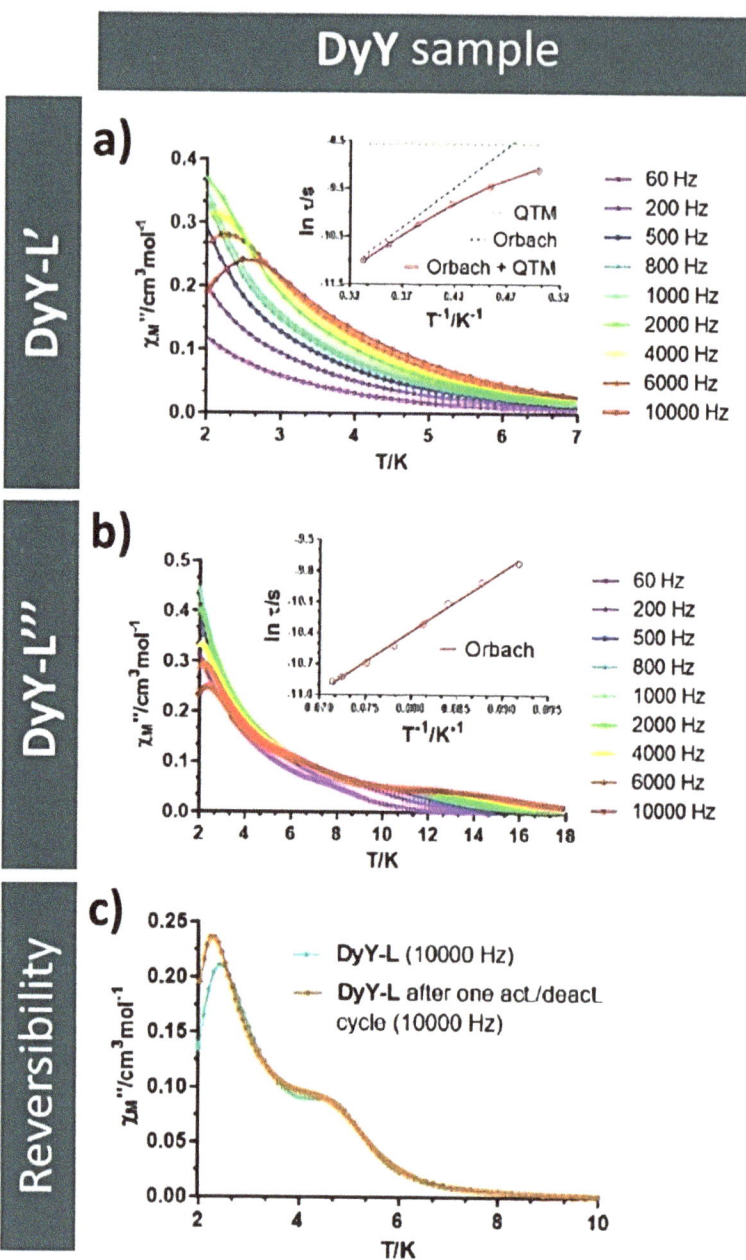

Figure 9. Thermal variation in out-of-phase $\chi_M T$ vs. T, under an applied field of 1000 Oe, for **22**, partially (**a**) and completely (**b**) dehydrated. Insets: Relaxation times vs. T and related best fits. (**c**) ac curves at 10,000 Hz for **22**, non-activated and completely dehydrated/rehydrated. Adapted with permissions from Ref. [47].

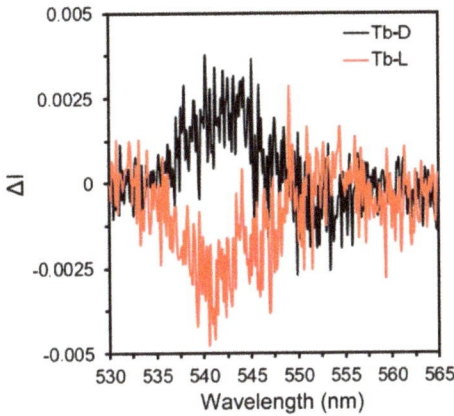

Figure 10. CPL spectra centered at the $^7F_5 \leftarrow {}^5D_4$ transition of **21** (Tb-L and Tb-D). Reprinted with permissions from Ref. [47].

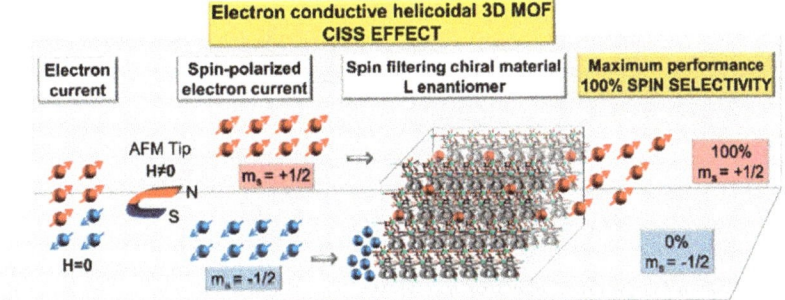

Figure 11. Schematic representation of the CISS effect in a helicoidal 3D MOF (**22**). Reprinted with permission from [50].

4. Summary and Outlook

Ln-MOFs have been demonstrated to be suitable platforms for the rational control of SMM behavior by the fine modulation of different parameters, such as synthetic protocols, involving temperature and the nature of the solvents, a careful choice of linkers and metal nodes, including preformed SMMs, and, most importantly, the coordination geometry, involving a rigid or flexible scaffold. The use of redox-active linkers has been shown to be a challenging strategy for achieving the fine-tuning of the SMM behavior through external stimuli and/or solvent exchange in redox-switchable and multi-stimuli-responsive SMM-MOFs. On the other hand, chiral linkers allow for exploiting physical phenomena such as CPL [44] and the CISS effect, observed for the first time in a 3D helicoidal Dy-MOF, which showed unprecedented spin polarization values. In this *scenario*, the combination of porosity with SMM and luminescent properties offers a route to building up multifunctionality as a powerful tool to further tune these properties through intermolecular interactions between guest molecules present in the pores and the 3D framework. It should be highlighted that a rational design of high-performance SMMs/SIMs, with improved energy barriers for the magnetization reversal, requires a careful choice of the coordination environment of lanthanide ions in terms of the ligand field. With this view, the linker donor atoms possessing high electron density should coordinate oblate Ln^{III} ions (Dy and Tb) at the axial positions, while prolate ions (Er and Yb) should be coordinated at the equatorial positions, in order to maximize their anisotropic electron density *(vide supra)*. As perspectives, the use of soft linkers, unexplored in the SMM-MOF field probably due to the hard character of the Ln^{III} ions, can be challenging due to the occurrence of strong exchange interactions observed in their molecular counterpart, a binuclear Dy^{III} complex bridged by sulfur atoms [29]. Furthermore, paramagnetic radical linkers should be developed, since these open-shell ligands have the potential to mediate strong magnetic interactions unless Ln^{III} centers bridged by these ligands exhibit short distances [51]. A further achievement is represented by photo-switchable MOFs, not reported up to now, since a change in magnetic properties with light could be promising in the construction of spintronic devices. In conclusion, these relevant findings pave the way to construct next-generation materials with improved performances for optical data storage and spintronic applications.

Author Contributions: N.A. and M.L.M. defined the topic and the main content of the review. F.M. and M.O. equally wrote parts of the initial draft, supervised, reviewed and completed by N.A. and M.L.M. All authors have read and agreed to the published version of the manuscript.

Funding: This research was funded in Italy by the Fondazione di Sardegna, Convenzione triennale tra la Fondazione di Sardegna e gli Atenei Sardi, Regione Sardegna, L.R. 7/2007 annualità 2020, through the SMAWRT project (CUP F75F21001260007).

Institutional Review Board Statement: Not applicable.

Informed Consent Statement: Not applicable.

Data Availability Statement: Not applicable.

Acknowledgments: The SMAWRT project (CUP F75F21001260007) is acknowledged for M.O.'s post-doctoral grant. The MUR (Ministry of Education, University, Research) and UNICA-UNISS Consortium PhD Course on Chemical Sciences and Technologies are also acknowledged for F.M.'s Ph.D. grant. The work in France was supported by the CNRS and the University of Angers.

Conflicts of Interest: The authors declare no conflict of interest. The funders had no role in the design of the study; in the writing of the manuscript; or in the decision to publish the results.

References

1. O'keeffe, M. Design of MOFs and intellectual content in reticular chemistry: A personal view. *Chem. Soc. Rev.* **2009**, *38*, 1215–1217. [CrossRef]
2. Yazaydin, A.Ö.; Snurr, R.Q.; Park, T.H.; Koh, K.; Liu, J.; LeVan, M.D.; Benin, A.I.; Jakubczak, P.; Lanuza, M.; Galloway, D.B.; et al. Screening of metal-organic frameworks for carbon dioxide capture from flue gas using a combined experimental and modeling approach. *J. Am. Chem. Soc.* **2009**, *131*, 18198–18199. [CrossRef] [PubMed]
3. Yaghi, O.M.; Li, G. Li Hailian Yaghi-Selective binding and removal of guests in a imcroporous metal-organic framework-Nature 1985. *Nature* **1995**, *378*, 703–706. [CrossRef]
4. Monni, N.; Oggianu, M.; Sahadevan, S.A.; Mercuri, M.L. Redox activity as a powerful strategy to tune magnetic and/or conducting properties in benzoquinone-based metal-organic frameworks. *Magnetochemistry* **2021**, *7*, 109. [CrossRef]
5. Sahadevan, S.A.; Abhervé, A.; Monni, N.; De Pipaón, C.S.; Galán-Mascarós, J.R.; Waerenborgh, J.C.; Vieira, B.J.C.; Auban-Senzier, P.; Pillet, S.; Bendeif, E.E.; et al. Conducting Anilate-Based Mixed-Valence Fe(II)Fe(III) Coordination Polymer: Small-Polaron Hopping Model for Oxalate-Type Fe(II)Fe(III) 2D Networks. *J. Am. Chem. Soc.* **2018**, *140*, 12611–12621. [CrossRef] [PubMed]
6. Sahadevan, S.A.; Monni, N.; Abhervé, A.; Marongiu, D.; Sarritzu, V.; Sestu, N.; Saba, M.; Mura, A.; Bongiovanni, G.; Cannas, C.; et al. Nanosheets of Two-Dimensional Neutral Coordination Polymers Based on Near-Infrared-Emitting Lanthanides and a Chlorocyananilate Ligand. *Chem. Mater.* **2018**, *30*, 6575–6586. [CrossRef]
7. Monni, N.; Andres-Garcia, E.; Caamaño, K.; García-López, V.; Clemente-Juan, J.M.; Giménez-Marqués, M.; Oggianu, M.; Cadoni, E.; Espallargas, G.M.; Clemente-León, M.; et al. A thermally/chemically robust and easily regenerable anilato-based ultramicroporous 3D MOF for CO2uptake and separation. *J. Mater. Chem. A* **2021**, *9*, 25189–25195. [CrossRef]
8. Oggianu, M.; Monni, N.; Mameli, V.; Cannas, C.; Sahadevan, S.A.; Mercuri, M.L. Designing magnetic nanomofs for biomedicine: Current trends and applications. *Magnetochemistry* **2020**, *6*, 39. [CrossRef]
9. Sahadevan, S.A.; Monni, N.; Abhervé, A.; Marongiu, D.; Saba, M.; Mura, A.; Bongiovanni, G.; Mameli, V.; Cannas, C.; et al. Heteroleptic NIR-Emitting YbIII/Anilate-Based Neutral Coordination Polymer Nanosheets for Solvent Sensing. *ACS Appl. Nano Mater.* **2020**, *3*, 94–104. [CrossRef]

10. Fernández, B.; Oyarzabal, I.; Fischer-Fodor, E.; MacAvei, S.; Sánchez, I.; Seco, J.M.; Gómez-Ruiz, S.; Rodríguez-Diéguez, A. Multifunctional applications of a dysprosium-based metal-organic chain with single-ion magnet behaviour. *Crystengcomm* **2016**, *18*, 8718–8721. [CrossRef]
11. Ashebr, T.G.; Li, H.; Ying, X.; Li, X.L.; Zhao, C.; Liu, S.; Tang, J. Emerging Trends on Designing High-Performance Dysprosium(III) Single-Molecule Magnets. *ACS Mater. Lett.* **2022**, *4*, 307–319. [CrossRef]
12. Sessoli, R.; Tsai, H.; Schake, A.R.; Wang, S.; Vincent, J.B.; Foiling, K.; Gatteschi, D.; Christou, G.; Hendrickson, D.N. High-Spin Molecules [Mn12O12(O2CR)16(H20)4]. *J.Am.Chem. Soc.* **1993**, *115*, 1804–1816. [CrossRef]
13. Zhu, Z.; Guo, M.; Li, X.L.; Tang, J. Molecular magnetism of lanthanide: Advances and perspectives. *Coord. Chem. Rev.* **2019**, *378*, 350–364. [CrossRef]
14. Zhang, P.; Guo, Y.N.; Tang, J. Recent advances in dysprosium-based single molecule magnets: Structural overview and synthetic strategies. *Coord. Chem. Rev.* **2013**, *257*, 1728–1763. [CrossRef]
15. Ungur, L.; Chibotaru, L.F. Strategies toward High-Temperature Lanthanide-Based Single-Molecule Magnets. *Inorg. Chem.* **2016**, *55*, 10043–10056. [CrossRef]
16. Benmansour, S.; Pintado-Zaldo, C.; Martínez-Ponce, J.; Hernández-Paredes, A.; Valero-Martínez, A.; Gómez-Benmansour, M.; Gómez-García, C.J. The Versatility of Ethylene Glycol to Tune the Dimensionality and Magnetic Properties in DyIII-Anilato-Based Single-Ion Magnets. *Cryst. Growth Des.* **2023**, *23*, 1269–1280. [CrossRef]
17. Espallargas, G.M.; Coronado, E. Magnetic functionalities in MOFs: From the framework to the pore. *Chem. Soc. Rev.* **2018**, *47*, 533–557. [CrossRef]
18. Zeng, M.; Ji, S.Y.; Wu, X.R.; Zhang, Y.Q.; Liu, C.M.; Kou, H.Z. Magnetooptical Properties of Lanthanide(III) Metal-Organic Frameworks Based on an Iridium(III) Metalloligand. *Inorg. Chem.* **2022**, *61*, 3097–3102. [CrossRef] [PubMed]
19. Gendron, F.; Pritchard, B.; Bolvin, H.; Autschbach, J. Single-ion 4f element magnetism: An ab-initio look at Ln(COT)$_2^-$. *Dalt. Trans.* **2015**, *44*, 19886–19900. [CrossRef] [PubMed]
20. Baldoví, J.J.; Coronado, E.; Gaita-Ariño, A.; Gamer, C.; Giménez-Marqués, M.; Espallargas, G.M. A SIM-MOF: Three-dimensional organisation of single-ion magnets with anion-exchange capabilities. *Chem. A Eur. J.* **2014**, *20*, 10695–10702. [CrossRef]
21. Topologies, M.O.F.; Stock, N.; Biswas, S. Synthesis of Metal-Organic Frameworks (MOFs): Routes to Various. *Chem. Rev.* **2012**, *2*, 933–969. [CrossRef]
22. Younis, S.A.; Bhardwaj, N.; Bhardwaj, S.K.; Kim, K.H.; Deep, A. Rare earth metal–organic frameworks (RE-MOFs): Synthesis, properties, and biomedical applications. *Coord. Chem. Rev.* **2021**, *429*, 213620. [CrossRef]
23. Safaei, M.; Foroughi, M.M.; Ebrahimpoor, N.; Jahani, S.; Omidi, A.; Khatami, M. A review on metal-organic frameworks: Synthesis and applications. *TrAC Trends Anal. Chem.* **2019**, *118*, 401–425. [CrossRef]
24. Rinehart, J.D.; Long, J.R. Exploiting single-ion anisotropy in the design of f-element single-molecule magnets. *Chem. Sci.* **2011**, *2*, 2078–2085. [CrossRef]
25. Dey, A.; Kalita, P.; Chandrasekhar, V. Lanthanide(III)-Based Single-Ion Magnets. *ACS Omega* **2018**, *3*, 9462–9475. [CrossRef]
26. Liu, J.L.; Chen, Y.C.; Tong, M.L. Symmetry strategies for high performance lanthanide-based single-molecule magnets. *Chem. Soc. Rev.* **2018**, *47*, 2431–2453. [CrossRef]
27. Kalinke, L.H.G.; Cangussu, D.; Mon, M.; Bruno, R.; Tiburcio, E.; Lloret, F.; Armentano, D.; Pardo, E.; Ferrando-Soria, J. Metal-Organic Frameworks as Playgrounds for Reticulate Single-Molecule Magnets. *Inorg. Chem.* **2019**, *58*, 14498–14506. [CrossRef]
28. Gould, C.A.; McClain, K.R.; Reta, D.; Kragskow, J.G.C.; Marchiori, D.A.; Lachman, E.; Choi, E.S.; Analytis, J.G.; Britt, R.D.; Chilton, N.F.; et al. Ultrahard magnetism from mixed-valence dilanthanide complexes with metal-metal bonding. *Science* **2022**, *375*, 198–202. [CrossRef] [PubMed]
29. Tuna, F.; Smith, C.A.; Bodensteiner, M.; Ungur, L.; Chibotaru, L.F.; McInnes, E.J.L.; Winpenny, R.E.P.; Collison, D.; Layfield, R.A. A High Anisotropy Barrier in a Sulfur-Bridged Organodysprosium Single-Molecule Magnet. *Angew. Chem.* **2012**, *124*, 7082–7086. [CrossRef]
30. Zhu, Z.; Li, X.L.; Liu, S.; Tang, J. External stimuli modulate the magnetic relaxation of lanthanide single-molecule magnets. *Inorg. Chem. Front.* **2020**, *7*, 3315–3326. [CrossRef]
31. Ma, Y.J.; Hu, J.X.; De Han, S.; Pan, J.; Li, J.H.; Wang, G.M. Manipulating On/Off Single-Molecule Magnet Behavior in a Dy(III)-Based Photochromic Complex. *J. Am. Chem. Soc.* **2020**, *142*, 2682–2689. [CrossRef]
32. Su, J.; Yuan, S.; Li, J.; Wang, H.Y.; Ge, J.Y.; Drake, H.F.; Leong, C.F.; Yu, F.; D'Alessandro, D.M.; Kurmoo, M.; et al. Rare-Earth Metal Tetrathiafulvalene Carboxylate Frameworks as Redox-Switchable Single-Molecule Magnets. *Chem. A Eur. J.* **2021**, *27*, 622–627. [CrossRef]
33. Zhang, X.; Vieru, V.; Feng, X.; Liu, J.L.; Zhang, Z.; Na, B.; Shi, W.; Wang, B.W.; Powell, A.K.; Chibotaru, L.F.; et al. Influence of Guest Exchange on the Magnetization Dynamics of Dilanthanide Single-Molecule-Magnet Nodes within a Metal-Organic Framework. *Angew. Chem. Int. Ed.* **2015**, *54*, 9861–9865. [CrossRef] [PubMed]
34. Liu, K.; Zhang, X.; Meng, X.; Shi, W.; Cheng, P.; Powell, A.K. Constraining the coordination geometries of lanthanide centers and magnetic building blocks in frameworks: A new strategy for molecular nanomagnets. *Chem. Soc. Rev.* **2016**, *45*, 2423–2439. [CrossRef]
35. Aratani, I.; Horii, Y.; Takajo, D.; Kotani, Y.; Osawa, H.; Kajiwara, T. Construction of a two-dimensional metal-organic framework with perpendicular magnetic anisotropy composed of single-molecule magnets. *J. Mater. Chem. C* **2023**, *11*, 2082–2088. [CrossRef]

36. Lin, P.H.; Burchell, T.J.; Clérac, R.; Murugesu, M. Dinuclear dysprosium(III) single-molecule magnets with a large anisotropic barrier. *Angew. Chem. Int. Ed.* **2008**, *47*, 8848–8851. [CrossRef] [PubMed]
37. Liu, K.; Li, H.; Zhang, X.; Shi, W.; Cheng, P. Constraining and Tuning the Coordination Geometry of a Lanthanide Ion in Metal-Organic Frameworks: Approach toward a Single-Molecule Magnet. *Inorg. Chem.* **2015**, *54*, 10224–10231. [CrossRef] [PubMed]
38. Wang, M.; Meng, X.; Song, F.; He, Y.; Shi, W.; Gao, H.; Tang, J.; Peng, C. Reversible structural transformation induced switchable single-molecule magnet behavior in lanthanide metal-organic frameworks. *Chem. Commun.* **2018**, *54*, 10183–10186. [CrossRef]
39. Xin, Y.; Wang, J.; Zychowicz, M.; Zakrzewski, J.J.; Nakabayashi, K.; Sieklucka, B.; Chorazy, S.; Ohkoshi, S.I. Dehydration-Hydration Switching of Single-Molecule Magnet Behavior and Visible Photoluminescence in a Cyanido-Bridged DyIIICoIII Framework. *J. Am. Chem. Soc.* **2019**, *141*, 18211–18220. [CrossRef]
40. Monni, N.; Baldoví, J.J.; García-López, V.; Oggianu, M.; Cadoni, E.; Quochi, F.; Clemente-León, M.; Mercuri, M.L.; Coronado, E. Reversible tuning of luminescence and magnetism in a structurally flexible erbium–anilato MOF. *Chem. Sci.* **2022**, *13*, 7419–7428. [CrossRef]
41. Xie, L.S.; Skorupskii, G.; Dincă, M. Electrically Conductive Metal-Organic Frameworks. *Chem. Rev.* **2020**, *120*, 8536–8580. [CrossRef] [PubMed]
42. Cador, O.; Le Guennic, B.; Pointillart, F. Electro-activity and magnetic switching in lanthanide-based single-molecule magnets. *Inorg. Chem. Front.* **2019**, *6*, 3398–3417. [CrossRef]
43. Echenique-Errandonea, E.; Mendes, R.F.; Figueira, F.; Choquesillo-Lazarte, D.; Beobide, G.; Cepeda, J.; Ananias, D.; Rodríguez-Diéguez, A.; Paz, F.A.A.; Seco, J.M. Multifunctional Lanthanide-Based Metal-Organic Frameworks Derived from 3-Amino-4-hydroxybenzoate: Single-Molecule Magnet Behavior, Luminescent Properties for Thermometry, and CO2 Adsorptive Capacity. *Inorg. Chem.* **2022**, *61*, 12977–12990. [CrossRef]
44. Riehl, J.P.; Richardson, F.S. Circularly Polarized Luminescence Spectroscopy. *Chem. Rev.* **1986**, *86*, 1–16. [CrossRef]
45. Brandt, J.R.; Salerno, F.; Fuchter, M.J. The added value of small-molecule chirality in technological applications. *Nat. Rev. Chem.* **2017**, *1*, 0045. [CrossRef]
46. Zinna, F.; Di Bari, L. Lanthanide Circularly Polarized Luminescence: Bases and Applications. *Chirality* **2015**, *27*, 1–13. [CrossRef] [PubMed]
47. Huizi-Rayo, U.; Zabala-Lekuona, A.; Terenzi, A.; Cruz, C.M.; Cuerva, J.M.; Rodríguez-Diéguez, A.; García, J.A.; Seco, J.M.; Sebastian, E.S.; Cepeda, J. Influence of thermally induced structural transformations on the magnetic and luminescence properties of tartrate-based chiral lanthanide organic-frameworks. *J. Mater. Chem. C* **2020**, *8*, 8243–8256. [CrossRef]
48. Naman, R.; Waldeck, D.H. Chiral-Induced Spin Selectivity Effectct. *J. Phys. Chem. Lett.* **2012**, *3*, 2178–2187. [CrossRef] [PubMed]
49. Naaman, R.; Paltiel, Y.; Waldeck, D.H. Chiral molecules and the electron spin. *Nat. Rev. Chem.* **2019**, *3*, 250–260. [CrossRef]
50. Huizi-Rayo, U.; Gutierrez, J.; Seco, J.M.; Mujica, V.; Diez-Perez, I.; Ugalde, J.M.; Tercjak, A.; Cepeda, J.; Sebastian, E.S. An Ideal Spin Filter: Long-Range, High-Spin Selectivity in Chiral Helicoidal 3-Dimensional Metal Organic Frameworks. *Nano Lett.* **2020**, *20*, 8476–8482. [CrossRef] [PubMed]
51. Demir, S.; Jeon, I.R.; Long, J.R.; Harris, T.D. Radical ligand-containing single-molecule magnets. *Coord. Chem. Rev.* **2015**, *289*, 149–176. [CrossRef]

Disclaimer/Publisher's Note: The statements, opinions and data contained in all publications are solely those of the individual author(s) and contributor(s) and not of MDPI and/or the editor(s). MDPI and/or the editor(s) disclaim responsibility for any injury to people or property resulting from any ideas, methods, instructions or products referred to in the content.

MDPI AG
Grosspeteranlage 5
4052 Basel
Switzerland
Tel.: +41 61 683 77 34

Magnetochemistry Editorial Office
E-mail: magnetochemistry@mdpi.com
www.mdpi.com/journal/magnetochemistry

Disclaimer/Publisher's Note: The title and front matter of this reprint are at the discretion of the . The publisher is not responsible for their content or any associated concerns. The statements, opinions and data contained in all individual articles are solely those of the individual Editors and contributors and not of MDPI. MDPI disclaims responsibility for any injury to people or property resulting from any ideas, methods, instructions or products referred to in the content.

www.ingramcontent.com/pod-product-compliance
Lightning Source LLC
LaVergne TN
LVHW070741100526
838202LV00013B/1279